NONLINEAR SYSTEMS

3

Control

NONLINEAR SYSTEMS

3

Control

Edited by

A.J. FOSSARD

ENSAE/CERT, Toulouse, France

D. NORMAND-CYROT

CNRS, Supélec, Gif-sur-Yvette, France

CHAPMAN & HALL

London · Weinheim · New York · Tokyo · Melbourne · Madras

Published by Chapman & Hall, 2-6 Boundary Row, London SEI 8HN, UK

Published with the support of the ministère de l'Enseignement supérieur et de la Recherche (MESR, France): Direction de l'information scientifique et technique et des bibliothèques (DISTNB).

Chapman & Hall, 2-6 Boundary Row, London SEI 8HN, UK

Chapman & Hall GmbH, Pappelallee 3, 69469 Weinheim, Germany

Chapman & Hall USA, 115 Fifth Avenue, New York NY 10003, USA

Chapman & Hall Japan, ITP-Japan, Kyowa Building, 3F, 2-2-1 Hirakawacho, Chiyoda-ku, Tokyo 102, Japan

Chapman & Hall Australia, 102 Dodds Street, South Melbourne, Victoria 3205, Australia

Chapman & Hall India, R. Seshadri, 32 Second Main Road, CIT East, Madras 600 035, India

---

Co-published by Masson Éditeur, 120 boulevard Saint-Germain, 75006 Paris, France

First English language edition 1997

English language translation revised by Mrs M.B. Groen-Garrer

Original French language edition: *Systèmes non linéaires*, coordonné par A.J. Fossard et D. Normand-Cyrot, © 1993, Masson, Paris.

Typeset in France by Publilog

Printed in Belgium by SNEL

ISBN 0 412 60010 2

A catalogue record for this book is available from the British Library

Preface

Under the aegis of the DRET and the AFCET, the "Nonlinear Group" (NLG) has, over the years, brought together researchers, with their different backgrounds, from specialized Schools of Engineering, Universities and the CNRS (National Scientific Research Center):

- from the Department of Studies and Research In Automatic Control of the Studies and Research Center of Toulouse associated to the Higher School of Aeronautics and Space ("Département d'Études et de Recherches en Automatique du Centre d'Études et de Recherches de Toulouse associé à l'École Supérieure d'Aéronautique et de l'Espace")
  DERA/CERT/ENSAE

- from the Automatic Control and Computer Science Center of the School of Mining Engineering in Paris ("Centre d'Automatique et d'Informatique de l'École des Mines de Paris")
  CAI/ENSMP

- from the Laboratory of Automatic Control of the State School of Engineering in Nantes ("Laboratoire d'Automatique de l'École Centrale de Nantes")
  LAN/ECN

- from the Laboratory of Automatic Control and System Analysis in Toulouse ("Laboratoire d'Automatique et d'Analyse des Systèmes à Toulouse")
  LAAS/CNRS

- from the Laboratory of Industrial Automatic Control of the INSA in Lyon ("Laboratoire d'Automatique Industrielle de l'INSA de Lyon")
  LAI/INSAL

- from the Laboratory of Automatic Control and Industrial Computer Science of the State School of Engineering in Lille ("Laboratoire d'Automatique et d'Informatique Industrielle de l'École Centrale de Lille")
  LAIL/CNRS

- from the University of Compiègne ("l'Université de Compiègne")
  HEUDIASYC/UTC

- from the Laboratory of Automatic Control in Grenoble ("Laboratoire d'Automatique de Grenoble")
  LAG/INPG/CNRS

- from the Laboratory of Automatic Control and Process Engineering of the University of Lyon I ("Laboratoire d'Automatique et de Génie des Procédés de l'Université Lyon I")
  LAGEP/CNRS

- from the Laboratory of Automatic Control and Industrial Micro-Computer Science of the University of Savoie ("Laboratoire d'Automatique et de Micro-Informatique Industrielle de l'Université de Savoie")
  LAMII/FAST

- from the Laboratory of Signals and Systems in Gif sur Yvette ("Laboratoire des Signaux et Systèmes à Gif sur Yvette")
  LSS/SUPELEC/CNRS

This group, created in 1993 under the initiative of A.J. Fossard, made at the time the following two observations:

• The first one was that substantial progress had been made these last two years in Nonlinear Automatic Control, in theory as well as in methodology. The number of defended thesis, the papers presented during different conferences, the number of published papers, all widely and publicly revealed the activity of this scientific area. Moreover, this research, which had been in progress for several years, had resulted in a certain number of applications, not only on experimental processes but also on real processes. Even if limited, they clearly translated both a desire for applicability on behalf of the researchers, and a certain maturity of the discipline.

• The second one was that this research effort had not been transferred with desired efficiency at the industrial level, whereas the industrial needs appeared to be more and more obvious, whatever they are related to the slowness of great changes or to the limitations of the controlling organization.

It seemed to us that we, as teachers and researchers, had to assume our responsibilities and should try:

- to have common ideas and a common language,
- to be clear and simple,
- to take into account the real needs, with increased willingness.

Structured by exchanges and reflections, this group therefore wished to make a certain inventory of the knowledge of "modern" automatic nonlinear control, to think about the applicability to new methods on long term and short term, to introduce them in a form which should be as tutorial as possible both in the initial training of engineers and for the necessary transfer to the research departments in industry.

It was largely assisted by the fact that it very soon received material and moral help of the DRET, which had initially been created as a working group of the AFCET. This help appeared to be deterministic in permitting the transformation of what was originally a simple working group into a concerted action, conducted by A.J. FOSSARD and D. NORMAND-CYROT. It is within the framework of this action that the objectives have been specified, both for their substance and form, and that the relationship with the industrial market bag been somewhat institutionalized.

This has not been an easy task. On the one hand because the mathematics which are most often used in nonlinear theory (Lie algebra, differential geometry, etc.) are not taught daily to the engineers, on the other hand because the differences in experience and sensitivity of the members of the group, who in a certain way constitute the richness of it, have created numerous problems for the adaptation to the research of a compromise between rigor and accessibility by the reader.

Each of the different themes treated in the following chapters, has been the object of numerous discussions, drafts, criticisms inside the research group of this difficult compromise. It is this long common work that is presented in the three different volumes, collected under the general theme:

"nonlinear systems"

Each of these volumes is related to a particular aspect:

- Volume 1 Modeling and estimation
- Volume 2 Stability and Stabilization
- Volume 3 Control of nonlinear systems

In this general preface it is out of the question to mention the specific prefaces of each volume, as the title by itself gives already an idea of the problems under study. We will find, however, in the index of each volume, a brief summary of the three volumes together, in order to enable the reader to have a quick view on the contents of the complete work.

Finally, the authors wish to thank all the persons who heave made it possible to work out this document. The DRET morally and materially supported the action of the group. Let us especially mention Mrs FARGEON and Mr ROUCHOUZE whose help and presence have been inestimable. The manufacturers collaborated on this work by their interest that they showed for the group, through their presence at the organized meetings in the framework of this work and through their remarks. The scientific automatic control society supplied with their research the foundations of the presented work, as is shown by the important bibliography at the end of each chapter.

We wish that those books, in accordance with the objectives of the group and thanks to a language which is understood by all of us, will be able to bring together the work of the researchers and the preoccupation of the manufacturers. In that case the objectives will have been completely carried out.

A.J. Fossard and D. Normand-Cyrot

Coordinators of the Concerted Action of the NG

Contents

Volume 1
Modeling and estimation

Volume 2
Stability and stabilization

Volume 3
Control

Introduction

Preoccupation with performance improvement in controlled systems always yields to more and more precise modeling. However, if a model is a good representation of the behavior of a system over a wide operating range, it is unfortunately most often nonlinear. Consequently the fundamental tools for control design commonly used within the linear field are obsolete and other methods are necessary.

Because of the complexity of nonlinear mathematical models, it is difficult to design and apply control laws. The theoretical and practical problems to be solved are still considerable. Nevertheless, some methods do exist that lead to interesting solutions. Throughout this book, we have tried to present their ins and outs as simply as possible, and simultaneously to display their limitations.

Such an overview of methods and applications is always incomplete since it reflects the competence and the experience of the various contributors. Furthermore, we have voluntarily excluded approaches that are not yet well established and also, those that are specific to a very specialized problem and whose use is very limited. Thus, our choice has been guided by the concern to present well-mastered techniques whose theoretical foundations are not subject to controversy and whose practical implementation on many various and real control problems is conceivable in the short term or even already established.

Faced with the difficulty of the nonlinear case, the natural approach for engineers is to look for a way to linearize the problem. This approach is justified in so far as it is possible after the linearization to use the gear of linear system control theory which now covers a wide range of problems.

Linearization of the model equations around an operating point has for a long time been the way used by engineers faced with a nonlinear control problem. This well-known quasi-linearization is well suited to regulation, with more or less large deviations from the above mentioned equilibrium point. The limitations of this local and approximate linearization have led to the appearance of new approaches that may be distinguished according to the following characteristics:

- local or global: approximation by the linearized model may be valid around a point, a surface, or globally that is to say in the whole state space,
- approximate or exact: according to the way the system behaves after linearization, exactly linearly or not,

- internal or external: (also called input-state and input–output) whether the behavior is linearized between input and state or between input and output.

All these aspects of the problem are discussed in the following chapters.

While quasi-linearization only uses a basic linear change of coordinates (state is replaced with its deviation from equilibrium), these new approaches use nonlinear changes of coordinates and also feedbacks (static or dynamic) that transform the behavior of the system to be controlled. Thus it is really a closed-loop which is required to achieve these linearizations. The following figure illustrates this new concept of the linearization for the internal and external cases:

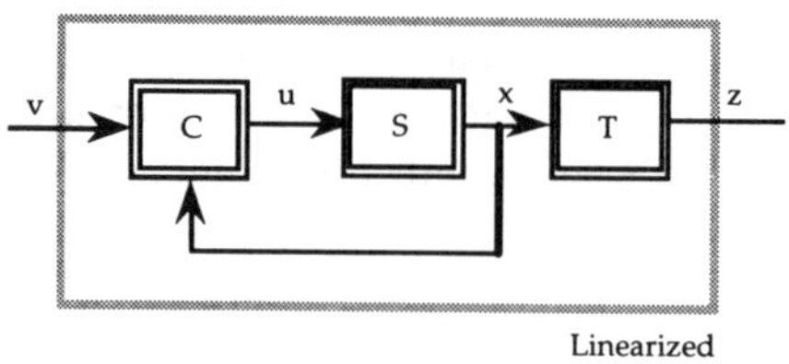

Internal linearization

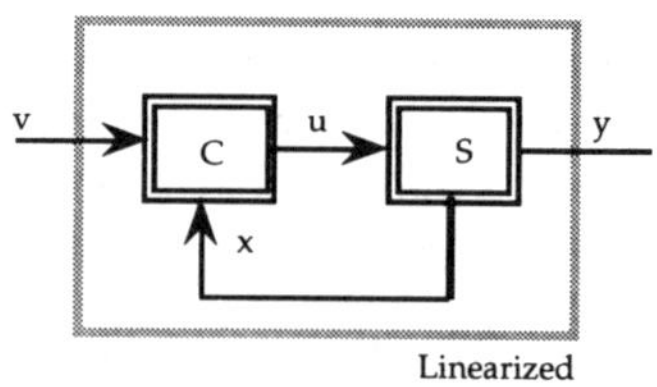

External linearization

The state x of the nonlinear system S is feedback to the controller C. In the first case, the closed-loop system output behaves linearly with respect to the new input v. In the second case, a change of state variables T demonstrates clearly a linear dependence between v and the new state z.

Many of the results presented here deal with these approaches of control via a linearization procedure since they are now fully developed and make up a coherent whole.

The first chapter shows how to pass from a linear law to a nonlinear one. A linear control law is computed from a locally linearized model at an equilibrium point; then it is changed into a nonlinear control law which is efficient around any equilibrium point.

Exact input–output linearization is dealt with in Chapter 2. This approach is rather simple at first glance but, in fact, many difficulties such as fixed mode existence may arise. These questions are clarified and tools are introduced that allow a precise analysis in order to achieve realistic control laws. Furthermore, the algorithmic point of view is developed in Chapter 3 in terms of the system graph. Connections with external decoupling are studied.

Chapter 4 presents in detail the input-state linearization. As suggested by the above figure, this approach may be considered as a search for the fictitious outputs $z = T(x)$ for an external linearization.

The discretization problem is approached in Chapter 5. How to implement a nonlinear control law? What precautions have to be taken?

Chapter 6 is devoted to the application of those control techniques that use a linearization step. We study the control of a chemical reactor, a pneumatic jack and a magnetic bearing.

Finally linearization is not always a requisite for control. The optimal control problem is studied in the final chapter and illustrates this point. Simplifications are possible and the introduction of the multiple time-scale concept allows to the separation of the dynamics. Working with dimensional reduced systems yields to solutions of the state feedback form.

Ph. Mouyon

CHAPTER 1

First-order control of nonlinear systems

PH. MOUYON

1.1 Introduction

The study of linearized models around steady state points in order to ascertain local stability is already old history. The use of these models for regulation goes back to the very beginning of automatic control. It is mainly justified by the easiness of the analysis of linear systems behavior.

Obviously, when the knowledge of a single linearized model around an operating point is not sufficient to satisfy control aims, one thinks of using several linearized models around different operating points. The set of all linearized models around all existing operating points forms what is called the *linear tangent model*. The knowledge of this model, since providing only partial information on the nonlinear system behavior, is often quite sufficient to design a control law satisfying common specifications. For example, the synthesis of flight-control laws and autopilots for aircrafts is commonly done from such linearized models, just as for rocket control laws or even regulation of numerous chemical processes.

Nevertheless, the use of several linearized models, and *a fortiori* of the tangent model, is not always an easy task. Law commutations often used are critical phases and now various alternatives based on the continuous integration of locally computed laws exist. This transition from a discrete integration (commutation) to a continuous one leads to a nonlinear control law which quicky adapts itself to operating point changes and consequently gives rise to better performances.

Within the framework of this approach of nonlinear systems control designed from the linear tangent model, which is called *control at first-order*, the fundamental notions are those of equilibrium, linearized models as well as some integration techniques on surfaces. The study of them leads us first to better specify the application field of control at first-order. We will also show why linear laws are sometimes so effective in controlling nonlinear systems. Finally, we will present an overview of control techniques at first-order. All this might well constitute the first step toward a gradual taking into account of nonlinear model complexity within the control law synthesis procedure.

It is worth noting that our aim throughout this chapter devoted to control at first-order is more to introduce new ideas in an understandable way and to throw some light on other older ones, than to show that they are well-founded in theory. Nevertheless, even a straightforward justification sometimes requires some mathematics. Here they are reduced to the minimum. Intentionally, we abandonned generality for the ease of reading. The reader inclined to go deeper into these questions is invited to refer to the works of Sontag (1987a; 1987b), Baumann and Rugh (1983; 1986), as well as Reboulet, Champetier and Mouyon (1986). A certain formalism is required in order to formulate general results, which seems not necessary here.

1.2 Basic notions

1.2.1 Nonlinear systems

1.2.1.1 State space

Generally, a system state may be described in several ways according to the set of physical variables being used. When the physical representation is changed, these variables are modified but the state remains the same. The state space M of a nonlinear system must be considered as a surface. The system state x has an intrinsic reality: it is a point on this surface. The state variables are the coordinates of this point within a surface parametrization. Physical representation changes are often nonlinear and thus the coordinates are curvilinear.

For example, let us consider a satellite. Choose any orthogonal frame which is fixed with respect to this solid. The satellite position around its gravity center (its attitude) at a time t is well defined by the rotation $R(t)$ which converts the frame at time $t = 0$, into the frame at time t. This rotation does not depend on the chosen frame: it is an intrinsic representation of the satellite orientation. Each satellite motion is comparable to a curve in the rotation space SO(3); it is in fact the application $t \to R(t)$. Various sets of coordinates are used to set up the satellite kinematic equations: Euler angles, quaternions. Satellite angular velocity is the tangent to the curve $t \to R(t)$. It is shown that there exists an antisymmetric operator $\Omega(t)$ such that:

$$\dot{R}(t) = \Omega(t) R(t) . \tag{1.1}$$

The satellite state is therefore a point on the surface SO(3) x Ant(3), where Ant(3) denotes the set of antisymmetric operators. This surface is called the *system state space.*

Generally speaking, there does not exist any global parametrization of the state space of a nonlinear system. That means that any parametrization presents singularities (think of Euler angles). Consequently, the state space is not an affine space but, in fact, what mathematicians call a manifold, generalizing the idea of surface.

1.2.1.2 State space model

If one changes the coordinate frame, the expression of kinematic and dynamic equations also changes, but system behavior remains the same. Thus, an intrinsic state space model must not depend on the frame. Rigorously speaking, we must consider that f in the expression

$$\dot{x} = f(x, u) \tag{1.2}$$

is a vector field parametrized by u, on the state manifold M. With any pair (x, u), this vector field associates a direction $f(x, u)$ tangent to the state manifold at point P with coordinates x on M. It is the direction in which the system is moving.

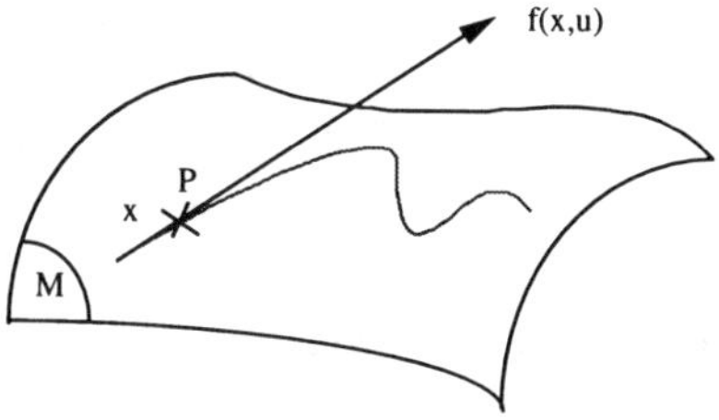

Figure 1.1: A trajectory on the state manifold.

For electrical or mechanical systems, f is a field of forces. The control u allows the modification of this field in order to obtain the desired behavior.

We still use expression (1.2) to represent the system's state model projected in a frame. In this case, f denotes the local expression of the vector field in the chosen frame and x, the state coordinates in the same frame. The state equation is a set of n ordinary nonlinear differential equations on $\mathbb{R}^n$.

1.2.2 *Linear tangent model*

1.2.2.1 The equilibrium set

Consider a nonlinear system of the general form (1.2), the set of equilibrium points $\mathcal{E}$ is the set of singular points for f, that is to say the set of points where f vanishes.

$$\mathcal{E} = \left\{ \left(x^0, u^0\right) \in M \times \mathbb{R}^m; f\left(x^0, u^0\right) = 0 \right\}. \tag{1.3}$$

Let n be the dimension of M, the equilibrium points are the solution of a set of n nonlinear equations with $n + m$ unknowns. Generally, they form an m-dimensional space, where m is the number of inputs.

What matters about control is the state value. This leads us to consider the projection of the set $\mathcal{E}$ on the state space M, denoted by $\mathcal{V}$.

$$\mathcal{V} = \left\{ x^0 \in M; \exists u^0; f\left(x^0, u^0\right) = 0 \right\}. \tag{1.4}$$

It is also generally an m-dimensional set, like $\mathcal{E}$.

We will come back to that notion of equilibrium in a more detailed manner later and we will also introduce partial and relative equilibriums.

1.2.2.2 Local linearization

In the neighborhood of an equilibrium point (x^0, u^0) of $\mathcal{E}$, we consider the approximation of the system behavior at first-order. The Taylor series expansion of f is

$$f(x, u) = f\left(x^0, u^0\right) + \frac{\partial f}{\partial x}\left(x^0, u^0\right)\left(x - x^0\right) + \frac{\partial f}{\partial u}\left(x^0, u^0\right)\left(u - u^0\right) + \sigma\left(x - x^0, u - u^0\right).$$

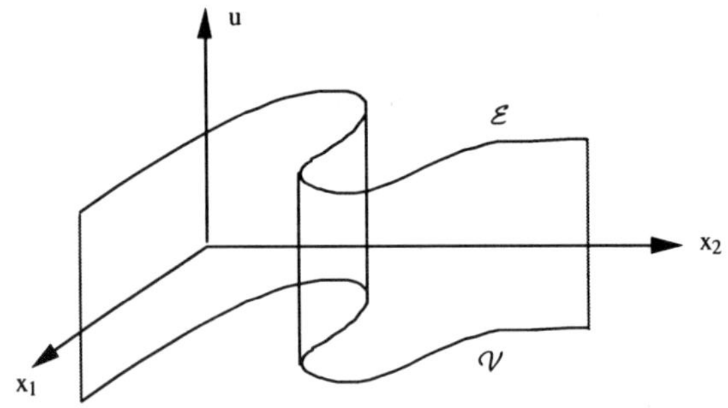

Figure 1.2: The equilibrium set and its projection.

Neglecting second order terms and setting:

$$\begin{aligned}\delta x &= x - x^0\\ \delta u &= u - u^0\end{aligned}$$

the state equation (1.2) becomes:

$$\delta \dot{x} = \frac{\partial f}{\partial x}\left(x^0, u^0\right)\delta x + \frac{\partial f}{\partial u}\left(x^0, u^0\right)\delta u. \tag{1.5}$$

It is the linearized model at point (x^0, u^0). It is a linear system. Its state space Tx^0M , is tangent to M at point x^0. Vectors $\delta\dot{x}$ and δx are tangent to M at x^0.

An important property of locally linearized models is that they intrinsically define a system if and only if linearization is done at an equilibrium point. This restriction is not always well-known because its effects appear only in very particular cases. The notion of relative equilibrium, which will be introduced later, enable us to enlarge the validity domain of local linearization.

In order to prove the above-mentioned property, consider a general system of the form (1.2). Under the coordinate's change $z = T(x)$, the state representation is modified. Let us compute how the locally linearized model is modified around any point (x^0, u^0). Linearization of the coordinate's change gives:

$$\delta z = \frac{\partial T}{\partial x}\left(x^0\right)\delta x.$$

Thus, on the new basis, the linearized model may be written as

$$\delta \dot{z} = \frac{\partial T}{\partial x}\left(x^0\right)\delta \dot{x} = \frac{\partial T}{\partial x}\left(x^0\right)\delta f\left(x, u\right).$$

Nevertheless, if the nonlinear system is first transformed and then linearized, we obtain a different result. As a matter of fact we have

$$\dot{z} = \frac{\partial T}{\partial x}\left(x\right)\dot{x} = \frac{\partial T}{\partial x}\left(x\right) f\left(x, u\right).$$

so

$$\delta \dot{z} = \delta\left(\frac{\partial T}{\partial x}\left(x\right)\right) f\left(x^0, u^0\right) + \frac{\partial T}{\partial x}\left(x^0\right)\delta f\left(x, u\right).$$

If $f(x^0, u^0) = 0$, that is to say if (x^0, u^0) is an equilibrium point, the two results are identical, but this is not true in the opposite case.

1.2.2.3 Linear tangent model: a definition

The set of linearized models around all the possible equilibrium points forms what one calls the *linear tangent model*. It is a nonlinear system, defined on $\mathcal{E}$ by:

$$\delta\dot{x} = F\left(x^0, u^0\right)\delta x + G\left(x^0, u^0\right)\delta u \tag{1.6}$$

setting

$$F\left(x^0, u^0\right) = \frac{\partial f}{\partial x}\left(x^0, u^0\right)$$
$$G\left(x^0, u^0\right) = \frac{\partial f}{\partial u}\left(x^0, u^0\right)$$

for all (x^0, u^0) satisfying $f(x^0, u^0) = 0$. One often assumes that a unique input value is associated with each equilibrium state. Then there exists a fonction φ_1 such that:

$$f\left(x^0, u^0\right) = 0 \Rightarrow u^0 = \varphi_1\left(x^0\right) \tag{1.7}$$

In this case, the projection of $\mathcal{E}$ on the state space is injective and the linear tangent model is defined on $\mathcal{V}$ by a state equation of the following form:

$$\delta\dot{x} = F\left(x^0\right)\delta x + G\left(x^0\right)\delta u \tag{1.8}$$

Some results on control at first-order, particularly those related to pseudolinearization, are much more easily explained with this special form of the linear tangent model.

1.2.2.4 Local controllability

In this chapter, we shall say that a system is locally controllable, or controllable at first-order, if its linear tangent model is controllable. We only consider nonlinear systems with this property. Note that, according to the classical terminology a system may be locally controllable without its linearized model being controllable: in that case, second-order terms provide controllability. Therefore, the controllability notion required here is quite strong. It is essential to the control synthesis proposed here, as is shown by the following counter example:

Let us return to satellite modeling. We have already stated the kinematic equation (1.1) describing the satellite attitude behavior. Physical laws yield the dynamic equation modeling the evolution of the instantaneous rotation vector with respect to an inertial frame. In the case of a satellite controlled by only two pairs of opposing gas jet actuators, it takes the following form:

$$\begin{cases} \dot{\omega}_1 &= a_1\omega_2\omega_3 + u_1 \\ \dot{\omega}_2 &= a_2\omega_1\omega_3 + u_2 \\ \dot{\omega}_3 &= a_3\omega_1\omega_2 \end{cases}$$

where a_i denote the inertial ratios and u_i the torques created by the gas jets, and ω_i the antisymmetric rotation matrix components:

$$\Omega(t) = \begin{pmatrix} 0 & \omega_3 & -\omega_2 \\ -\omega_3 & 0 & \omega_1 \\ \omega_2 & -\omega_1 & 0 \end{pmatrix}$$

This system is still locally controllable in the neighborhood of an equilibrium state with only two inputs. This noteworthy fact allows us to consider the continuation of pointing maneuvers in the presence of one faulty actuator [12]. Techniques of control at first-order are not adapted to these cases when controllability is achieved thanks to second-order terms. About a rest point ($\omega_i = 0$), the linearized model is not controllable since it can be written:

$$\begin{cases} \delta\dot{\omega}_1 = \delta u_1 \\ \delta\dot{\omega}_2 = \delta u_2 \\ \delta\dot{\omega}_3 = 0 \end{cases}$$

Rotation around the inertial axis whose actuator has broken down is uncontrollable at first-order. Fortunately, the linear tangent model very often retains enough of the controllability properties of the system. Controllability at first-order is a necessary and sufficient condition for the existence of control laws satisfying asymptotic stability with arbitrarily chosen time response. If not, system control is often a very difficult task.

1.2.2.5 Controllability structure

The controllability structure is represented by the way the inputs act on the system states. It is summed up by the controllability indices of the linear tangent model. Since this model depends on the equilibrium point, the indices are not be always constant along $\mathcal{E}$. For analytic systems, they are almost everywhere constant on each connex component of $\mathcal{E}$. But even isolated variations are annoying. Around these points, the control structure must be changed in order to avoid possible singular behaviors.

A more tricky problem arises when the controllability structure is different from, and located outside, the set of equilibrium points. In this case the use of the linear tangent model, even for a local regulation around an equilibrium point, does not make sense. This problem is encountered with some distillation column models.

A necessary condition for control at first-order to be efficient is that its linear tangent model is controllable and that its controllability structure is invariant in a neighborhood of the equilibrium set. That means that a certain structural robustness is required.

1.2.3 A pseudolinear system: the pneumatic jack

1.2.3.1 Introduction

Pneumatic jacks are commonly used in robotics. In practice, the implemented control laws are often linear ones although the modeling is nonlinear. And they are efficient! All the studies show that the contribution of a nonlinear control law is only perceptible in an extreme working condition: that is to say with a jack at long last. Moreover, according to the users, in order to improve performances one cannot do better than buying a higher quality and speedier jack. To try to take into account unknown time-varying nonlinearities by complicating the control law is hazardous.

An explanation of the efficiency of linear control laws for pneumatic jack control is the following: in spite of an obvious nonlinear modeling, the temporal behavior of these jacks is almost linear all over the commonly used domain. This will now be checked in the case of a jack with single effect and with return spring.

1.2.3.2 Modeling

We consider a very simple jack model, nevertheless it depicts all fundamental characteristics of any jack. The principle diagram is shown in figure 1.3.

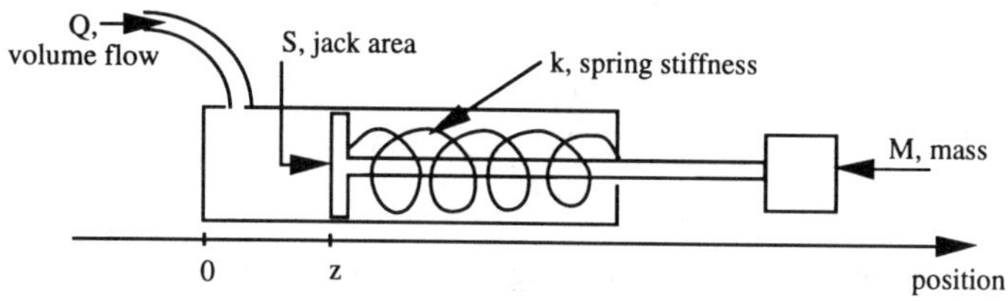

Figure 1.3: Single effect pneumatic jack.

For this system, physical laws easily yield state equations. The first law of mechanics is written as:

$$\overrightarrow{F} = M\vec{\gamma}, \qquad \text{or} \qquad M\ddot{z} = PS - kz$$

where P stands for jack chamber pressure. Supposing isothermal behavior, the ideal gas law gives:

$$\frac{PV}{n} = RT = \text{constant}, \qquad \text{or} \qquad \dot{P}V + P\dot{V} - PV\frac{\dot{n}}{n} = 0$$

where $V = Sz$ denotes the chamber volume, and n the number of moles in that chamber. Air mass conservation amounts to:

$$Q = \left(\frac{V}{n}\right)\dot{n} \qquad \text{and so} \qquad \dot{P} = -\frac{P}{V}\left(\dot{V} - Q\right)$$

We can now easily set out the state space equations of this simplified model (some assumptions have been made such as isothermal transformations, no leak and no friction).

$$\begin{cases} \dot{x}_1 &= x_2 \\ \dot{x}_2 &= -\frac{k}{M}x_1 + \frac{S}{M}x_3 \\ \dot{x}_3 &= -\frac{x_3}{x_1}\left(x_2 - \frac{u}{S}\right) \end{cases}$$

They have the form $\dot{x} = f(x, u)$ where $x_1 = z$ is the position, $x_2 = \dot{z}$ the velocity, $x_3 = P$ the pressure and $u = Q$ the volume flow.

1.2.3.3 Linearization

The set $\mathcal{E}$ of equilibrium points for the jack is derived from $f(x^0, u^0) = 0$, that is to say:

$$\begin{cases} x_2^0 &= 0 \\ kx_1^0 &= Sx_3^0 \\ u^0 &= Sx_2^0 = 0 \end{cases}$$

Therefore

$$\mathcal{E} = \left\{\left(x^0, u^0\right) = \left(x_1^0, 0, \frac{k}{S}x_1^0, 0\right); x_1^0 \in \mathbb{R}\right\}$$

We study small variations around any equilibrium point (x^0, u^0). To do this, let us introduce the linear tangent model

$$\begin{cases} \delta\dot{x}_1 &= \delta x_2 \\ \delta\dot{x}_2 &= -\dfrac{k}{M}\delta x_1 + \dfrac{S}{M}\delta x_3 \\ \delta\dot{x}_3 &= -\dfrac{x_3^0}{x_1^0}\delta\left(x_2 - \dfrac{u}{S}\right) - \delta\left(\dfrac{x_3}{x_1}\right)\cdot\left(x_2^0 - \dfrac{u^0}{S}\right) \end{cases}$$

Note that we carry out this calculus considering the"δ" operator as a derivative, formally applied to the nonlinear model. Taking into account the equilibrium, we obtain:

$$\begin{cases} \delta\dot{x}_1 = \delta x_2 \\ \delta\dot{x}_2 = -\dfrac{k}{M}\delta x_1 + \dfrac{S}{M}\delta x_3 \\ \delta\dot{x}_3 = -\dfrac{k}{S}\left(\delta x_2 - \dfrac{1}{S}\delta u\right) \end{cases}$$

As regards pneumatic jacks, the linear tangent model is a good local approximation of the nonlinear model, since it retains the system controllability structure.

1.2.3.4 Control

The most important fact is that this linear tangent model does not depend on the equilibrium point (x^0, u^0). In other words, the local behavior is, at first-order, independent of the operating point. This explains why linear control laws with constant gains are efficient.

The linearized model is not only valid in a neighborhood of an operating point, but also in a neighborhood of the whole equilibrium set $\mathcal{E}$. We call such a system pseudolinear.

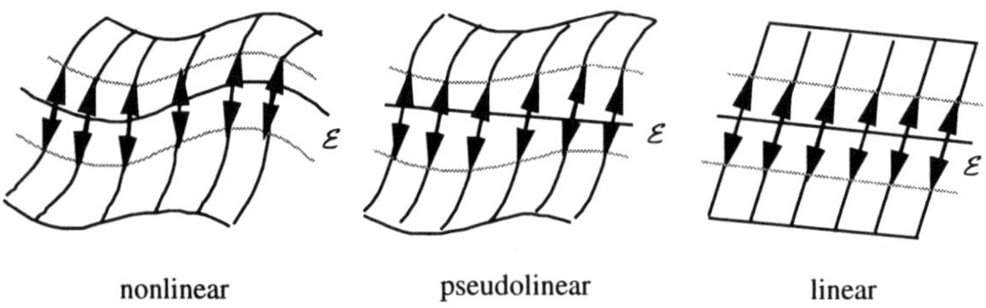

Figure 1.4: From linear to nonlinear systems.

In the above figure 1.4, double arrows represent linearized models.When the linearized model changes with the equilibrium point, the double arrow orientation also changes. In the pseudolinear case, all double arrows are parallel. A "tube" exists, containing the equilibrium set, on which a unique linear model (the linear tangent model) is a good approximation of the system behavior.

Under these conditions, one easily understands why a linear constant gain law is efficient. Define

$$\delta u = k_1\delta x_1 + k_2\delta x_2 + k_3\delta x_3$$

The linearized closed-loop dynamic is kept invariant along $\mathcal{E}$ since gains are constant. Therefore, damping ratio and response time are fixed by the three gains k_i independently of the operating point. Transition from one rest point to another will be carried out accurately provided that the dynamic of the reference rest point is slower than the linear tangent model. As a matter of fact, the jack will then be able to follow the reference. For that reason and other practical considerations, one generally uses a slope type reference.

The control law actually applied to the jack is obtained by integrating the above-mentioned linearized control law. Constant gain structure is preserved. The reference allows one to fix the equilibrium point to reach $\left(x_1^c, 0, \frac{k}{S}x_1^c, 0\right)$.

$$u = k_1\left(x_1 - x_1^c\right) + k_2 x_2 + k_3\left(x_3 - \frac{k}{S}x_1^c\right)$$

In order not to introduce static error it is essential that equilibrium conditions are taken into account when integrating the control law.

1.3 Equilibrium set

1.3.1 Structure

Under local controllability assumption, the set $\mathcal{E}$ of equilibrium points defined by (1.3) is an m-dimensional surface of $M \times \mathbb{R}^m$ (more precisely an embedded submanifold). A necessary and sufficient condition for controllability at first-order is that for any real λ

$$\operatorname{rank}\left[\frac{\partial f}{\partial x}\left(x^0, u^0\right) - \lambda I, \frac{\partial f}{\partial u}\left(x^0, u^0\right)\right] = n \tag{1.9}$$

This must be true especially with $\lambda = 0$ which proves independence of the n equations defining the equilibrium $f(x^0, u^0) = 0$.

The set $\mathcal{V}$ of possible equilibrium states (1.4) is the projection of $\mathcal{E}$ on the state space M. Singularities may appear during this projection with the consequence that $\mathcal{V}$ is not always as regular as $\mathcal{E}$.

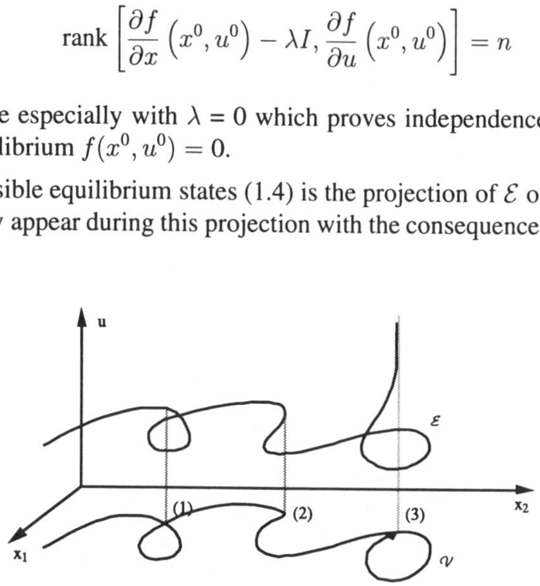

Figure 1.5: Equilibrium singularities.

Among the three kinds of singularities represented in figure 1.5, the last two appear when there exists a value of u_0 (finite in the second case, or infinite in the third case) which gives rise to rank degenerancy of the matrix $G(x^0, u^0) = \partial f / \partial u(x^0, u^0)$.Therefore, provided that controllability structure invariance property is fulfilled they do not appear. The first kind of singularities is not introduced if the projection of $\mathcal{E}$ on $\mathcal{V}$ is injective. In that case, $\mathcal{V}$ is an embedded submanifold of M.

1.3.2 Feedback invariance

We show now that the set $\mathcal{V}$ of equilibrium states is invariant under regular feedback, which emphasizes the importance of taking into account equilibrium constraints for control design.

1.3.2.1 Static state feedback

Under static state feedback $u = \phi(x, v)$, the system state equation becomes:

$$\dot{x} = g(x, v) = f(x, \phi(x, v))$$

Obviously, any equilibrium state x^0 of the closed-loop is always an equilibrium state of the open-loop system. Just take $u^0 = \phi(x^0, v^0)$ with v^0 satisfying $g(x^0, v^0) = 0$. The converse is also true if the feedback is regular on $\mathcal{E}$, that is to say there exists $\phi^\#$ such that:

$$\phi\left(x^0, \phi^\#\left(x^0, u^0\right)\right) = u^0 \qquad \text{or} \qquad \det\left|\frac{\partial \phi}{\partial v}\left(x^0, v^0\right)\right| \neq 0$$

In fact in that case to any open-loop equilibrium point x^0 corresponds the control $v^0 = \phi^\#(x^0, u^0)$ where u^0 satisfies $f(x^0, u^0) = 0$.

1.3.2.2 Dynamic state feedback

Once more it is very easy to show that closed-loop equilibrium states are open-loop equilibrium states. To establish the converse, and thus prove the invariance of $\mathcal{V}$, one needs to assume the controller to be statically right invertible on $\mathcal{E}$. Consider a controller of the form:

$$\begin{aligned} \dot{\omega} &= R(\omega, x, v) \\ u &= \phi(\omega, x, v) \end{aligned}$$

The hypothesis means that there exist $R^\#$ and $\phi^\#$ such that at any equilibrium point (x^0, u^0) of $\mathcal{E}$, we have:

$$\begin{aligned} R\left(R^\#\left(x^0, u^0\right), x^0, \phi^\#\left(x^0, u^0\right)\right) &= 0 \\ \phi\left(R^\#\left(x^0, u^0\right), x^0, \phi^\#\left(x^0, u^0\right)\right) &= u^0 \end{aligned}$$

It is therefore sufficient to choose $\omega^0 = R^\#(x^0, u^0)$ and $v^0 = \phi^\#(x^0, u^0)$ in order to show that x^0 is still an equilibrium state of the closed-loop.

1.3.3 *Partial equilibrium*

A characterization of partial equilibriums is that only a reduced number of equilibrium constraints is satisfied. An example is an aircraft in flight at cruising speed: its position obviously changes though all other state variables are constant. More generally partial equilibrium notion may be defined for systems with a "triangular" or "cascade" structure. Consider state equations of the form:

$$\begin{cases} \dot{x}_1 &= f_1(x_1, u_1) \\ \dot{x}_2 &= f_2(x_1, x_2, u_1, u_2) \end{cases}$$

If the equilibrium constraint is only satisfied for the first equation we have a partial equilibrium. This situation is closely related to multiple time-scale systems. The main interest of triangular systems is that this structure allows a multiple step control law design, with a reduced number of states and thus less complexity at each step.

The concept of partial equilibrium is quite misleading since various cascade decompositions of a given system may exist depending on the chosen state variables and varying under feedback. Let us go back to the aircraft example. A simplified model of the longitudinal flight dynamic is (George *et al.*, 1969):

$$\begin{cases} m\dot{V} &= P - \frac{1}{2}\rho S V^2 C_{xD} - mg\sin\gamma \\ mV\dot{\gamma} &= \frac{1}{2}\rho S V^2 C_L - mg\cos\gamma \\ \dot{h} &= V\sin\gamma \\ \dot{x} &= V\cos\gamma \end{cases}$$

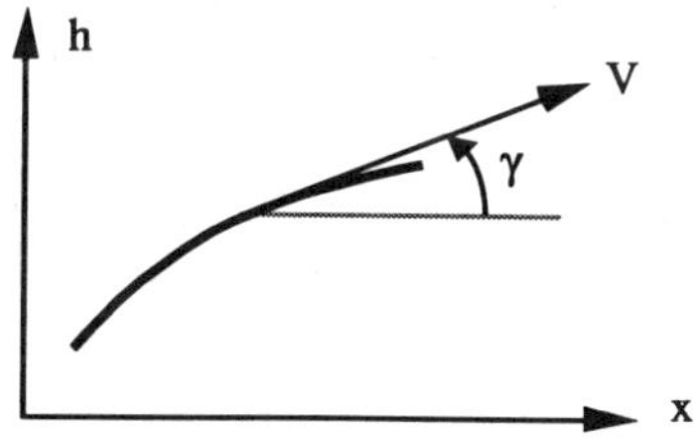

Figure 1.6: longitudinal trajectory.

where P is the thrust, γ the trajectory slope, h the altitude, ρ the atmospheric pressure, V the speed and x the position. The lifting coefficient C_L depends on the angular position of the elevator and the drag coefficient C_D is deduced from C_L.

The system is triangular since the last kinematic equation is decoupled from others. Partial equilibrium does not represent the aircraft at a standstill but the aircraft in level flight (constant speed, nul slope). The equation describing the evolution of the altitude cannot be isolated in a second subsystem because of the air volume mass ρ which depends on h, and affects the aerodynamic efforts (lift and drag) through coefficient ρV^2.

On the other hand, if thrust is used to regulate at a constant value an equivalent speed V_e defined by:

$$\rho V^2 = \rho_0 V_e^2$$

where ρ_0 is a reference volume mass, then the altitude h no longer affects the first two equations. The corresponding partial equilibrium is characterised by a nonzero constant slope γ and a speed V inversely proportional to $\sqrt{\rho}$, which increases with altitude to cancel lift diminution induced by air volume mass decrease.

This example shows that cascade subsystem decomposition may be modified by control.

1.3.4 Equilibrium along a trajectory

A common approach to the control of machines such as spacecrafts, ships, missiles and robots is a hierarchical one. First of all guidance, consists in defining a reference trajectory and computing the reference control which allows us to follow it. Then there is the piloting function: this is a regulation that controls differences by comparison with a reference.

We have seen that local linearization only makes sense in the neighborhood of equilibrium points. Therefore, it is important to show that piloting is a control around an equilibrium.

Consider a mechanical system, such as a manipulator robot. Its nominal function mode is obviously not the equilibrium which, here, is at rest. The robot must convey loads from one point to another and it should work at constant speed or, still better, in accordance with a pre-defined speed profile. The state model always has the following form:

$$\begin{cases} \dot{x}_1 = x_2 \\ \dot{x}_2 = f(x_1, x_2, u) \end{cases}$$

where positions are denoted by x_1, velocities by x_2 and applied torques and forces by u. It generally does not have a triangular form because of x_1, which explicitly appears in the second equation. This may be due to inertial variations, gravity efforts depending on robot configuration, or efforts induced by springs proportional to positions.

Consider a given reference trajectory $(x^r(t), u^r(t))$. The reference control maintains the reference state if no perturbation occurs. So the pair (x^r, u^r) satisfies the system state equation. Reference and system state equations may be written as:

$$\begin{cases} \dot{x}_1^r = x_2^r \\ \dot{x}_2^r = f(x_1^r, x_2^r, u^r) \\ \vdots \\ \Delta\dot{x}_1 = \Delta x_2 \\ \Delta\dot{x}_2 = f(x_1^r + \Delta x_1, x_2^r + \Delta x_2, u^r + \Delta u) - f(x_1^r, x_2^r, u^r) \end{cases}$$

setting $\Delta x_i = x_i - x_i^r$ and $\Delta u = u - u^r$. The equilibrium along the reference trajectory is equivalent to the second subsystem equilibrium. It is a partial equilibrium and the linear tangent model of this subsystem can be used for regulation (piloting) design. Reference variables x^r and u^r, act on this linearized model and may be viewed as time-varying parameters. The control law is then not only a function of differences between states and references but also of references themselves (Sontag, 1987).

1.4 Control at first-order

1.4.1 Generalities

1.4.1.1 Definition

A control law at first-order is designed by a two-step approach. First of all, a local control law is computed for each operating point using linear systems control theory (pole assignment, optimal control, ...) with linearized models. To achieve this stage the only required knowledge is the linear tangent model, *i.e.* the first-order dynamical system behavior. The second step consists in searching for a nonlinear control law whose linearization gives the linear local control laws again. This integration is by far the most difficult phase. At this point, one has to take into account equilibrium constraints, *i.e.* the knowledge of the static system behavior. Generally, integration will be feasible only through that restriction to the equilibrium set.

1.4.1.2 Application field

Generally speaking, the standard approach to control in engineering consists of a regulation along a preplanned reference. Different operating modes have to be distinguished according to the reference type

- regulation: the reference is known and constant,
- transition: the reference is known and piecewise constant with possible drift sections,
- reference following: the reference is the output of a known dynamical system,
- tracking: the reference is not known in advance.

Among these four control types, only the last one requires a perfect knowledge of the nonlinear model. In the other cases, an exact knowledge of the system is obviously favorable but not essential. Besides, it leads to an improvement of the closed-loop performances only if one knows how to use it, which is not always an obvious task. For instance, the conditions under which control laws via exact feedback linearization exist are very stringent and, even when applicable, the method may result in control laws that are not implementable due to the existence of actuator constraints which were necessarily ignored in the synthesis. It is a common problem in the computed torques' approach in robotics. The question is to know where and how to take into account system nonlinearities in the design procedure. We think that the knowledge of the linear tangent model associated with equilibrium characteristics is very often sufficient for satisfying control objectives.

As regards regulation, the aim is to maintain an equilibrium. Without any perturbations and measurement errors, control would be constant. System states move in a limited area around the operating point. The use of the linear tangent model is fully justified if it retains the essential controllability properties of the system. It is the case, for example, with airplane autopilots or ship steering systems.

In transition mode, the system which is regulated at an equilibrium point must be transferred to another equilibrium point where it will again be regulated (case of engine speed transition from idling speed to full speed). The most commonly required constraints on transition are: minimal response time and no overshoot according to reference. Using the linearized model to achieve transition accurately is a more difficult task since the system may deviate a long way from the equilibrium, especially when one tries to minimize the transition duration. A first solution is to restrict the reference speed variation so that it remains slower than the linearized model dynamic. Another way to tackle this problem, when transition speed is essential, is to consider partial equilibrium and the corresponding linearization.

Tracking is the most general operating mode. A reference trajectory is defined as a time function in the state space, and we oblige the system to follow it. If the reference is admissible, *i.e.* if there exists a control which solves the problem in the absence of any perturbations, equilibrium along the reference trajectory may be considered. Then the procedure of the regulation law design relies on the linearized model along that equilibrium (Sontag, 1983). However, in that case, the choice of the reference generator (whose outputs are fed as open-loop controls to the plant) is perhaps the most difficult problem.

1.4.2 Extension of local controls

When control aims fit into the use of a linearized model (regulation, transition and reference following cases) and if the linearized model variations are rather representative of the essential nonlinear behavior, then the use of control at first-order may be considered. Nevertheless the knowledge supplied by the linear tangent model and equilibrium conditions, though limited, is not easily integrated into the control design.

1.4.2.1 Robust control

One way to solve the problem is robust control. A unique linear control law with constant gain is obtained from one locally linearized model. The linearized model variations are viewed as perturbations, whereas the control law has to exhibit a certain robustness with regards to those perturbations, at least in stability.

From this point of view, the most famous example is undoubtedly provided by rotorcraft flight control laws. The linear tangent model of a helicopter is strongly coupled between longitudinal and lateral motions. Furthermore, these nonlinear coupling terms are not well-known and much depend on speed.

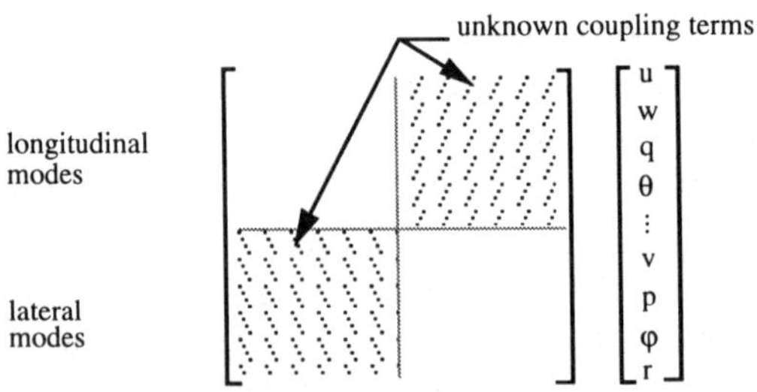

Figure 1.7: Helicopter linearized structure.

Robust control laws, such as proportional-derivative, are derived by neglecting all these nonlinearities. It is worth noting they ensure stability over the whole flight envelope (Samblancat, 1991).

$$\begin{aligned}\theta_{\text{longi}} &= k_q q + k_\theta \theta \\ \theta_{\text{lat}} &= k_p p + k_\varphi \varphi \\ \theta_{\text{tail}} &= k_r \left(r - r_c\right)\end{aligned}$$

Note also the reference term r_c which enables coordinated turns (partial equilibrium). It is the only speed-dependent term. Its expression is carried out writing the equilibrium between lateral lift and centrifugal force.

$$u r_c = g \operatorname{tg} \varphi$$

Finally, we point out that this straightforward flight control structure is not valid for aircraft control since their flight envelope is much larger in altitude as well as speed. Linearized model variations are so important that gain scheduling becomes necessary.

1.4.2.2 Gain scheduling or discrete integration

When the use of only one linearized model does not lead to an efficient control law, we must consider several local linearizations around different operating point $\left(x_i^0, u_i^0\right)$. Suppose a control law is designed for each linearized model.

$$\delta u_i = K_i \left(x_i^0, u_i^0\right) \delta x_i + \delta v$$

Then two questions are to be solved: estimate the current equilibrium point $\left(x^0, u^0\right)$, and deal with the commutations between laws according to the variations of that point.

So the control gain K is variable. It is a function of the various gains K_i computed at points $\left(x_i^0, u_i^0\right)$ but it also depends on the estimated value of the current equilibrium point $\left(x^0, u^0\right)$ compared with these points.

The gains of an aircraft flight control law change with the conventional speed V_c and the altitude h. It seems that this double dependency is necessary in order to obtain satisfactory performances in the whole flight envelope and achieve robustness according to parametric variations such as location modifications of the center of gravity. In that case, the current equilibrium point is simply measured since on-board sensors supply V_c and h.

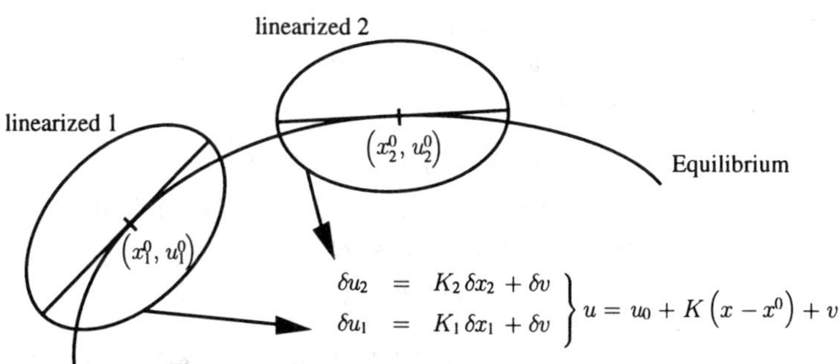

Figure 1.8: Discrete integration of local laws.

Figure 1.9 depicts the control system structure with emphasis on linearized model adaptation. It shows the two fundamental components: the current operating point estimator and the gain scheduling system.

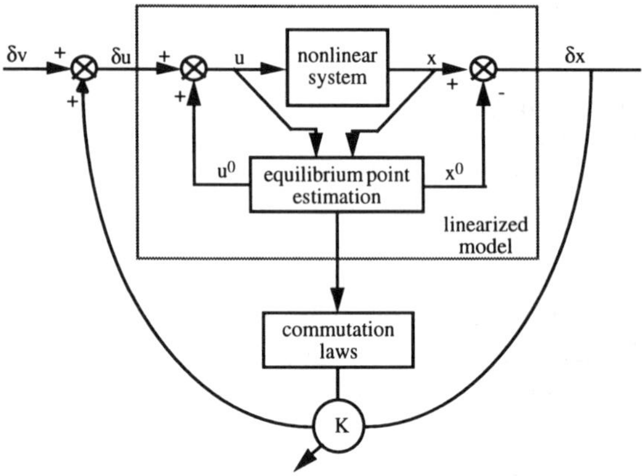

Figure 1.9: Linearized model adaptation.

Several procedures have been proposed to adjust these two components. Current equilibrium may be estimated through law pass filtering of some variables wich are representative for the equilibrium of the system. Bayesian estimation techniques have also been proposed. Several Kalman filters are used simultaneously (one for each linearized model); the one that yields the smallest *a posteriori* prediction error is assumed to correspond to the right model and thus gives the current equilibrium point. Finally, some advocate the use of the generalized likelihood maximum principle to determine the most likely linearized model. Drawbacks of the last two approaches are the implementation complexity as well as the slow time response. From that point of view, multi-model techniques should be more efficient, but their use in a closed-loop has been considered only recently.

As regards gain scheduling, note that they are generally smoothed in order to avoid control discontinuity instead of merely commuted when changing the linearized model.

1.4.2.3 Continuous integration

Now, suppose we have at our disposal not only several linearized models around some equilibrium points but also the analytical expression of the linear tangent model according to the current operating point (x^0, u^0):

$$\delta\dot{x} = F\left(x^0, u^0\right)\delta x + G\left(x^0, u^0\right)\delta u$$

Then a local control law δu can be designed which is still explicitly a function of the operating point. In the case of a state feedback we have:

$$\delta u = -K\left(x^0, u^0\right)\delta x + \delta v \tag{1.10}$$

The question to be answered is under what conditions this law may be integrated "along the set of operating points" (and how to do this) in order to obtain a nonlinear control law

$$u = \phi(x, v) \tag{1.11}$$

whose linearization around any equilibrium (x^0, u^0) restores (1.10). The main interest of that continuous integration of the local laws is as follows. Gain scheduling according to the equilibrium point is now automatic and continuous, without any need for estimation or commutation.

In fact, we have to solve first-order partial derivative equations on the equilibrium surface. This is obvious if x and u dependent terms are brought together.

$$\delta v = \delta u + K\left(x^0, u^0\right) \delta x \tag{1.12}$$

So we are searching a function $S(x, u)$ such that at any equilibrium point the following is true:

$$\left\{ \begin{array}{lcl} \dfrac{\partial S}{\partial u}\left(x^0, u^0\right) & = & I \\ \dfrac{\partial S}{\partial x}\left(x^0, u^0\right) & = & K\left(x^0, u^0\right) \end{array} \right. \tag{1.13}$$

We emphasize the fact that these equations must be satisfied only on the set $\mathcal{E}$. It is the key point to demonstrate the existence of a solution and also to find it. This solution will have the form:

$$v = S(x, u) \tag{1.14}$$

To reach actually get the required form (1.11), we also impose the additional condition

$$\det \left| \frac{\partial S}{\partial u}\left(x^0, u^0\right) \right| \neq 0$$

One recognizes the feedback regularity condition which is required for the invariance of the equilibrium state set.

If the local control law assigns the linearized model dynamic independently of the equilibrium point, then this kind of nonlinear control yields a closed-loop system whose behavior is approximately linear. The dynamic is always the same, whatever the equilibrium point, provided that the variations are small enough. The system will always respond more or less in the same way to small variations of the reference v. This is very much appreciated by the user, in the case of a human supervision. Moreover, if the reference is a ramp with a slope which is small enough regarding to the linearized model dynamic, the effective equilibrium point will "slip" along the equilibrium set.

Let us finally note that one should sometimes maintain a certain dependence of the dynamic according to the equilibrium. For instance, when regulating the heading Ψ of a ship one should respect its natural behavior, otherwise it may result in excessive control values. The trajectory of a ship is practically independent of its speed V, because the gyration radius R, though varying with the angle of the rudder, depends on the speed to a very small degree. So, for a given rudder angle R one has $V = R\dot{\Psi}$ and thus the dynamic of $\Psi(t/V)$ can be independent of the operating point but not that of $\Psi(t)$.

Depending on the type of local control laws δu (state feedback, dynamic or static feedback) and also on the structure imposed by control aims (invariance of poles, zeros, eigenvectors) it will be more or less easy to integrate these laws to get a unique nonlinear law. The remainder of this chapter is devoted to the exposition of solutions in some particular cases:

- pseudo-compensation, which yields a closed-loop dynamic invariant along the equilibrium set; and
- pseudo-linearization which completely fixes the linear tangent model.

However, first we illustrate our talk with a purely academic example which nevertheless allows a good understanding of the ins and outs of each approach to control at first-order.

1.4.3 *Example*

In order to introduce calculus apparatus relative to the different nonlinear system control laws proposed later, we consider the state system:

$$\begin{cases} \dot{x}_1 &= x_2 + u\cos x_1 \\ \dot{x}_2 &= u \end{cases} \qquad y = x_1$$

Its equilibrium set is defined by:

$$\mathcal{E} = \left\{ \left(x^0, u^0\right) \; ; \; x_2^0 = u^0 = 0 \right\} \qquad \mathcal{V} = \left\{ x^0 \; ; \; x_2^0 = 0 \right\}$$

and so, its linear tangent model is:

$$\begin{cases} \delta\dot{x}_1 &= \delta x_2 + \cos x_1^0 \delta u \\ \delta\dot{x}_2 &= \delta u \end{cases}$$

1.4.3.1 Pseudo-compensation by static state feedback

A linearized state feedback control law is designed as:

$$\delta u = -k_1\left(x_1^0\right)\delta x_1 - k_2\left(x_1^0\right)\delta x_2 + \delta v$$

to fix the closed-loop dynamic of the linear tangent model at $s^2+2\xi\omega_0 s+\omega_0^2$ (independently of the operating point), we easily obtain:

$$\begin{aligned} k_1\left(x_1^0\right) &= \omega_0^2 \\ k_2\left(x_1^0\right) &= 2\xi\omega_0 - \omega_0^2\cos x_1^0 \end{aligned}$$

Integration of the term $k_1\left(x_1^0\right)\delta x_1$ gives $k_1 x_1$ since k_1 is constant. The term $k_2\left(x_1^0\right)\delta x_2$ is also easily integrated, because x_2 is zero at the equilibrium. Thus we find the nonlinear control law:

$$u = -\omega_0^2 x_1 - \left(2\xi\omega_0 - \omega_0^2\cos x_1\right)x_2 + v$$

Comparison with linear control is very interesting. The only nonlinear term is the one in $\cos x_1$, centered around zero. So the nonlinear system under consideration may be viewed as "close" to the linear system.

$$\begin{cases} \dot{x}_1 &= x_2 \\ \dot{x}_2 &= u \end{cases}$$

However, for this system, pole assignment is merely obtained with:

$$u = -\omega_0^2 x_1 - 2\xi\omega_0 x_2 + v$$

Thus the extra "effort" required to ensure pole invariance is small (we only add $x_2\omega_0^2 \cos x_1$).

1.4.3.2 Robust control

On the other hand, imagine now that we intend to design a robust constant gain linear control law:

$$u = -k_1 x_1 - k_2 x_2 + v$$

Suppose the robustness is measured by the maximum relative variation of the damping ratio of the closed-loop linearized model. For instance, the required robustness corresponds to a 10% limited margin of tolerance for these variations. Closed-loop poles are given by $s^2 + s(k_2 + k_1 \cos x_1) + k_1 = 0$. Since k_1 is a constant, the frequency $\omega_0 = \sqrt{k_1}$ remains contant. But the damping ratio changes since it is given by:

$$\xi_0 = \frac{k_2 + k_1 \cos x_1}{2\sqrt{k_1}}$$

The maximum value of its relative variations around the central value corresponding to $\cos x_1 = 0$ is k_1/k_2. Then the robustness constraint is satisfied if $k_2 \geq 10k_1$. It follows that very often the control "effort" required to achieve robustness will be much more important than the one required for pseudo-compensation in the previous case. The damping ratio is necessarily important (we need $\xi_0 \geq 9\sqrt{k_1/2}$) and there are high gains.

1.4.3.3 Pseudo-compensation by output feedback

In the case of output feedback, we will use linear tangent transfer functions. At first-order, the output $y = x_1$ satisfies:

$$\delta y = \frac{\cos x_1^0 s + 1}{s^2} \delta u$$

The dynamic controller of the form (1.15) hereafter allows us to fix the closed-loop eigenvalues at specific values. In that way we realize a dynamic output feedback pseudo-compensation.

$$\delta u = \left(a\left(x_1^0\right) + \frac{b\left(x_1^0\right)}{s + c\left(x_1^0\right)} \right) \delta\left(y - v\right) \tag{1.15}$$

Notice that even if pole invariance is achieved, there may exist, inside the linear tangent transfer function, zeros that vary with the operating point.

A priori, the three coefficients a, b and c are equilibrium-dependent. Nevertheless, it is always possible to find the nonlinear controller whose linearized transfer function is this one. As a matter of fact it is sufficient to set:

$$\left\{ \begin{array}{rcl} \dot{z} & = & -c\left(x_1\right) z + b\left(x_1\right)\left(y - v\right) \\ u & = & a\left(x_1\right)\left(y - v\right) + z \end{array} \right. \tag{1.16}$$

Linearization of this controller will give (1.15) if z and $y - v$ are equal to zero at the equilibrium. This is true because at the equilibrium point we have:

$$u = \left(a\left(x_1^0\right) + \frac{b\left(x_1^0\right)}{c\left(x_1^0\right)} \right)(y - v)$$

The static gain of the controller appears in a factor. It is not zero, otherwise the controller pole would cancel the system zero. Now, since u is zero at the equilibrium, $y - v$ is also zero and the same holds for z. Consequently, linearization of (1.16) yields indeed the expected equation (1.15).

The principle of dynamic controller integration is quite simple. The problems we encounter are related to the use of freedom degrees regarding the choice of the controller structure. Another approach proposed by Baumann and Rugh (1986) is to use an observer at first-order.

1.4.3.4 Pseudo-linearization

The last approach of control at first-order we talk about in this chapter is pseudo-linearization. In this case, not only the dynamic must be made invariant along the equilibrium set but also the whole linear tangent model must be fixed, with respect to an appropriate coordinate frame. We are looking for a state feedback whose inverse is:

$$v = S(x, u)$$

and for a change of coordinates

$$z_1 = T_1(x) \qquad\qquad z_2 = T_2(x)$$

such that the linear tangent model becomes constant. Without loss of generality, we can choose the canonical form:

$$\begin{cases} \delta \dot{z}_1 = \delta z_2 \\ \delta \dot{z}_2 = \delta v \end{cases}$$

(Later we will set $v = -\omega_0^2 z_1 - 2\xi\omega_0 z_2 + w$ to fix the dynamic as previously). So the following must be true:

$$\begin{cases} \dfrac{\partial T_1}{\partial x}\delta \dot{x} & = & \delta T_2(x) \\ \dfrac{\partial T_2}{\partial x}\delta \dot{x} & = & \delta S(x, u) \end{cases}$$

or, equivalently

$$\begin{cases} \dfrac{\partial T_1}{\partial x_1}\left(\delta x_2 + \cos x_1^0 \delta u\right) + \dfrac{\partial T_1}{\partial x_2}\delta u & = & \delta T_2(x) \\ \dfrac{\partial T_2}{\partial x_1}\left(\delta x_2 + \cos x_1^0 \delta u\right) + \dfrac{\partial T_2}{\partial x_2}\delta u & = & \delta S(x, u) \end{cases}$$

Necessarily, the δu-dependent terms of the first equation cancel each other out, and then:

$$\begin{cases} \dfrac{\partial T_1}{\partial x_1}\cos x_1^0 + \dfrac{\partial T_1}{\partial x_2} = 0 \\ \delta T_2(x) = \dfrac{\partial T_1}{\partial x_1}\delta x_2 \\ \delta S(x,u) = \dfrac{\partial T_2}{\partial x_1}\left(\delta x_2 + \cos x_1^0 \delta u\right) + \dfrac{\partial T_2}{\partial x_2}\delta u \end{cases}$$

From the last two equations, one gets T_2 and S as functions of T_1. Then, T_1 is computed from the first equation. To solve it, we write down the characteristic system (see Appendix A, for the integration of homogeneous first-order partial derivative equations):

$$\frac{dx_1}{\cos x_1^0} = dx_2$$

This equation only needs to be verified at the equilibrium (x_2 nul) and so a solution is:

$$x_1 - x_2 \cos x_1 = \text{cste}.$$

Now, set

$$T_1(x) = x_1 - x_2 \cos x_1$$

and other transformations are easily deduced:

$$\begin{aligned} T_2(x) &= x_2 = z_2 \\ S(x,u) &= u = v \end{aligned}$$

If the dynamic is chosen fixed as before

$$u = -\omega_0^2(x_1 - x_2 cos x_1) - 2\xi\omega_0 x_2 + w$$

In this particular case, we find again the same law as for pseudo-compensation by state feedback.

1.5 Some methods

Depending on the goals of the control at first-order (invariance of poles or of the whole linear tangent model) and also depending on the structure used (state or output feedback, static or dynamic feedback, change of coordinates), the continuous integration techniques of the local laws will differ. We now present the procedure in different cases.

1.5.1 Pseudo-compensation by state feedback

The control design that consists in fixing the linearized model poles to specified values independent of the equilibrium point is called pseudo-compensation. In that case, stability is achieved during slow transitions (regarding this dynamic).

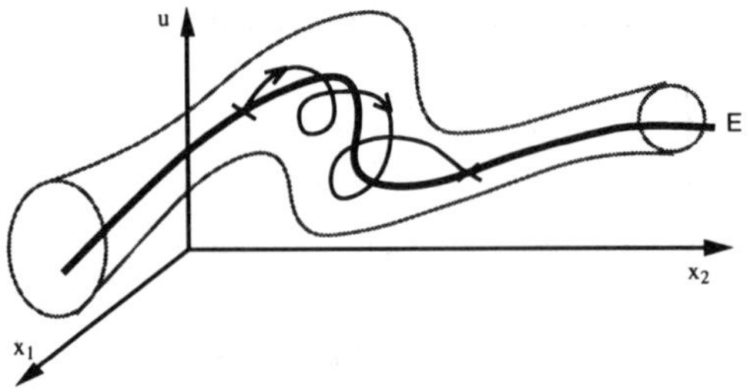

Figure 1.10: Stability domain around $\mathcal{E}$.

There exists a stability domain around $\mathcal{E}$ (shown with a "tube" in figure 1.10) and transitions are trajectories from one point of $\mathcal{E}$ to another, that do not leave this domain.

Here, we are interested in the case where the local laws that achieve the pole assignment are static state feedbacks. Successively, we examine the single-input case and the multi-input case.

1.5.1.1 Single-input systems

First of all, we check that the linearized model of such systems only depends on x^0. If local controllability is true then the gradient $\partial f/\partial u(x^0, u^0)$ is nonzero. So we can state u^0 as a function of x^0, at the equilibrium. We have established (1.7) which is recalled here

$$u^0 = \varphi_1(x_0)$$

Obviously, the linear tangent model (1.8) still depends only on x^0.

$$\delta\dot{x} = F\left(x^0\right)\delta x + G\left(x^0\right)\delta u$$

The local control law has the form:

$$\delta u = -K\left(x^0\right)\delta x + \delta v \tag{1.17}$$

We now show that there always exists a function $\phi(x)$ such that the nonlinear control law

$$u = \phi(x) + v$$

carries out the integration of the local laws (1.17) above. The demonstration is a constructive one.

We begin by "straightening up" the equilibrium set. That is to say, we search a coordinate frame $(s_1, s_2, \ldots, s_n)$ such that the equilibrium corresponds to $s_2 = \cdots = s_n = 0$. The independency of equilibrium equations (1.9) implies there exist, besides φ_1 , $n-1$ other functions $\varphi_i(x)$ such that $\mathcal{E}$ is defined by:

$$\varphi_1\left(x^0\right) = u^0 \qquad \varphi_2\left(x^0\right) = \cdots = \varphi_n\left(x^0\right) = 0$$

Then, the coordinate change[1]

$$s_i = \varphi_i(x) \qquad i = 1, \ldots, n$$

[1] If φ_1 is zero, another function independent of the $n-1$ other functions must be chosen. The remainder of the proof is the same.

achieves the straightening up of $\mathcal{V}$ since within this frame it is given by $s_2 = \cdots = s_n = 0$. Let Ψ be the inverse of φ. With the new coordinates the local control laws are written as:

$$\delta u = -K\left(\psi\left(s^0\right)\right)\frac{\partial \Psi}{\partial s}\left(s^0\right)\delta s + \delta v$$

this expression has the form:

$$\delta u = \alpha\left(s_1^0\right)\delta s + \delta v = \sum_{i=1}^{n}\alpha_i\left(s_1^0\right)\delta s_i + \delta v$$

since only the first coordinate $s - 1$ is nonzero at the equilibrium.Then, the integration is straightforward and

$$u = \int_0^{s_1}\alpha_1(\tau)\,\mathrm{d}\tau + \alpha_2(s_1)\,s_2 + \cdots + \alpha_n(s_1)\,s_n + v$$

because if we linearize the undesirable terms $\dot{\alpha}_i(s_1)\,s_i\delta s_1$ are all zero at the equilibrium. So pseudo-compensation by state feedback is always possible.

When faced with a difficult integration, one can introduce a coefficient $R(x^0)$ modifying the static gain

$$\delta u = -K(x^0)\delta x + R(x^0)\delta v \tag{1.18}$$

This extra degree of freedom may be used as an integrating factor for the differential form $K(x^0)\delta x + \delta u$.

Example 1

The approach is applied to the control of an inverted pendulum. This system is made up of an inverted pendulum put on a cart moving along a track. The control is the force applied to the cart.

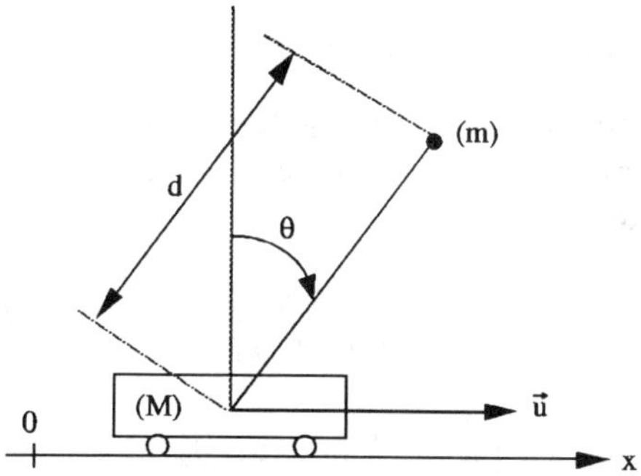

Figure 1.11: inverted pendulum.

The state equations modeling the behavior of the pendulum are

$$\begin{cases} \dot{x}_1 &= x_2 \\ \dot{x}_2 &= \dfrac{1}{d\left(1 - a\cos^2 x_1\right)}\left(-ad\sin x_1\cos x_1 x_2^2 + g\sin x_1 - \dfrac{a}{m}\cos x_1 u\right) \end{cases}$$

with $x_1 = \theta$ and $a = m/(m+M)$. M is the cart mass, m the pendulum mass and d its length. The equilibrium points are given by:

$$x_2^0 = 0 \qquad \text{and} \qquad u^0 = \frac{mg}{a}\operatorname{tg} x_1^0$$

The linear tangent model can be parametrized in terms of x_1^0, and may be written as:

$$\begin{cases} \delta\dot{x}_1 &= \delta x_2 \\ \delta\dot{x}_2 &= \dfrac{1}{d\left(1 - a\cos^2 x_1^0\right)}\left(\dfrac{g}{\cos x_1^0}\delta x_1 - \dfrac{a}{m}\cos x_1^0 \delta u\right) \end{cases}$$

bringing together δx_1-dependent terms.

In closed-loop, with the following local control law

$$\delta u = k_1 \delta x_1 + k_2 \delta x_2 + r\delta v$$

the linear tangent model poles satisfy

$$s^2 + s\frac{a\cos x_1^0 k_2}{md\left(1 - a\cos^2 x_1^0\right)} - \frac{1}{d\left(1 - a\cos^2 x_1^0\right)}\left(\frac{g}{\cos x_1^0} - \frac{a\cos x_1^0 k_1}{m}\right) = 0$$

To fix the characteristic polynomial at $s^2 + 2\xi\omega_0 s + \omega_0^2$ one can choose:

$$\begin{aligned} k_2 &= \frac{2\xi\omega_0 md\left(1 - a\cos^2 x_1^0\right)}{a\cos x_1^0} \\ k_1 &= \frac{m\,d\,\omega_0^2}{a\cos x_1^0} - m\,d\,\omega_0^2\cos x_1^0 + \frac{mg}{a\cos^2 x_1^0} \end{aligned}$$

The integration of

$$\delta v = \frac{1}{r}\left(\delta u - k_1\delta x_1 - k_2\delta x_2\right)$$

is done one term after the other. The last one is quite easy to integrate since x_2 is zero at the equilibrium. Integration gives $-k_2 x_2/r$. For the δu-dependent term, we introduce the equilibrium condition so as to obtain:

$$\delta v = \frac{1}{r}\left[\delta\left(u - \frac{mg}{a}\operatorname{tg} x_1\right) + \left(\frac{mg}{a\cos^2 x_1^0} - k_1\right)\delta x_1 - k_2\delta x_2\right]$$

Then the first term is integrated in $\frac{1}{r}\left(u - \frac{mg}{a}\operatorname{tg} x_1\right)$ since $u^0 - \frac{mg}{a}\operatorname{tg} x_1^0 = 0$. It remains to integrate

$$\frac{1}{r}\left(m\,d\,\omega_0^2\cos x_1^0 - \frac{m\,d\,\omega_0^2}{a\cos x_1^0}\right)\delta x_1.$$

To facilitate this last calculation, we choose the integrating factor r equal to $\cos x_1^0$ rather than one. In the end, we obtain

$$v = \left(u - \frac{mg}{a}\operatorname{tg} x_1\right)/\cos x_1 + m\,d\,\omega_0^2\left(x_1 - \frac{1}{a}\operatorname{tg} x_1\right) - \frac{2\xi\omega_0 m\,d\left(1 - a\cos^2 x_1\right)}{a\cos^2 x_1}x_2$$

This equation allows the computation of u as a function of v and the state. Another solution may be found by choosing the integrating factor equal to:

$$r = \frac{a\cos^2 x_1^0 - 1}{a\cos x_1^0},$$

In that case, the δx_1-dependent term becomes $m\,d\omega_0^2\delta x_1$ and may be even more easily integrated. Then the corresponding control law is defined by

$$v = \frac{a\cos x_1}{a\cos^2 x_1 - 1}\left(u - \frac{mg}{a}\operatorname{tg} x_1\right) + m\,d\omega_0^2 x_1 + 2\xi\omega_0 m\,dx_2$$

A straightforward calculation shows that this law fixes the dynamic independently of the equilibrium point but also pseudo-linearizes the system, *i.e.* its linearized model becomes entirely independent of the equilibrium.

1.5.1.2 Multi-input systems

The case of multi-input systems is somewhat more difficult. We will not go into details since that is not in the spirit of this didactic exposition. The reader who wants to examine the question more in details is invited to study Mouyon, 1986. Generally speaking[2] local control laws are always integrable, provided that an input matrix is introduced. We must consider the most general form of regular linear state feedback, that uses an invertible matrix R.

$$R\left(x^0, u^0\right)\delta v = \delta u + K\left(x^0, u^0\right)\delta x \tag{1.19}$$

The gain K is designed to satisfy control objectives, but the input matrix R is free. Several degrees of freedom are introduced by that means. They are used as integrating factors to solve (1.19) and find the nonlinear control law.

$$v = S(x, u) \tag{1.20}$$

It is always possible to compute u in terms of v and x since

$$\frac{\partial S}{\partial u}\left(x^0, u^0\right) = R^{-1}\left(x^0, u^0\right)$$

is invertible

As for the single-input case, the computation of S needs to straighten up the equilibrium set $\mathcal{V}$, and this may be done thanks to assumptions of local controllability and linearized model structure invariance.

1.5.2 Pseudo-compensation by output feedback

1.5.2.1 Observation at first-order

Design of output feedback control may be done in a straightforward manner or through the introduction of an observer that estimates the state. The principle of this latter approach

[2]In the very special case where at least two poles must be assigned at the origin, unintegrable laws may exist.

is adaptable from the linear to the nonlinear case at first-order. So consider a nonlinear system

$$\begin{cases} \dot{x} &= f(x,u) \\ y &= h(x) \end{cases}$$

and its linear tangent model

$$\begin{cases} \delta\dot{x} &= F\left(x^0,u^0\right)\delta x + G\left(x^0,u^0\right)\delta u \\ \delta y &= H\left(x^0\right)\delta x \end{cases}$$

A linearized observer is chosen in the form

$$\delta\dot{\hat{x}} = F\left(x^0,u^0\right)\delta\hat{x} + G\left(x^0,u^0\right)\delta u + L\left(x^0,u^0\right)\left(\delta y - H\left(x^0\right)\delta\hat{x}\right)$$

If the gain L ensures observer convergence around any equilibrium point, then we can subsitute x^0 by $\hat{x}^0$. Since the innovation is equal to zero at the equilibrium, integration of the linearized observer is easily done and results in:

$$\dot{\hat{x}} = f(\hat{x},u) + L(x,u)(y - h(\hat{x}))$$

or rather:

$$\dot{\hat{x}} = f(\hat{x},u) + L(\hat{x},u)(y - h(\hat{x}))$$

This nonlinear observer, called *observer at first-order*, has a classical structure with nevertheless an observer gain which is a function of the state and the input. It appears that the integration guidelines are, at least formally, much simpler in that case than for state feedback design. As a matter of fact, we have seen that the control laws do not keep the structure of a product between gain and state.

1.5.2.2 Parametrization of tangent transfer functions

Design of output feedback in the frequency domain first requires the computation of the tangent transfer function:

$$\delta y/\delta u = H(sI - F)^{-1}G = S\left(x^0,u^0\right)(s)$$

Secondly, a dynamic controller is designed, which provides accurate responses about any operating point. Generally, this controller will also depend on the equilibrium point; it will be denoted by $C(x^0,u^0)(s)$. The question now is to find a controller (certainly nonlinear) that integrates all these local laws. We restrict ourselves to the case, simple but nevertheless demonstrative of the main ideas of the approach, where the system is single-input with a controllable and observable linear tangent model. The integration process requires the parametrization of the system tangent tranfer function S as a function only of the measured variables. It is a trivial matter to see that S can be parametrized almost everywhere either in terms of u^0 or y^0. If the linearized model has no pole at the origin then F is full rank n. Then the implicit function theorem states that the application $x \to f(x,u)$ admits an inverse. Now $f(x,u)$ is equal to zero at the equilibrium and thus we can express x^0 as a function of u^0. In other words, if there is no true integration in the system then the equilibrium state is determined by the input value at equilibrium. Obviously, in that case the tangent transfer function may be parametrized by u^0 only. Conversely, if the system has a pole at the origin then there is no zero at the origin (otherwise the pole would be uncontrollable or unobservable). So

$$\text{rank}\begin{pmatrix} F & G \\ H & 0 \end{pmatrix} = n+1$$

and the application $x \rightarrow [f(x,u), h(x)]^T$ has an inverse. At the equilibrium, x^0 is a function of y^0 and u^0. If a perturbation acts on the system, some arguments show that parametrization by u^0 and y^0 still exists if this perturbation is statically detectable, *i.e.* the linearized transfer function between the perturbation and the output has no zero at the origin.

1.5.2.3 Dynamic controller

The tangent transfer function is parametrized by measured variables u^0 and y^0. It is the same for the tangent controller $C(u^0, y^0)(s)$. Suppose we use a unitary feedback structure. The nonlinear controller to be found is denoted C.

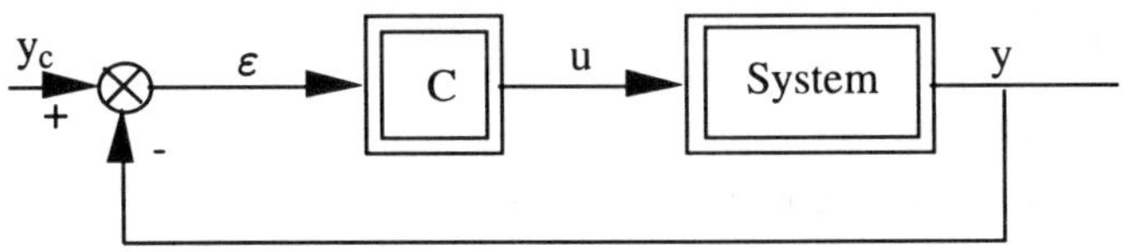

Figure 1.12: pseudo compensation by output feedback.

Assume that an integral control is used within the tangent controller. We now show that the integration is quite straightforward. $C(u^0, y^0)(s)$ is expressed in the form:

$$\frac{\delta u}{\delta \varepsilon} = k_1 + k_2 \frac{1 + \beta_1 s + \cdots + \beta_{n-1} s^{n-1}}{\alpha_1 s + \cdots + \alpha_{n-1} s^{n-1} + s^n}$$

An equivalent state space model is (controllability canonical form):

$$\begin{cases} \delta \dot{z}_1 &= \delta z_2 \\ \vdots & \\ \delta \dot{z}_{n-1} &= \delta z_n \\ \delta \dot{z}_n &= -\alpha_1 \left(u^0, y^0\right) \delta z_2 - \cdots - \alpha_{n-1}\left(u^0, y^0\right) \delta z_n + k_2 \left(u^0, y^0\right) \delta \varepsilon \end{cases}$$

$$\delta u = \delta z_1 + \beta_1 \left(u^0, y^0\right) \delta z_2 + \cdots + \beta_{n-1}\left(u^0, y^0\right) \delta z_n + k_1 \left(u^0, y^0\right) \delta \varepsilon$$

At the equilibrium we will have $z_2 = \cdots = z_n = 0$, for C. We only need to write:

$$\begin{cases} \dot{z}_1 &= z_2 \\ \vdots & \\ \dot{z}_{n-1} &= z_n \\ \dot{z}_n &= -\alpha_1 (u, y) z_2 - \cdots - \alpha_{n-1} (u, y) z_n + k_2 (u, y) (y - y_c) \end{cases}$$

$$u = z_1 + \beta_1 (u, y) z_2 + \cdots + \beta_{n-1} (u, y) z_n + k_1 (u, y) (y - y_c)$$

Note that the integral effect is maintained since at equilibrium we still have $y = y_c$.

If there is no integral term in the linearized controller, the integration is more complicated. When the controller is parametrized by u^0 and proper, *i.e.*

$$\frac{\delta u}{\delta \varepsilon} = k_1 \frac{1 + \beta_1 s + \cdots + \beta_{n-1} s^{n-1}}{\alpha_0 + \alpha_1 s + \cdots + \alpha_{n-1} s^{n-1} + s^n}$$

we write the state space model as:

$$\begin{cases} \delta \dot{z}_1 &= \delta z_2 \\ \vdots & \\ \delta \dot{z}_{n-1} &= \delta z_n \\ \delta \dot{z}_n &= -\alpha_0 \left(u^0\right) \delta z_1 - \cdots - \alpha_{n-1} \left(u^0\right) \delta z_n + k_1 \left(u^0\right) \delta \varepsilon \end{cases}$$

$$\delta u = \delta z_1 + \beta_1 \left(u^0\right) \delta z_2 + \cdots + \beta_{n-1} \left(u^0\right) \delta z_n$$

The nonlinear controller must have the form:

$$\begin{cases} \dot{z}_1 &= z_2 \\ \vdots & \\ \dot{z}_{n-1} &= z_n \\ \dot{z}_n &= -\Psi(z_1) - \alpha_1(z_1) z_2 - \cdots - \alpha_{n-1}(z_1) z_n + k_1(z_1) \varepsilon \end{cases}$$

$$u = z_1 + \beta_1(z_1) z_2 + \cdots + \beta_{n-1}(z_1) z_n$$

where Ψ is to be defined such that the linearization around any equilibrium point reproduces the previous linearized model. Ψ must satisfy

$$-\Psi'(z_1) + k_1'(z_1) \varepsilon = -\alpha_0(z_1)$$

Since $\varepsilon = \Psi(z_1)/k_1(z_1)$ at the equilibrium, this condition becomes $(\Psi/k_1)' = \alpha_0/k - 1$, and it is fulfilled if we choose:

$$\Psi(z_1) = k_1(z_1) \int \alpha_0(z_1)/k_1(z_1)\, \mathrm{d}z_1$$

Notice that z_1 is equal to u at the equilibrium.

We have presented a few solutions to the integration of tangent controllers in some particular simple cases. The general case seems to be very difficult to solve since the controller equilibrium, along which integration should be done, is not known *a priori*. We now give a realistic example of the use of this approach to the speed control of an asynchronous induction machine.

Example 2

An asynchronous induction machine is composed of a stator whose coils induce a rotating electromagnetic field, and of a rotor which tries to follow that field in order to minimize flux variations (according to the law of Lenz). Because of energy loss (a torque is produced), the rotor is not able to revolve as fast as the field (the engine is asynchronous).

A state space modeling of such a system, accurate over a quite large operating domain, is (De Fornel *et al.*, 1983):

$$\begin{cases} \dot{x}_1 &= -\frac{x_1}{\tau} + (u_1 - x_3)\, x_2 \\ \dot{x}_2 &= -\frac{x_2}{\tau} - (u_1 - x_3)\, x_1 + \frac{k}{\tau} u_2 \\ \dot{x}_3 &= -\frac{x_3}{T} + K\,(g u_2 x_1 - C_r) \end{cases}$$

τ, k, T, K and g are constant. u_1 denotes the angular velocity of the electromagnetic field, u_2 the strength of that field.

The first two equations describe the behavior of the flux ϕ through the rotor: x_1 and x_2 are the components of the flux

$$\phi = x_1^2 + x_2^2$$

The third equation shows the torques acting on the rotor: x_3 is the angular velocity of the rotor, *i.e.* the engine speed.

The resistive torque applied to the machine is denoted C_r. The control aims are the regulation of the flux to a constant reference ϕ_c, and piloting the speed (transitions from one speed to another are often requested). The states x_1 and x_2 are measured (they depend on current characteristics in the stator coils) but x_3 is generally not measured since speed measurement of a rotating piece is expensive and tricky.

It is easy to regulate the flux since it satisfies the differential equation

$$\dot{\phi} = -\frac{2}{\tau}\phi + \frac{2k}{\tau} u_2 x_2$$

The time constant $(\tau/2)$ is often sufficiently fast and it is not necessary to accelerate the flux dynamic. We just choose:

$$\dot{\phi} = -\frac{2}{\tau}(\phi - \phi_c)$$

which leads to:

$$u_2 = \frac{\phi_c}{k x_2}$$

and the input-output linearization of the flux dynamic is realized. The most difficult problem is that of the speed control. Let us recall that this speed is not measured. A double loop control structure is adopted. The first loop is fast and will control the electromagnetic torque

$$C = g u_2 x_1$$

The second loop is slower and will control the speed. This structure has two advantages. On the one hand, the control design is simplified, on the other hand, the inner loop provides an estimation of the speed which is essential to the outer loop. As a matter of fact, we have:

$$C = \frac{g \phi_c}{k} \frac{x_1}{x_2}$$

and thus, if ϕ and C are maintained constant, x_1 and x_2 are also constant. Then the equilibrium of C allows the estimation of x_3 by setting:

$$\hat{x}_3 = u_1 - \frac{1}{\tau}\frac{k}{g\phi_c} C$$

Calculus will be further simplified if we suppose ϕ equal to the constant reference ϕ_c. This means that we consider a partial equilibrium of the system which is legitimated by the fact that u_2 maintains this equilibrium. The speed return term will be neglected in the torque equation.

Let us see how to control the electromagnetic torque. Its dynamic is described by:

$$\dot{C} = \frac{g\phi_c}{kx_2^2}\left[(u_1 - x_3)\,\phi_c - \frac{k}{\tau g}C\right]$$

Now x_3 satisfies approximatively:

$$\dot{x}_3 = K\,(C - C_r)$$

Then the tangent linear transfer functions may be written as:

$$\begin{cases} s\,\delta C &= \dfrac{g\phi_c}{kx_2^2}\left[\phi_c\,\delta u_1 - \phi_c\,\delta x_3 - \dfrac{k}{\tau g}\,\delta C\right] \\ s\,\delta x_3 &= K\,[\delta C - \delta C_r] \end{cases}$$

that is:

$$\delta C = \frac{1}{as^2 + bs + K}\,(s\,\delta u_1 + K\delta C_r)$$

with

$$a = \frac{kx_2^2}{g\phi_c^2} \qquad \text{and} \qquad b = \frac{k}{\tau g\phi_c}$$

The linearized dynamic is fixed using a proportional integral controller, with varying gains.

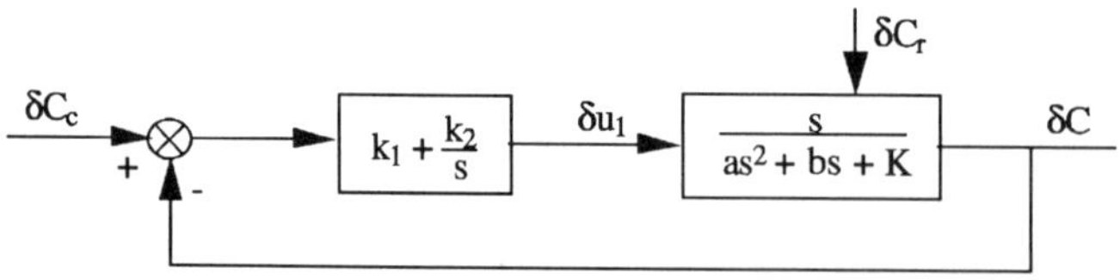

Figure 1.13: Electromagnetic torque control.

Since coefficients a and b depend on ϕ_c and x_2, it will be the same for k_1 and k_2. To obtain the closed-loop dynamic equal to $s^2 + 2\xi\omega_0 s + \omega_0^2 = 0$ it is sufficient to pose:

$$\begin{aligned} k_1\,(\phi_c, x_2) &= 2\xi\omega_0\frac{kx_2^2}{g\phi_c^2} - \frac{k}{\tau g\phi_c} \\ k_2\,(\phi_c, x_2) &= \omega_0^2\frac{kx_2^2}{g\phi_c^2} - K \end{aligned}$$

The linear tangent controller is then

$$\begin{cases} \delta\dot{z}_1 &= k_2\,\delta\,(C_c - C) \\ \delta u_1 &= k_1\,\delta\,(C_c - C) + \delta z_1 \end{cases}$$

Though k_1 and k_2 depend on x_1 and x_2, it can be integrated in a straightforward manner, since $C = C_c$ at the equilibrium. We get:

$$\begin{cases} \dot{z}_1 &= k_2(\phi_c, x_2) \cdot \left(C_c - \dfrac{g\phi_c}{k} \dfrac{x_1}{x_2} \right) \\ u_1 &= k_1(\phi_c, x_2) \cdot \left(C_c - \dfrac{g\phi_c}{k} \dfrac{x_1}{x_2} \right) + z_1 \end{cases}$$

The torque control is assumed to be very fast, and thus C may be assimilated to C_c. Under that assumption $\hat{x}_3$ is equal to x_3. As the torque equation is linear, a proportional integral controller with a constant gain will be used for speed control.

This control design has been completely validated by simulation experiments. It provides an excellent invariance of time responses both for acceleration and deceleration as for large variations of the load. Commonly used linear control laws do not lead to such a symmetry of responses around an equilibrium point (positive and negative load variations) nor to invariance of the response over a wide operating range.

1.5.3 Pseudo-linearization

1.5.3.1 Principle

To complete this survey about control at first-order, we now present the pseudo-linearization technique which approximately linearizes the state equation. The aim here is to obtain invariance, along the equilibrium set, of the whole linear tangent model, and not only invariance of its poles. To achieve this ambitious goal, a coordinate change

$$z = T(x) \tag{1.21}$$

may be used, in addition to the state feedback (1.20):

$$v = S(x, u)$$

The linear tangent model of the closed-loop system expressed in the new coordinates must be independent of the equilibrium point. It will have the form:

$$\delta\dot{z} = A\,\delta z + B\,\delta v$$

with A and B constant. A pseudo-linearized system is still nonlinear but its non linearities are of the first-order around any equilibrium point; its state equation in the (z, v) space may be written as:

$$\dot{z} = Az + Bv + \varepsilon(z, v)$$

with

$$\varepsilon\left(z^0, v^0\right) = 0 \qquad \text{and} \qquad d\varepsilon\left(z^0, v^0\right) = 0$$

Consequently, its behavior is quite well approximated by the linear terms if one remains close enough to the equilibrium.

While the pseudo-linearization is to be regarded as a linearization in a neighborhood of a surface, the exact input-state linearization by diffeomorphism and feedback is more ambitious since it attemps to obtain a linear system

$$\dot{z} = Az + Bv$$

Thus stability of the closed-loop is guaranteed in the whole state space and tracking is possible. However, assumptions under which exact input-state linearization may be applied are much more restrictive than for pseudo-linearization. Law complexity is also often increased in the latter case.

It has been shown that any nonlinear system satisfying the conditions of the local controllability and structural robustness is pseudo-linearizable if the number of inputs does not exceed two. Beyond this, a certain vector field involutivity condition is required that is not generically satisfied, as is always the case for exact input-state linearization. In that case, state feedback pseudo-linearization cannot be solved. Nevertheless, there always exists a solution if dynamic state feedback is available.

Finally, the main drawbacks of this approach are also those of the exact input state linearization: linearized state variables have no physical meaning, range and dynamic of the input is not controlled, laws are often complicated, singularities may exist. Its main advantage is, when available, a wider validity domain allowing tracking.

1.5.3.2 Single-input systems

We show in this simple case how to compute the transformations S and T. The local controllability assumption permits us to express u^0 as a function of x^0, and the linear tangent model may be written in the form (1.8) recalled below

$$\delta\dot{x} = F\left(x^0\right)\delta x + G\left(x^0\right)\delta u$$

First, we choose a controllability canonical form for the pair (A, B). This is not restrictive since by using other transformations it will be possible to get any matrix A and B. Thus we want:

$$\left\{\begin{array}{lcl} \delta\dot{z}_1 & = & \delta z_2 \\ \vdots & & \\ \delta\dot{z}_{n-1} & = & \delta z_n \\ \delta\dot{z}_n & = & \delta v \end{array}\right. \tag{1.22}$$

That is

$$\left\{\begin{array}{lcl} \dfrac{\partial T_1}{\partial x}(F\delta x + G\delta u) & = & \dfrac{\partial T_2}{\partial x}\delta x \\ \vdots & & \\ \dfrac{\partial T_{n-1}}{\partial x}(F\delta x + G\delta u) & = & \dfrac{\partial T_n}{\partial x}\delta x \\ \dfrac{\partial T_n}{\partial x}(F\delta x + G\delta u) & = & \dfrac{\partial S}{\partial x}\delta x + \dfrac{\partial S}{\partial u}\delta u \end{array}\right.$$

This must be true whatever δx and δu are, and thus:

$$\left\{\begin{array}{lll} \frac{\partial T_2}{\partial x} & = & \frac{\partial T_1}{\partial x} F \\ \vdots & & \\ \frac{\partial T_n}{\partial x} & = & \frac{\partial T_{n-1}}{\partial x} F \\ \frac{\partial S}{\partial x} & = & \frac{\partial T_n}{\partial x} F \end{array}\right. \quad \text{and} \quad \left\{\begin{array}{lll} \frac{\partial T_1}{\partial x} G & = & 0 \\ \vdots & & \\ \frac{\partial T_{n-1}}{\partial x} G & = & 0 \\ \frac{\partial T_n}{\partial x} G & = & \frac{\partial S}{\partial u} \end{array}\right.$$

We obtain a set of $n-1$ first-order homogeneous partial differential equations that must be solved for T_1 at equilibrium.

$$\frac{\partial T_1}{\partial x}\left[G\left(x^0\right), F\left(x^0\right) G\left(x^0\right), \ldots, F^{n-2}\left(x^0\right) G\left(x^0\right)\right]=0 \tag{1.23}$$

Other equations give gradients of T_i for $i=2,\ldots,n$ always at equilibrium:

$$\frac{\partial T_i}{\partial x}=\frac{\partial T_1}{\partial x} F^{i-1}\left(x^0\right) \qquad i=2,\ldots,n \tag{1.24}$$

and the last one fixes the gradient of S:

$$\left(\frac{\partial S}{\partial x}, \frac{\partial S}{\partial u}\right)=\frac{\partial T_1}{\partial x} F^{n-1}\left(x^0\right)\left(F\left(x^0\right), G\left(x^0\right)\right) \tag{1.25}$$

Let us note that by (1.23) the gradient direction of T_1 is fixed. As a matter of fact, according to the local controllability assumption $G, FG, \ldots, F^{n-1}G$ are independent. Then the rank of $[G, FG, \ldots, F^{n-2}G]$ is $n-1$. Denote E as that direction. Solving (1.23) is equivalent to finding k such that

$$\frac{\partial T_1}{\partial x}=k\left(x^0\right) E\left(x^0\right) \tag{1.26}$$

where k is a scalar function. As previously, this equation needs to be satisfied only at equilibrium.

To solve (1.23) or (1.26), several approaches exist. We can set up the corresponding characteristic system and integrate taking into account equilibrium conditions. This is done when there are only few equations to solve, as in the example dealt with at 1.5.3.3. The imprecise sentence "take into account equilibrium conditions" means that we will straighten up the equilibrium set implicitly. A second approach involves explicitly straightening up the equilibrium set, as has been done for output feedback pseudo-compensation. The coordinate change:

$$s_i=\varphi_i(x) \qquad i=1,\ldots,n$$

is such that $\mathcal{V}$ is defined by $s_2=\cdots=s_n=0$ (it is an embedded submanifold). Equation (1.26) to be solved for T_1 is changed into

$$\frac{\partial T_1}{\partial s}=l\left(s_1^0\right) \alpha\left(s_1^0\right)$$

because other components of s are equal to zero at equilibrium. l is a scalar function and α the gradient direction. Now a solution is easily obtained

$$T_1(s) = \int_0^{s_1} l(\tau)\,\alpha_1(\tau)\,\mathrm{d}\tau + l(s_1)\,\alpha_2(s_1)\,s_2 + \cdots + l(s_1)\,\alpha_n(s_1)\,s_n$$

The integrating factor l may be chosen to simplify the integration. For instance with $l = 1/\alpha_1$, we get:

$$T_1(s) = s_1 + \frac{\alpha_2(s_1)}{\alpha_1(s_1)}s_2 + \ldots + \frac{\alpha_n(s_1)}{\alpha_1(s_1)}s_n$$

Integration of other equations (1.24) defining the T_i may be done in the same way, but integrating factors cannot be used since the gradients are completely fixed. As regards (1.25), we find with the new coordinates:

$$\left(\frac{\partial S}{\partial s}, \frac{\partial S}{\partial u}\right) = \left(\beta\left(s_1^0\right), \gamma\left(s_1^0\right)\right)$$

We search a solution of the form

$$S(s,u) = \int_0^{s_1} \beta_1(\tau)\,\mathrm{d}\tau + \beta_2(s_1)\,s_2 + \cdots + \beta_n(s_1)\,s_n + \gamma(s_1)\,u + h(s_1)$$

Linearizing and equalizing δs_1 dependent terms, show that h must satisfy:

$$\beta_1\left(s_1^0\right) + \gamma'\left(s_1^0\right)u^0 + h'\left(s_1^0\right) = \beta_1\left(s_1^0\right)$$

But, according to the controllability assumption, at the equilibrium, u^0 can be expressed as a function of s_1^0, $u^0 = \psi\left(s_1^0\right)$. Therefore:

$$h(s_1) = -\int_0^{s_1} \gamma'(\tau)\,\Psi(\tau)\,\mathrm{d}\tau$$

We conclude with some complementary remarks. If there exists a nonzero solution T_1 to (1.23), it can be shown that the transformation $z = T(x)$ is a welldefined change of coordinates. This ensures that the whole system state has been pseudo-linearized. Furthermore, the state feedback is inversible and u may be stated as a function of x and v.

Suppose the opposite is true, *i.e.* the T_i are not independent. Then the Jacobian matrix of T is not inversible. Consider a nonzero linear combination of its rows

$$\sum_{i=1}^{n} a_i \frac{\partial T_i}{\partial x} = 0$$

Let k be the greatest index i such that a_i is nonzero. We have:

$$\sum_{i=1}^{k} a_i \frac{\partial T_i}{\partial x} = \sum_{i=1}^{k} a_i \frac{\partial T_1}{\partial x} F^{i-1} = 0 \qquad \text{with} \qquad a_k \neq 0$$

Multiplying by $F^{n-k}G$ leads to

$$\sum_{i=1}^{k} a_i \frac{\partial T_1}{\partial x} F^{n-k+i-1} G = 0$$

Now if $i < k$, then $n+i-k-1 \leq n-2$ and thus according to the definition of T_1 (1.23),

$$\frac{\partial T_1}{\partial x} F^{n-k+i-1} G = 0 \qquad \text{if } i < k$$

The linear combination reduces to only one term, and we have

$$\frac{\partial T_1}{\partial x} F^{n-1} G = 0$$

Consequently, the gradient of T_1 is orthogonal to n vectors $G, FG, \ldots, F^{n-1}G$ which are independent according to the local controllability assumption. It must be zero which contradicts our hypothesis. In the same way the gradient

$$\frac{\partial S}{\partial u} = \frac{\partial T_1}{\partial x} F^{n-1} G$$

cannot be zero and the feedback $v = S(x,u)$ is regular.

1.5.3.3 Multi-input systems

The theoretical solution of the problem is more complicated in this case. There does not always exist a solution. We must consider the "projections" of the partial derivative equations to be solved in the tangent space $\mathcal{V}$ and its orthogonal. On this latter space, integration is easy but, on the tangent space, an involutivity condition involving the "projected" fields is required to integrate. If the tangent space dimension does not exceed two $(m \leq 2)$, the projection leads on to a unique vector field which is necessarily involutive, on to a pair of vector fields which form an involutive distribution since it lines up the tangent space to $\mathcal{V}$.

Under assumptions of controllability and structural robustness of the linear tangent model, it has been shown that a system is pseudo-linearizable if and only if the distributions of vector fields

$$\left\{G, FG, \ldots, F^{k_i-2}G\right\} \cap T\mathcal{V}$$

are involutive. $T\mathcal{V}$ denotes the tangent space to $\mathcal{V}$, and the k_i are the controllability indices of (F, G). The rank of these distributions is given by:

$$\text{rank}\left[\left\{G, FG, \ldots, F^{k_i-2}G\right\} \cap T\mathcal{V}\right] = \text{card}\left\{j/k_j < k_i\right\}$$

They are all rank one distributions if $k_1 = k_2 = \ldots = k_{m-1} \geq k_m$. Now we can easily introduce integrators on the inputs so that the indices of the linearized model satisfy this relation. Since rank one distributions are always involutive, we have shown that it is always possible to pseudo-linearize a system which is locally controllable and whose controllability structure is constant in a neighborhood of the equilibrium set.

We will not go deeper into such theoretical considerations and we will simply illustrate how to deal with a multivariable case using an example. We consider the system state equation

$$\begin{cases} \dot{x}_1 &= u_2 \\ \dot{x}_2 &= u_1 \\ \dot{x}_3 &= x_1 x_3 + u_3 \\ \dot{x}_4 &= x_1^2 - 2x_2 + x_2 u_2 \end{cases}$$

The set of equilibrium points is given by:

$$\mathcal{E} = \left\{(x,u)\,;u_1 = u_2 = 0, u_3 = -x_1x_3, x_1^2 = 2x_2\right\} \qquad \text{and} \qquad \mathcal{V} = \left\{x; x_1^2 = 2x_2\right\}$$

Its linearized model can be written as:

$$\begin{cases} \delta\dot{x}_1 &= \delta u_2 \\ \delta\dot{x}_2 &= \delta u_1 \\ \delta\dot{x}_3 &= x_3\delta x_1 + x_1\delta x_3 + \delta u_3 \\ \delta\dot{x}_4 &= 2x_1\delta x_1 - 2\delta x_2 + x_2\delta u_2 \end{cases} \qquad \text{where} \qquad F = \begin{pmatrix} 0 & 0 & 0 & 0 \\ 0 & 0 & 0 & 0 \\ x_3^0 & 0 & x_1^0 & 0 \\ 2x_1^0 & -2 & 0 & 0 \end{pmatrix}$$

$$\text{and} \qquad G = \begin{pmatrix} 0 & 1 & 0 \\ 1 & 0 & 0 \\ 0 & 0 & 1 \\ 0 & x_2^0 & 0 \end{pmatrix}$$

It is controllable with controllability indices $(2, 1, 1)$. Since $\mathcal{V}$ is defined by $x_1^2 = 2x_2$, its perpendicular is $(x_1, -1, 0, 0)$ and its tangent plane is:

$$T\mathcal{V} = \text{span}\begin{pmatrix} 1 & 0 & 0 \\ x_1 & 0 & 0 \\ 0 & 1 & 0 \\ 0 & 0 & 1 \end{pmatrix}$$

Pseudo-linearization is feasible if $G \cap T\mathcal{V}$ is involutive. We have:

$$G \cap T\mathcal{V} = \text{span}\begin{pmatrix} 0 & 1 & 0 \\ 1 & 0 & 0 \\ 0 & 0 & 1 \\ 0 & x_2 & 0 \end{pmatrix} \cap \text{span}\begin{pmatrix} 1 & 0 & 0 \\ x_1 & 0 & 0 \\ 0 & 1 & 0 \\ 0 & 0 & 1 \end{pmatrix} = \text{span}\begin{pmatrix} 0 & 1 \\ 0 & x_1 \\ 1 & 0 \\ 0 & x_2 \end{pmatrix} = \text{span}\,(f_1, f_2)$$

The Lie bracket of these two vector fields $[f_1, f_2]$ is equal to zero and involutivity of $G \cap T\mathcal{V}$ is fulfilled. We can pseudo-linearize. Since controllability indices are $(2, 1, 1)$, we choose the canonical form

$$\begin{cases} \delta\dot{z}_1 &= \delta z_2 \\ \delta\dot{z}_2 &= \delta v_1 \\ \delta\dot{z}_3 &= \delta v_2 \\ \delta\dot{z}_4 &= \delta v_3 \end{cases}$$

We search $T(x)$ and $S(x,u)$ satisfying:

$$\begin{aligned}
\frac{\partial T_1}{\partial x}(F\delta x + G\delta u) &= \frac{\partial T_1}{\partial x}\delta x \\
\frac{\partial T_2}{\partial x}(F\delta x + G\delta u) &= \frac{\partial S_1}{\partial x}\delta x + \frac{\partial S_1}{\partial u}\delta u \\
\frac{\partial T_3}{\partial x}(F\delta x + G\delta u) &= \frac{\partial S_2}{\partial x}\delta x + \frac{\partial S_2}{\partial u}\delta u \\
\frac{\partial T_4}{\partial x}(F\delta x + G\delta u) &= \frac{\partial S_3}{\partial x}\delta x + \frac{\partial S_3}{\partial u}\delta u
\end{aligned}$$

The first transformation, T_1, must be such that $\delta \dot{z}_1$ does not depend on the input control. So, at equilibrium

$$\frac{\partial T_1}{\partial x}G = 0$$

i.e.

$$\frac{\partial T_1}{\partial x_2} = 0 \qquad \frac{\partial T_1}{\partial x_1} + x_2^0 \frac{\partial T_1}{\partial x_4} = 0 \qquad \frac{\partial T_1}{\partial x_3} = 0$$

An existence of a solution to these equations is guaranteed by the involutivity of $G \cap T\mathcal{V}$. The characteristic system corresponding to the second equation is

$$\mathrm{d}x_1 = \frac{\mathrm{d}x_4}{x_2^0}$$

now $x_2^0 = \left(x_1^0\right)^2/2$, and a solution is

$$T_1(x) = \frac{1}{6}x_1^3 - x_4$$

Next, it remains to integrate other transformations. T_2 is defined by

$$\frac{\partial T_2}{\partial x} = \frac{\partial T_1}{\partial x}F = (-2x_1, 2, 0, 0)$$

and we can choose

$$T_2(x) = -x_1^2 + 2x_2$$

For T_3 and T_4 we choose functions independent of the previous ones, *i.e.*, such that $\det \dfrac{\partial T}{\partial x} \neq 0$

$$\begin{aligned}
T_3(x) &= x_1 \\
T_4(x) &= x_3
\end{aligned}$$

Finally, the state feedback is computed from

$$\left(\frac{\partial S_1}{\partial x}, \frac{\partial S_1}{\partial u}\right) = \frac{\partial T_1}{\partial x}F(F,G) = (0,0,0,0,2,-2x_1,0)$$

which yields

$$S_1(x,u) = 2u_1 - 2x_1u_2$$

because u_2 is equal to zero at equilibrium. Other control loops are easily found.

$$\begin{aligned}
\left(\frac{\partial S_2}{\partial x}, \frac{\partial S_2}{\partial u}\right) &= \frac{\partial T_3}{\partial x}(F,G) \quad \Rightarrow \quad S_2(x,u) = u_2 \\
\left(\frac{\partial S_3}{\partial x}, \frac{\partial S_3}{\partial u}\right) &= \frac{\partial T_4}{\partial x}(F,G) \quad \Rightarrow \quad S_3(x,u) = x_1x_3 + u_3
\end{aligned}$$

We check that the transformations are nonsingular and the feedback is regular. We have on $\mathcal{V}$:

$$\left|\frac{\partial T}{\partial x}\right| = -2 \qquad \left|\frac{\partial S}{\partial u}\right| = 2$$

Using these transformations the original system is brought to:

$$\begin{cases}
\dot{z}_1 = z_2 - \frac{1}{2}v_2z_2 \\
\dot{z}_2 = v_1 \\
\dot{z}_3 = v_2 \\
\dot{z}_4 = v_3
\end{cases}$$

Note that the nonlinear term is a second-order term around any equilibrium point: pseudo-linearization is well achieved.

1.6 Conclusion

Control at first-order of nonlinear systems is based on the use of linearized models along the equilibrium set. When the controllability structure of these models is invariant and if they are controllable, one can try to integrate the control laws which are computed at some equilibrium points, in order to obtain a unique nonlinear control law.

In this chapter, we have tried to show how this integration may be done analytically when the local control laws are parametrized by the operating point. A few solutions have been explained in some particular cases: pole assignment by state feedback, by output feedback, transformation to a canonical form.

The trick which often solves the integration problems with which we are faced is to reckon with equilibrium conditions. The user of such approaches will have to solve homogeneous first-order partial differential equations, whose local solutions may often be easily found using these equilibrium conditions.

The potential application field of these approaches is quite large. We insisted on the invariance of performances according to the operating point which is an important feature often required.

1.7 Bibliography

[1] SONTAG, E.D. (1987), Nonlinear control via equilinearization, *Proc. IEEE, CDC*, Los Angeles.

[2] SONTAG, E.D. (1987), *Equilinearization : A simplified deviation and experimental results*, Proc. Conf. Info. Sci. and Systems, John Hokins, University Press.

[3] SONTAG, E.D. (1987), *Controllability and linearized regulation IEEE Trans. on Autom. Control,* **32**, 10.

[4] RUGH, W.J. (1983), Linearization about constant operating points: on input-output viewpoint, *Proc. IEEE Conf. Decision Contr.*, San Antonio.

[5] BAUMANN, W.T. and RUGH, W.J. (1986), Feedback control of nonlinear systems by extended linearization *IEEE Trans. Autom. Contr.*, **31**.

[6] REBOULET, C. and CHAMPETIER, C. (1984), A new method for linearizing non-linear systems: the pseudolinearization, *Int. J. Contr.*, **40**, 631–638.

[7] REBOULET, C., MOUYON, PH. and CHAMPETIER, C. (1986), About the local linearization of nonlinear systems, *Nonlinear control and systems theory*, M. Fliess and M. Hazewinkel eds, D. Reidel Publishing Company, Dordrecht.

[8] MOUYON, PH. (1986), *Pseudolinéarisation et pseudocompensation des systèmes non linéaires*, Doctorate thesis of the ENSAE.

[9] GEORGE, L., VERNET, J.F. and WANNER, J.C. (1969), *La mécanique du vol : performances des avions et des engins*, Dunod.

[10] SAMBLANCAT, CH. (1991), *Commande robuste multivariable : applications à l'hélicoptère*, Doctorate thesis of the ENSAE.

[11] DE FORNEL, B., REBOULET, C., and BOIDIN, M. (1983), Speed control by microprocessor for an induction machine fed by a static convecter, *3rd IFAC Symposium* Lausanne.

[12] KRISHNAN, H., REYHANOGLU, M., and MC CLAMROCH, H. (1994), Attitude stabilization of a rigid spacecraft using two control torques, *Automatica*, **30**, 6, 1023–1027.

CHAPTER 2

Input–Output linearization

J. DESCUSSE

2.1 Introduction

2.1.1 Problem setting, motivations, links with quasilinearization and pseudolinearization

The control of a nonlinear system is difficult to carry out, mainly because the mathematical models, which are supposed to describe the real world, are complex and not easily analyzed. The determination and the implementation of control laws require tremendous endeavors; keep in mind, how difficult it is to calculate optimal controls, such as "minimum time", even with academic models.

For engineers, linearizing a nonlinear system is an appealing approach since, after linearization, a complete range of tools is available, from the linear system theory.

The underlying questions are then:

– Which type of linearization: approximate or exact?

– For which goal: local or aggregate?

– Internal (input-state) or external (input–output)?

The approximate linearization, around an equilibrium point, in other words the quasi-linearization, well-known to all engineers, is well suited for regulation problems, with a running point more or less far from the considered equilibrium point, according to the degree of approximation.

The exact linearization is well suited for transitions of the equilibrium point and for tracking given trajectories.

Whether approximate or exact, linearization often requires transformations like (static or dynamic) state feedback, which has to be measured, and diffeomorphism or immersion.

From the approximate to the exact solutions, we use the following order:

– the internal linearization with a degree of approximation r ($r \geq 1$); the case $r = 1$ does not differ from the well-kown linearization in the tangent bundle,

– the internal pseudolinearization, which makes the linearized tangent system independent from the equilibrium point (Chapter 1),

– the exact internal linearization (Chapter 4),

– the exact external linearization, which is the purpose of this chapter.

With every application, we must keep in mind the limitations of these different methods, and especially the loss of information for the approximate methods and the problem of singularities for the exact ones.

The exact external linearization (e.e.l.) is closely related to noninteracting control also named "input–output decoupling" (i.o.d.), because solving i.o.d. always provides a solution to e.e.l. On the other hand, the only tractable solutions for e.e.l. are those of i.o.d. This is the reason why we shall pay strong attention to i.o.d.

2.1.2 Input–output decoupling (i.o.d.)

The Input–Output Decoupling Problem is of paramount importance for automaticians. Since the beginning of multivariable control, it has enthused a huge number of researchers, throughout the world. Far from being an academic issue, it is spurred on by practical reasons: input–output decoupling enables an easy synthesis of the aggregate control system, but it also answers to system security constraints by splitting the control chains into independent sets.

In the nonlinear case, there have been only a few contributions, mainly because of mathematical difficulties. An exhaustive bibliography on this topic can be found in the well-known book by Isidori (1989).

Although a quite thorough knowledge is available for the nonlinear Input–Output Decoupling theory, its transfer to users facing practical questions remains an open problem. They come up against difficulties occurring due to intricate mathematical formalism. In the present state of the art, one can claim that, for some classes of nonlinear systems, a whole apparatus, theoretically useable, is at the disposal of any user. We aim a didactic presentation of this apparatus but we also intend to show its limits in view of practical implementation. All the mathematical developments will be illustrated by simple examples (so that the reader may choose the appropriate tool for his mathematical problem).

At the present time, the main difficulties encompass the computational procedures as well as the understanding of insufficiently analyzed structural properties (fixed and unobservable dynamics). They often cause instabilities, the study of which remains open. These issues will extensively be developed in the remainder of this chapter.

2.2 Problem setting: Mathematical model and class of feedback laws

Definition 2.1

Let a system Σ with m inputs and p outputs be given. Σ is said to be decoupled if it is split into p independent mono-output subsystems, in the sense that the inputs of the subsystem i do not affect the output y_j and conversely that the inputs of the subsystem i only affect the output y_i.

Usually, a physical system is never decoupled in the above sense. It is therefore interesting to search for transformations, such as "feedbacks", the action of which make decoupling possible.

In this chapter, two types of feedback will be considered:

1. static state feedback
2. dynamic compensator.

The class of systems Σ under consideration deals with affine nonlinear systems, modeled as follows:

$$\begin{aligned} \dot{x} &= f(x) + \sum_{i \in \underline{m}} g_i(x) u_i \\ y &= h(x) \end{aligned}$$

In brief, hereunder the triple (h, f, g) will denote, such a dynamic system.

We assume that:

- the vector fields $f(x)$ and $g_i(x)$ are analytic or C^∞,
- the state x belongs to $\mathbb{R}^n$,
- h is an analytic or C^∞ submersion with rank p (at each point, the rank of h equals the number of outputs p). Locally, h is given by $[h_1(x), \ldots, h_p(x)]$ with $h_i(x)$ from $\mathbb{R}^n$ to $\mathbb{R}$.

2.3 The solution of the problem using static state feedback

We want to find a static state feedback: $u = \alpha(x) + \beta(x) \cdot v$, such that the closed-loop system $(h, f + g\alpha, g\beta)$ is decoupled.

The feedback is said to be regular if $\beta(x)$ is a regular matrix.

At first, we shall examine the simplest case of square systems $(p = m)$.

2.3.1 The situation for square systems

2.3.1.1 An introductive example

Let us consider the following example in order to introduce the general procedure of input–output decoupling given below by theorem 2.1.

Example 1

Let us consider the system Σ described by

$$\begin{pmatrix} \dot{x}_1 \\ \dot{x}_2 \\ \dot{x}_3 \\ \dot{x}_4 \end{pmatrix} = \begin{pmatrix} x_1 x_4 \\ x_3(1 - x_4) \\ 0 \\ 0 \end{pmatrix} + \begin{pmatrix} 0 & 0 \\ 0 & 0 \\ 0 & x_1 \\ 1 & 0 \end{pmatrix} \begin{pmatrix} u_1 \\ u_2 \end{pmatrix}$$

$$\begin{aligned} y_1 &= x_1 \\ y_2 &= x_2 \end{aligned}$$

Differentiating y_1 and y_2 with respect to the time t gives

$$\begin{aligned}\dot{y}_1 &= x_1 x_4\\ \dot{y}_2 &= x_3(1-x_4)\end{aligned}$$

These derivatives do not explicitly depend on the input u explicitly. Differentiating once more, we find

$$\begin{aligned}\ddot{y}_1 &= x_1 x_4^2 + x_1 u_1\\ \ddot{y}_2 &= -x_3 u_1 + x_1(1-x_4)u_2\end{aligned}$$

As the input u now appears explicitly, let us consider the matrix $\Delta(x)$, multiplying the inputs u_1 and u_2:

$$\Delta(x) = \begin{pmatrix} x_1 & 0\\ -x_3 & x_1(1-x_4)\end{pmatrix}$$

On $M_0 = \mathbb{R}^4 - \{x_1(1-x_4) = 0\}$, its rank, noted as rank $\Delta(x)$, equals 2. It is thus invertible on M_0. Its inverse is:

$$\Delta^{-1}(x) = \begin{pmatrix} \dfrac{1}{x_1} & 0\\ \dfrac{x_3}{x_1^2(1-x_4)} & \dfrac{1}{x_1(1-x_4)}\end{pmatrix}$$

Let $\Delta_0(x)$ be the vector defined by those terms which, in the above expressions of $\ddot{y}_1$ and $\ddot{y}_2$, do not depend on u.

$$\Delta_0(x) = \begin{pmatrix} x_1 x_4^2\\ 0\end{pmatrix}$$

Then, let us consider the feedback $u = \alpha(x) + \beta(x)\cdot v$, with

$$\begin{aligned}\alpha(x) &= -\Delta(x)^{-1}\Delta_0(x)\\ \beta(x) &= \Delta(x)^{-1}\end{aligned}$$

We have $$\alpha(x) = -\begin{pmatrix} \dfrac{1}{x_1} & 0\\ \dfrac{x_3}{x_1^2(1-x_4)} & \dfrac{1}{x_1(1-x_4)}\end{pmatrix}\cdot\begin{pmatrix} x_1 x_4^2\\ 0\end{pmatrix} = -\begin{pmatrix} x_4^2\\ \dfrac{x_3 x_4^2}{x_1(1-x_4)}\end{pmatrix}$$

After closing the loop, the state equations become

$$\dot{x} = f(x) + g(x)\cdot[\alpha(x) + \beta(x)\cdot v]$$

Let

$$\dot{x} = \tilde{f}(x) + \tilde{g}(x)\cdot v(t)$$

with

$$\begin{aligned}\tilde{f}(x) &= f(x) + g(x)\cdot\alpha(x)\\ \tilde{g}(x) &= g(x)\cdot\beta(x)\end{aligned}$$

Let:
$$\begin{pmatrix} \dot{x}_1 \\ \dot{x}_2 \\ \dot{x}_3 \\ \dot{x}_4 \end{pmatrix} = \begin{pmatrix} x_1 x_4 \\ x_3(1-x_4) \\ \dfrac{-x_3 x_4^2}{(1-x_4)} \\ -x_4^2 \end{pmatrix} + \begin{pmatrix} 0 & 0 \\ 0 & 0 \\ \dfrac{x_3}{x_1(1-x_4)} & \dfrac{1}{(1-x_4)} \\ \dfrac{1}{x_1} & 0 \end{pmatrix} \cdot \begin{pmatrix} v_1 \\ v_2 \end{pmatrix}$$

Now
$$\begin{aligned} \dot{y}_1 &= x_1 x_4 \\ \ddot{y}_1 &= \dot{x}_1 x_4 + x_1 \dot{x}_4 \\ &= x_1 x_4^2 + x_1 \left(-x_4^2 + \frac{v_1}{x_1}\right) \end{aligned}$$

thus
$$\ddot{y}_1 = v_1$$

and,
$$\begin{aligned} \dot{y}_2 &= x_3(1-x_4) \\ \ddot{y}_2 &= \dot{x}_3(1-x_4) - x_3 \dot{x}_4 \\ &= -x_4^2 x_3 + \frac{x_3}{x_1} v_1 + v_2 + x_3 x_4^2 - v_1 \frac{x_3}{x_1} \end{aligned}$$

thus
$$\ddot{y}_2 = v_2$$

We get two independent chains of cascaded integrators:

$$\begin{aligned} \ddot{y}_1 &= v_1 \\ \ddot{y}_2 &= v_2 \end{aligned}$$

The closed-loop system is thus decoupled. Moreover, the state feedback law considered yields a closed-loop system, with a linear input–output behavior, on M_0, described by:

$$\ddot{y}_i = v_i \quad \forall i \in \underline{2}$$

The procedure followed for the above example can be generalized.

2.3.1.2 The general result

Let the system Σ be defined by:
$$\begin{aligned} \dot{x} &= f(x) + \sum_{i \in \underline{m}} g_i(x) u_i \\ y &= h(x) \end{aligned}$$

Let $(\rho_1, \ldots, \rho_m)$ be the set of row infinite zeros of the system (h, f, g). Recall that they are defined as follows:

$$\rho_i := \left(\inf l \in \mathbb{N}, \text{ such that, } \exists j \in \underline{m}, L_{g_j} L_f^{l-1} h_i \neq 0\right)$$

(L_f denotes the Lie derivative with respect to the vector field f.)

Recall that ρ_i corresponds to the first derivative of y_i where the control input u appears explicitly:

$$y^{(\rho_i)} = L_f^{\rho_i}(x) + L_g L_f^{\rho_i - 1} h(x) \cdot u$$

the multiplicative term of u denoting the concatenation of the terms $L_{g_j} L_f^{\rho_i - 1} h(x), \forall j \in \underline{m}$

Let $\Delta(x)$ be the matrix defined by:

$$\Delta(x) = \begin{pmatrix} L_{g_1} \cdot L_f^{\rho_1 - 1} \cdot h_1(x) & \dots & L_{g_m} \cdot L_f^{\rho_1 - 1} \cdot h_1(x) \\ \dots & \dots & \dots \\ L_{g_1} \cdot L_f^{\rho_m - 1} \cdot h_m(x) & \dots & L_{g_m} \cdot L_f^{\rho_m - 1} \cdot h_m(x) \end{pmatrix}$$

This $\Delta(x)$ matrix is named the "decoupling matrix" of the system.

Let $\Delta_0(x)$ be the vector defined by:

$$\Delta_0(x) = \begin{pmatrix} L_f^{\rho_1} \cdot h_1(x) \\ \dots \\ L_f^{\rho_m} \cdot h_m(x) \end{pmatrix}$$

Theorem 2.1

(i) The system Σ can be decoupled statically on a submanifold M_0 of $\mathbb{R}^n$ if and only if

$$\text{rank } \Delta(x) = m \qquad \forall x \in M_0 \qquad (m = p : \textit{number of outputs})$$

(ii) If this condition for $\Delta(x)$ is fulfilled, the static state feedback defined by:

$$u = \alpha(x) + \beta(x) \cdot v,$$

with

$$\begin{aligned} \alpha(x) &= -\Delta(x)^{-1} \cdot \Delta_0(x) \\ \beta(x) &= \Delta(x)^{-1} \end{aligned}$$

decouples the system Σ, on M_0.

(iii) Moreover, the closed-loop system $(h, f + g\alpha, g\beta)$ has a linear input–output behavior described by

$$y_i^{(\rho_i)} = v_i \quad \forall i \in \underline{m}$$

The proof of this theorem, and especially the sufficient condition, can be found in Isidori (1989). As an exercise, the reader can work it out. As in example 1, differentiate y_i until the input u appears explicitly. This operation will show $\Delta(x)$ and $\Delta_0(x)$, in a natural way. The result follows immediately.

The dynamic behavior of a system composed of cascaded integrators is obviously not admissible in practice. It is therefore necessary to modify it using a second feedback, linear if possible. The calculation of this second feedback needs requires to put the decoupled system in a canonical form which is called decoupled.

Before tackling this point, which is the main point in this contribution and crucial for any application, we shall first examine the case of non square systems.

2.3.2 *The situation for non square systems*

Now we suppose that Σ has p outputs and m inputs, with $p < m$. This case encounters tremendous difficulties like in the linear situation, because the problem can admit a solution with non regular feedbacks. For nonlinear systems, no efficient solution of this type is available up to now. Nevertheless, one may propose the following result.

Theorem 2.2

(i) The system Σ can be decoupled on M_0 of $\mathbb{R}^n$ using static state feedback if

$$\operatorname{rank}\Delta(x) = p \quad \forall x \in M_0$$

(ii) The above condition for $\Delta(x)$ being fulfilled, the static state feedback defined by $u = \alpha(x) + \beta(x) \cdot v$ with

$$\begin{aligned} \alpha(x) &= -\Delta(x)^{\dagger}\Delta_0(x) \\ \beta(x) &= \Delta(x)^{\dagger} \end{aligned}$$

decouples the system Σ; $\Delta^{\dagger}$ denotes the pseudo-inverse of Δ.

(iii) Moreover, the input–output behavior of the closed-loop system $(h, f+g\alpha, g\beta)$ is linear and described by

$$y_i^{(\rho_i)} = v_i \quad \forall i \in \underline{p}$$

Note that the use of this theorem, for nonsquare systems, leads to a decreasing number of independent inputs between the open-loop and the closed-loop systems.

2.4 The decoupled canonical form

The decoupling feedback built up from $\Delta(x)$ and $\Delta_0(x)$ cannot be a good candidate since it yields chains of cascaded integrators. One has to use another feedback to get a correct dynamical behavior for the closed-loop system.

The underlying issues are relative to the class of decoupling feedbacks and to the stability analysis, which need to put the system in a decoupled canonical form. As in the linear case, the latter will have an essential role in the full synthesis of a decoupled system.

2.4.1 First qualitative approach

Theorem 2.3

There exists a local change of coordinates $\Phi : x \longrightarrow z$ such that in the new coordinates, the closed-loop system $(h, f + g\alpha, g\beta)$ is written canonically as:

$$\dot{z}^i = \begin{pmatrix} 0 & 1 & \dots & \dots & \dots \\ 0 & 0 & 1 & \dots & \dots \\ \dots & \dots & \dots & \dots & \dots \\ 0 & \dots & \dots & \dots & 1 \\ 0 & \dots & \dots & \dots & 0 \end{pmatrix} \cdot z^i + \begin{pmatrix} 0 \\ \dots \\ \dots \\ 0 \\ 1 \end{pmatrix} \cdot v_i \quad \forall i \in \underline{m}$$

$$\begin{aligned} \dot{z}^{m+1} &= f_{m+1}(z) + \sum_{i \in \underline{m}} g_i^{m+1}(z) \cdot v_i \\ y_i &= [1, 0, \dots, 0] z^i \end{aligned}$$

with $\dim z^i = \rho_i$ *and* $\dim z^{m+1} = n - \sum_{i \in \underline{m}} \rho_i$

Below, we shall only report, for the diffeomorphism Φ, the mainstream informations. All the relative details can be found in Isidori (1989).

Let us consider the mapping $\Phi_i : x \;\rightarrow\; z^i$ defined by

$$z^i = \Phi_i(x) = \text{ col }[h_i(x), \ldots, \ldots, L_f^{\rho_i - 1} h_i(x)]$$

Let the vector

$$\Phi^\circ = \begin{pmatrix} \Phi_1 \\ \cdots \\ \Phi_m \end{pmatrix} = \begin{pmatrix} \text{col }[h_1(x), \ldots\ldots, L_f^{\rho_1 - 1} h_1(x)] \\ \cdots \\ \text{col }[h_m(x), \ldots\ldots, L_f^{\rho_m - 1} h_m(x)] \end{pmatrix}$$

The mapping Φ° fulfils the important following property: rank $\Phi^\circ = \sum_{i \in \underline{m}} \rho_i$.

Two situations may occur: $\sum_{i \in \underline{m}} \rho_i = n$ or $\sum_{i \in \underline{m}} \rho_i < n$

If $\sum_{i \in \underline{m}} \rho_i = n$

the sum of the ρ_i equals the dimension of the state space exactly and in this case, the set of functions $\Phi_i^k(x) = L_f^{k-1} h_i(x)$ with $1 <= k <= \rho_i$ and $1 <= i <= m$ defines the change of coordinates completely (*i.e.*, $\Phi = \Phi^\circ$ is a diffeomorphism). As an exercise, the reader can study again example 1.

Conclusion

In this case, the closed-loop system characterized by the m decoupled subsystems is clearly *linear, controllable* and *observable.*

If $\sum_{i \in \underline{m}} \rho_i < n$

we must complete the set of functions $\Phi_i^k(x)$ with an arbitrary function $\Phi_{m+1}(x)$ such that the rank of $\Phi = (\Phi_1, \ldots, \Phi_m, \Phi_{m+1})$ equals n. Usually the calculation of f_{m+1} and of g_i^{m+1} is difficult to perform, because it requires the calculation of Φ^{-1}.

Note that, in this second case, *when* $z = 0$ *is an equilibrium point of the decoupled system,* the "zero dynamics" of Σ is given by the dynamical behavior of the subsystem

$$\dot{z}^{m+1} = f_{m+1}(0, \ldots, 0, z^{m+1})$$

The zero dynamics play a key role for the stability of the decoupled system.

The reader will make a link between this concept and the zeros of linear systems. For the latter concept, the book by Wonham (1989) is a useful reference.

2.4.2 *The decoupling control laws*

Taking into account the above canonical form, it is clear that the feedbacks which maintain decoupling for the decoupled integrators system $(\tilde{f}, \tilde{g})$ look like

$$v_i = \gamma_i(z^i) + \delta_i(z^i) \cdot w_i$$

or as functions of the state x

$$v_i = \gamma_i(\Phi_i(x)) + \delta_i(\Phi_i(x)) \cdot w_i$$

We shall see in section 2.4.4, through an example, that this type of control law is only a subclass of the whole set of decoupling control laws.

2.4.3 An illustrative example:

Let the system Σ be defined by

$$\begin{aligned} f(x) &= \begin{pmatrix} 0 \\ -x_1^2 \\ (x_1^2 - 1)x_3 + \mathrm{e}^{-x_2}(x_2 + x_1x_4 + x_4) \\ x_1 \end{pmatrix} \\ g(x) &= \begin{pmatrix} 0 & 1 \\ 1 & -x_4 \\ -x_3 & x_3x_4 \\ 0 & 0 \end{pmatrix} \\ h(x) &= \begin{pmatrix} x_2 + x_1x_4 \\ x_1 \end{pmatrix} \end{aligned}$$

A simple calculation leads to

$$\dot{y} = \begin{pmatrix} u_1 \\ u_2 \end{pmatrix}$$

This system is thus decoupled. Note that decoupling induces two unobservable dynamics, since the sum of ρ_i equals 2.

The procedure followed in section 2.4.1 gives

$$\Phi_1(x) = x_2 + x_1x_4 \qquad \text{and} \qquad \Phi_2(x) = x_1$$

Now, we must complete these functions to obtain a diffeomorphism. A wide choice is available. Undoubtedly, the simplest one to choose is

$$\Phi_3(x) = \begin{pmatrix} x_3 \\ x_4 \end{pmatrix}$$

Through this example, the reader immediately understands the huge difficulty which occurs when the situation is non academic and it is difficult to "draw" the "best solution"!

Thus, the diffeomorphism is written as $z = \Phi(x)$, where

$$\Phi(x) = \begin{pmatrix} x_2 + x_1 x_4 \\ x_1 \\ x_3 \\ x_4 \end{pmatrix}$$

In general, the calculation of its inverse is extremely difficult. Here, things go well. We have

$$\Phi^{-1}(z) = \begin{pmatrix} z_2 \\ z_1 - z_2 z_4 \\ z_3 \\ z_4 \end{pmatrix}$$

Under the action of this diffeomorphism, the system Σ is turned into

$$f(z) = \begin{pmatrix} 0 \\ 0 \\ (z_2^2 - 1)z_3 + e^{-(z_1 - z_2 z_4)}(z_1 + z_4) \\ z_2 \end{pmatrix}$$

$$g(z) = \begin{pmatrix} 1 & 0 \\ 0 & 1 \\ -z_3 & z_3 z_4 \\ 0 & 0 \end{pmatrix}$$

$$h(z) = \begin{pmatrix} z_1 \\ z_2 \end{pmatrix}$$

In accordance with the notation of section 2.4.1, we have

$$f_3(z) = \begin{pmatrix} (z_2^2 - 1)z_3 + e^{-(z_1 - z_2 z_4)}(z_1 + z_4) \\ z_2 \end{pmatrix}$$

$$g_3(z) = \begin{pmatrix} -z_3 & z_3 z_4 \\ 0 & 0 \end{pmatrix}$$

The "zero dynamics" are thus given by

$$\begin{pmatrix} \dot{z}_3 \\ \dot{z}_4 \end{pmatrix} = f_3(0, 0, z_3, z_4) = \begin{pmatrix} -z_3 + z_4 \\ 0 \end{pmatrix}$$

These dynamics are linear; one pole equals zero and the other one equals -1.

According to section 2.4.2, each subsystem can be controlled and stabilized by control laws such as $v_i = \gamma_i(\Phi_i(x)) + \delta_i(\Phi_i(x)) \cdot w_i$. However, the zero dynamics being unstable, the internal stability of the decoupled system cannot be achieved through this type of solution.

Hereafter, we shall see that decoupling and also internally stabilizing solutions can be achieved.

To proceed, we shall use a general method, analogous to the method followed in the linear case. Contrarily to the above method, it helps to detect the partially unobservable dynamics (in other words, unobservable for some particular feedbacks). The corresponding change of coordinates will be of the type $\Psi : z \rightarrow x$, z denoting the new coordinates.

2.4.4 An illustrative example: continued

We shall again consider the above example and (we shall) proceed as in the linear case. The reader will understand the procedure, without difficulty! All the mathematical justifications will be given in section 2.4.5.

Let the regular controllability distributions be defined as

$$R_i^* = \{ad_f^q g_i \; ; \; q \in \mathbb{N}\}_{\text{L.A.}} \qquad i \in \underline{2}$$

where

$$ad_f^q g_i =: [ad_f^{q-1}, g_i]; q \geq 1 \qquad \text{with} \qquad ad_f g_i =: [f, g_i] =: \left(\frac{\partial g_i}{\partial x}\right) \cdot f - \left(\frac{\partial f}{\partial x}\right) \cdot g_i$$

the abbreviation $\{\}_{\text{L.A.}}$ meaning Lie Algebra generated by the set of elements between $\{\}$. "Regular" means that R_i^* is g-invariant, in other words that:

$$[g_j, R_i^*] \subset R_i^* \qquad \forall j \in \underline{2}.$$

In the linear case, these distributions are nothing else but the controllability subspaces (Wonham, 1989).

After calculation, we have:

$$R_1^* = \text{sp}\left\{\begin{pmatrix} 0 & 0 \\ 1 & 0 \\ -x_3 & -\mathrm{e}^{-x_2} \\ 0 & 0 \end{pmatrix}\right\} = \text{sp}\{(g_1, ad_f g_1)\}$$

$$R_2^* = \text{sp}\left\{\begin{pmatrix} 1 & 0 & 0 \\ -x_4 & x_1 & 0 \\ x_3 x_4 & -x_1 x_3 & \mathrm{e}^{-x2} \\ 0 & -1 & 0 \end{pmatrix}\right\} = \text{sp}\{(g_2, ad_f g_2, adf^2 g_2)\}$$

Calculating $R_1^* \cap R_2^*$ we have:

$$R_1^* \cap R_2^* = \mathrm{sp}\left\{\begin{pmatrix} 0 \\ 0 \\ e^{-x_2} \\ 0 \end{pmatrix}\right\}$$

Let us now consider the following sequence of inclusions.

$$R_1^* \cap R_2^* \subset R_1^* \cap R_2^* + R_1^* \subset R_1^* \cap R_2^* + R_1^* + R_2^* = T\,\mathbb{R}^4$$

Taking into account the above calculus, we can write

$$T\mathbb{R}^4 = \mathrm{sp}\left\{\underbrace{\underbrace{\underbrace{\begin{matrix} X_1 & X_2 \\ \begin{pmatrix} 0 \\ 0 \\ e^{-x_2} \\ 0 \end{pmatrix} & \begin{matrix} 0 \\ 1 \\ -x_3 \\ 0 \end{matrix}\end{matrix}}_{R_1^* \cap R_2^*} \begin{matrix} X_3 \\ 1 \\ -x_4 \\ x_3x_4 \\ 0 \end{matrix}}_{R_1^*} \begin{matrix} X_4 \\ 0 \\ x_1 \\ -x_1x_3 \\ -1 \end{matrix}}_{R_1^* + R_2^*}\right\}$$

Let us now use the whole mathematical apparatus given in appendix B and more precisely theorem B.1.

First, we rectify the vector field X_1. This operation is left to the reader, as an exercise. The reader is kindly requested to do the exercises detailed in appendix B! Taking for the time the variable z_3, we find that the diffeomorphism that rectifies the vector field is

$$\begin{aligned} x_1 &= z_1 \\ x_2 &= z_2 \\ x_3 &= z_3\,e^{-z_2} \\ x_4 &= z_4 \end{aligned}$$

Under its action, the above basis of $T\mathbb{R}^4$ is written as

$$T\,\mathbb{R}^4 = \mathrm{sp}\left\{\begin{matrix}\begin{pmatrix} 0 & 0 & 1 & 0 \\ 0 & 1 & -z_4 & z_1 \\ 1 & 0 & 0 & 0 \\ 0 & 0 & 0 & -1 \end{pmatrix} \\ X_1 \quad X_2 \quad X_3 \quad X_4 \end{matrix}\right\}$$

Let us use the procedure followed to prove the Frobenius theorem (See appendix B.) It amounts to turn X_3 and X_4 respectively into:

$$Y_3 = X_3 - X_3(z_2)X_2$$

and

$$Y_4 = X_4 - X_4(z_2)X_2$$

Doing this, it is clear that we fully rectify the distribution $R_1^* + R_2^*$.

Then, apply the above diffeomorphism to the original system. We get:

$$f(z) = \begin{pmatrix} 0 \\ -z_1^2 \\ -z_3 + z_2 + z_1 z_4 + z_4 \\ z_1 \end{pmatrix}$$

$$g(z) = \begin{pmatrix} 0 & 1 \\ 1 & -z_4 \\ 0 & 0 \\ 0 & 0 \end{pmatrix}$$

$$h(z) = \begin{pmatrix} z_2 + z_1 z_4 \\ z_1 \end{pmatrix}$$

By means of renumbering, we get the following form. Let

$$\eta_0 = z_3, \eta_1 = z_2, \eta_2 = z_4, \eta_3 = z_1$$

Then, we have:

$$\begin{pmatrix} \dot{\eta}_0 \\ \dot{\eta}_1 \\ \dot{\eta}_2 \\ \dot{\eta}_3 \end{pmatrix} = \begin{pmatrix} -\eta_0 + \eta_1 + \eta_2\eta_3 + \eta_2 \\ -\eta_3^2 \\ \eta_3 \\ 0 \end{pmatrix} + \begin{pmatrix} 0 & 0 \\ 1 & -\eta_2 \\ 0 & 0 \\ 0 & 1 \end{pmatrix} u$$

and

$$y = \begin{pmatrix} \eta_1 + \eta_2\eta_3 \\ \eta_3 \end{pmatrix}$$

Let us now consider the reduced system

$$\begin{pmatrix} \dot{\eta}_1 \\ \dot{\eta}_2 \\ \dot{\eta}_3 \end{pmatrix} = \underbrace{\begin{pmatrix} -\eta_3^2 \\ \eta_3 \\ 0 \end{pmatrix}}_{f'} + \begin{pmatrix} 1 & -\eta_2 \\ 0 & 0 \\ 0 & 1 \end{pmatrix} u$$

$$\qquad\qquad\qquad\qquad g_1' \quad g_2'$$

In accordance with the notation of appendix B, we have

$$L_{01} = \text{sp}\left\{\begin{pmatrix} 1 \\ 0 \\ 0 \end{pmatrix}\right\} = \text{sp}\{g'_1\}$$

and

$$L_{02} = \text{sp}\left\{\begin{pmatrix} -\eta_2 & \eta_3 \\ 0 & -1 \\ 1 & 0 \end{pmatrix}\right\} = \text{sp}\{g'_1, ad_{f'}g'_2\}$$

The distributions L_{01} and L_{02} fulfil the conditions of theorem B.5, therefore, they can be rectified separately. Note that both vector fields g'_1 and $ad_{f'}g'_2$ commute. It thus suffices to apply theorem B.2 on the coordinate frame fields. Here, we shall detail the calculation. First, we calculate the flow generated by the first vector field. Let

$$\begin{pmatrix} \dot{\eta}_1 \\ \dot{\eta}_2 \\ \dot{\eta}_3 \end{pmatrix} = \begin{pmatrix} -\eta_2 \\ 0 \\ 1 \end{pmatrix} \quad \Longrightarrow \quad \Phi_t^1(c_1, c_2, c_3) = \begin{pmatrix} -c_2 t + c_1 \\ c_2 \\ t + c_3 \end{pmatrix}$$

We now calculate the flow generated by the second vector field. Let

$$\begin{pmatrix} \dot{\eta}_1 \\ \dot{\eta}_2 \\ \dot{\eta}_3 \end{pmatrix} = \begin{pmatrix} \eta_3 \\ 0 \\ -1 \end{pmatrix} \quad \Longrightarrow \quad \Phi_t^2(c'_1, c'_2, c'_3) = \begin{pmatrix} -c'_3 t + c'_1 \\ c'_2 - t \\ c'_3 \end{pmatrix}$$

Now calculate

$$\begin{aligned} \Phi_{\xi_2}^2(\Phi_{\xi_1}^1(0,0,0)) &= \Phi_{\xi_2}^2(\xi_1, 0, \xi_3) \\ &= \begin{pmatrix} \xi_2\xi_3 + \xi_1 \\ -\xi_2 \\ \xi_3 \end{pmatrix} \end{aligned}$$

As an exercise, the reader will verify that the change of coordinates

$$\begin{pmatrix} \eta_1 \\ \eta_2 \\ \eta_3 \end{pmatrix} = \begin{pmatrix} \xi_2\xi_3 + \xi_1 \\ -\xi_2 \\ \xi_3 \end{pmatrix}$$

in fact rectifies the above vector fields.

Now, apply the change of coordinates

$$\begin{pmatrix} \eta_0 \\ \eta_1 \\ \eta_2 \\ \eta_3 \end{pmatrix} = \begin{pmatrix} \xi_4 \\ \xi_2\xi_3 + \xi_1 \\ -\xi_2 \\ \xi_3 \end{pmatrix}$$

After calculation, we find

$$\begin{pmatrix} \dot{\xi}_1 \\ \dot{\xi}_2 \\ \dot{\xi}_3 \\ \dot{\xi}_4 \end{pmatrix} = \begin{pmatrix} 0 \\ -\xi_3 \\ 0 \\ \xi_1 - \xi_2 - \xi_4 \end{pmatrix} + \begin{pmatrix} v_1 \\ 0 \\ v_2 \\ 0 \end{pmatrix}$$

$$y = \begin{pmatrix} \xi_1 \\ \xi_3 \end{pmatrix}$$

The system is linear. It has a stable fixed dynamic, for any decoupling feedback,

$$\dot{\xi}_4 = -\xi_4$$

It also has another fixed dynamic but only for some decoupling feedbacks. It is related to the subsystem

$$\begin{aligned} \dot{\xi}_2 &= -\xi_3 \\ \dot{\xi}_3 &= u_2 \end{aligned}$$

which has one pole and zero equal to zero. Using a feedback such as $u_2 = -a_1\xi_2 - a_2\xi_3 + v_2$, we get for the transfer function between y_2 and v_2,

$$\frac{s}{s^2 + a_2 s - a_1}$$

which can always be stabilized with a proper choice of a_1 and a_2.

The class of decoupling feedbacks, functions of the state ξ is given by

$$\begin{aligned} u_1 &= -a_0\xi_1 + v_1 \\ u_2 &= -a_1\xi_2 - a_2\xi_3 + v_2 \end{aligned}$$

Finally, we must give its expression as a function of the state x. For this purpose, the reader will verify that the diffeomorphism $\Psi : \xi \rightarrow x$ turns the original system into the system given above. In order to provide the general class of the decoupling feedbacks, as a function of the state x, we have to calculate Ψ^{-1}. Usually, this operation is difficult to perform and sometimes even impossible. Here things go well: Ψ^{-1} is given by

$$\begin{pmatrix} \xi_1 \\ \xi_2 \\ \xi_3 \\ \xi_4 \end{pmatrix} = \begin{pmatrix} x_2 + x_1 x_4 \\ -x_4 \\ x_1 \\ x_3\, e^{x2} \end{pmatrix}$$

Thus, the class of decoupling feedbacks as a function of the state x, is given by,

$$\begin{aligned} u_1 &= -a_0(x_2 + x_1 x_4) + v_1 \\ u_2 &= a_1 x_4 - a_2 x_1 + v_2 \end{aligned}$$

The above example introduced the problem of the calculation of the decoupled canonical form. Through this example, the reader can understand the crucial difference between Φ and Ψ, and also the interest of calculating Ψ to get the "best" decoupled canonical form.

The general procedure for calculating Ψ now follows.

2.4.5 A general approach (determination of Ψ)

The goal of this section is to point out the existence of a local coordinate frame in which the decoupled system can be written in a "decoupled canonical form". For calculating the latter, the key elements are the lemmas and theorems from differentiel geometry given in appendix B.

Hereafter, we shall assume that the necessary and sufficient condition for decoupling (given in §2.3) is fulfilled, and we shall use the feedback $u = \alpha(x) + \beta(x) \cdot v$ with $\alpha(x) = -\Delta(x)^{-1} \cdot \Delta_0(x)$ and $\beta(x) = \Delta(x)^{-1}$.

The decoupled system will be noted as:

$$\begin{aligned} \dot{x}(t) &= \tilde{f}(x) + \sum_{i \in \underline{m}} \tilde{g}_i(x) \cdot v_i(t) \\ y_i(t) &= h_i(x) \end{aligned}$$

We wish to build a local coordinate frame $z = (z_1, \ldots, z_n)$ under the action of which the decoupled system looks like:

$$\tilde{f}(z) = \begin{pmatrix} \tilde{f}^1(z^1) \\ \vdots \\ \tilde{f}^m(z^m) \\ \tilde{f}^{m+1}(z) \end{pmatrix} \quad \tilde{g}(z) = \begin{pmatrix} \tilde{g}^1(z^1) & 0 & 0 & 0 \\ 0 & \tilde{g}^2(z^2) & . & . \\ . & . & . & . \\ . & . & . & . \\ . & . & . & \tilde{g}^m(z^m) \\ \tilde{g}_1^{m+1}(z) & \tilde{g}_2^{m+1}(z) & . & \tilde{g}_m^{m+1}(z) \end{pmatrix}$$

where the z^i are independent subsets of the coordinates z_j, with a proper numbering.

Definition 2.2

Let the regular controllability distributions be defined by:

$$R_i^* = \{ad_{\tilde{f}}^q, \tilde{g}_i, q \in \mathbb{N}\}_{\text{L.A.}} \qquad i \in \underline{m}$$

the abbreviation $\{\}_{\text{L.A.}}$ meaning Lie Algebra generated by the set of elements between $\{\}$.

Definition 2.3

We call radical of a set of distributions $R_1, \ldots, R_m$, the distribution:

$$\check{R} = \cap_{i \in \underline{m}} \check{R}_i \text{ with } \check{R}_i = \sum_{j \neq i} R_j \qquad j \in \underline{m}$$

Remark 2.1

In this section, like in the linear case, the concept of a radical will play a key role. The main idea to get the "decoupled canonical form" is to point out the set of dynamics of the decoupled system related to the radical $\check{R}^*$ of the set of distributions $R_i^*, \forall i \in \underline{m}$. In the above example, $\check{R}^*$ is not different from the intersection $R_1^* \cap R_2^*$.

Lemma 2.1 (Bournonville, 1984)

If R_i^* and $(R_i^* \cap G)$ have fixed dimensions, G denotes the distibution generated by the columns of g. If the decoupling condition is fulfilled, then: $\sum_{i \in \underline{m}} R_i^* = T\mathbb{R}^n$

Remark 2.2

As the feedback is regular, the same property holds for the decoupled system, if the system fulfils the strong accessibility condition (See appendix B).

Lemma 2.2 (Bournonville, 1984)

If R_i^* and $(R_i^* \cap G)$ have fixed dimensions and if the decoupling condition is fulfilled, then:

$$\begin{aligned} \left[\tilde{f}, \check{R}^*\right] &\subset \check{R}^* \\ \left[\tilde{g}_i, \check{R}^*\right] &\subset \check{R}^* \qquad \forall i \in \underline{m}. \end{aligned}$$

Proposition 2.1 (Bournonville, 1984)

(i) The radical $\check{R}^*$ of the set R_i^* is involutive.

(ii) $\check{R}^* + \sum_{i \in \underline{l}} R_i^*$ is involutive $\qquad \forall l \in \underline{m}$

(iii) The radical $\check{R}^*$ of the set R_i^* verifies $\check{R}^* \subset (\cap_{i \in \underline{m}} \mathrm{Ker}(\mathrm{d}h_i))$

Theorem 2.4 (Fundamental theorem)

Let the system

$$\begin{aligned} \dot{x}(t) &= \tilde{f}(x) + \sum_{i \in \underline{m}} \tilde{g}_i(x) \cdot v_i(t) \\ y_i(t) &= h_i(t) \end{aligned}$$

such that

(i) R_i^* et $(R_i^* \cap G)$ have fixed dimensions

(ii) the decoupling condition is fulfilled.

Then, there exists a local coordinate frame in which the decoupled system appears under the above "decoupled canonical form".

We do not provide here the detailed proof of this theorem. We ask the interested reader to check the thesis of F. Bournonville (1984) or the paper by Nijmeijer and Schumacher (1986). Indeed, in the present work, our purpose does not aim to underscore all the refinements of the proof, but to exhibit clearly the right procedure, to point out the efficient way leading to the "decoupled canonical form".

The goal of the method, of which we now sketch the main lines, aims to draw the dynamics of the decoupled system related to the radical $\check{R}^*$ of the set R_i^*. It is obvious that

$$\check{R}^* \subset (R_1^* + \check{R}^*) \subset \cdots \subset (R_1^* + R_2^* + \cdots + R_m^* + \check{R}^*)$$

However, we know that $\forall I \in 1, \ldots, m$, $\sum_{j \in \underline{I}} R_j^* + \check{R}^*$ is involutive. Lemma B.1 allows us to assert that, in a neighborhood of each $x \in M_0$, there exists a local coordinate frame $(x_1, \ldots, x_n)$ such that:

$$\check{R}^* = \text{sp}\left\{\frac{\partial}{\partial x_1}; \ldots; \frac{\partial}{\partial x_{\mu_o}}\right\}$$

$$\sum_{j \in \underline{i}} R_j^* + \check{R}^* = \text{sp}\left\{\frac{\partial}{\partial x_1}; \ldots; \frac{\partial}{\partial x_{\mu_o}}; \frac{\partial}{\partial x_{\mu_o+1}}; \ldots; \frac{\partial}{\partial x_{\mu_i}}\right\}$$

To simplify the notation, write:

$$\begin{aligned} 0 &= \mu_{-1} \\ d_o &= \mu_o \\ d_i &= \mu_i - \mu_{i-1}\text{ , for } i \in \underline{m} \end{aligned}$$

As a convention, denote by:

x^o the d_o first coordinates $(x_1, \ldots, x_{d_o})$ of x.

x^1 the d_1 following coordinates $(x_{d_o+1}, \ldots, x_{d_o+d_1})$ of x.

$\vdots$

x^m the d_m last coordinates of x.

In this local coordinate frame (**dropping, in what follows, the ˜ above f and above g, to simplify the notation**) the system is written as

$$\begin{pmatrix} \dot{x}^0 \\ \dot{x}^1 \\ \vdots \\ \dot{x}^{m-1} \\ \dot{x}^m \end{pmatrix} = \begin{pmatrix} f^0(x^0, x^1, \ldots, x^m) \\ f^1(x^1, \ldots, x^m) \\ f^2(x^2, \ldots, x^m) \\ \vdots \\ f^{m-1}(x^{m-1}, x^m) \\ f^m(x^m) \end{pmatrix} + \sum_{i \in \underline{m}} \cdot \begin{pmatrix} g_i^0(x^0, x^1, \ldots, x^m) \\ g_i^1(x^1, \ldots, x^m) \\ g_i^2(x^2, \ldots, x^m) \\ \vdots \\ g_i^{m-1}(x^{m-1}, x^m) \\ g_i^m(x^m) \end{pmatrix} v_i(t)$$

Now, since $\check{R}^* \subset \cap_{i \in \underline{m}} \text{Ker}(dh_i)$, we can, without loss of generality from an input–output viewpoint, write the system modulo the radical $\check{R}^*$. Then, we get the following system:

$$\begin{pmatrix} \dot{x}^1 \\ \vdots \\ \dot{x}^{m-1} \\ \dot{x}^m \end{pmatrix} = \begin{pmatrix} f^1(x^1, \ldots, x^m) \\ f^2(x^2, \ldots, x^m) \\ \vdots \\ f^{m-1}(x^{m-1}, x^m) \\ f^m(x^m) \end{pmatrix} + \sum_{i \in \underline{m}} \cdot \begin{pmatrix} g_i^1(x^1, \ldots, x^m) \\ g_i^2(x^2, \ldots, x^m) \\ \vdots \\ g_i^{m-1}(x^{m-1}, x^m) \\ g_i^m(x^m) \end{pmatrix} v_i(t)$$

which verifies the conditions of theorem B.5.

So, we can check that there exists a local coordinate frame $(z_1, \ldots, z_n - \mu_0)$ in which the system admits a parallel decomposition.

To simplify the notation, write:

z^1 the coordinates $(z_1, \ldots, z_{\mu_1-\mu_o})$

$\vdots$

z^i the coordinates $(z_{(\mu_i-\mu_0)+1}, \ldots, z_{\mu_i+1-\mu_0}) \quad i \in \underline{m}$

and when $\check{R}^* \neq 0$, we write $z^{m+1} = x^0$

In this local coordinate frame, the decoupled system takes the following form:

$$\begin{pmatrix} \dot{z}^1 \\ \vdots \\ \dot{z}^m \\ \dot{z}^{m+1} \end{pmatrix} = \begin{pmatrix} f^1(z^1) \\ \vdots \\ f^m(z^m) \\ f^{m+1}(z) \end{pmatrix} + \begin{pmatrix} g^1(z^1) & 0 & 0 & 0 \\ 0 & g^2(z^2) & . & . \\ . & . & . & . \\ . & . & . & . \\ . & . & . & g^m(z^m) \\ g_1^{m+1}(z) & g_2^{m+1}(z) & . & g_m^{m+1}(z) \end{pmatrix} \cdot \begin{pmatrix} v_1 \\ v_2 \\ \vdots \\ v_m \end{pmatrix}$$

Also, since the strong accessibility property holds (Bournonville, 1984),

$$R_i^* = \text{sp}\left\{\frac{\partial}{\partial z^i}\right\}$$

modulo $\check{R}^*$, for $i \in \underline{m}$. By construction, we have:

$$R_i^* \subset \cap_{i\in\underline{m}} \text{Ker}(\mathrm{d}h_j) \quad i \neq j \tag{I}$$

We know that

$$\check{R}^* = \text{sp}\left\{\frac{\partial}{\partial z^{m+1}}\right\} \subset \cap_{i\in\underline{m}}\text{Ker}\,(\mathrm{d}h_i) \tag{II}$$

From (I) and (II) it follows that

$$\text{sp}\left\{\frac{\partial}{\partial z^j}; \frac{\partial}{\partial z^{m+1}}\right\} \subset \cap_{i\in\underline{m}}\text{Ker}\,(\mathrm{d}h_i) \qquad i \neq j$$

and so: $\dfrac{\partial h_i}{\partial z^j} = 0 \;\; i \neq j \quad i, j \in \underline{m} \quad \Longrightarrow \quad h_i(z) = h_i(z^i) \qquad i \in \underline{m}$

Conclusion

We have just built a local coordinate frame $z = (z_1, \ldots, z_n)$ that splits up the decoupled system in a suitable "decoupled canonical form".

Notation

Let $\nabla_i^* = \sup I\{f_i, g_i, \text{Ker}\,(\mathrm{d}h_i)\}$ the supremal (f_i, g_i) invariant distribution included in $\text{Ker}\,(\mathrm{d}h_i)$.

Analyze in depth the subsystem

$$\begin{aligned} \dot{z}^i &= f_i(z^i) + g_i(z^i) \cdot v_i \\ y_i &= h_i(z^i) \end{aligned} \tag{III}$$

Two situations can occur: $\rho_i = \dim(z^i)$ or $\rho_i < \dim(z^i)$

If $\rho_i = \dim(z^i)$ there are no unobservable dynamics and the coordinate frame (z^i) can always be chosen such that

$$\dot{z}^i = \underbrace{\begin{pmatrix} 0 & 1 & . & . & . \\ 0 & 0 & 1 & . & . \\ . & . & . & . & . \\ 0 & . & . & . & 1 \\ 0 & . & . & . & 0 \end{pmatrix}}_{\rho_i * \rho_i} \cdot z^i + \begin{pmatrix} 0 \\ \vdots \\ 0 \\ 1 \end{pmatrix} \cdot v_i \tag{IV}$$

If $\rho_i < \dim(z^i)$ unobservable dynamics exist for the feedback (α, β)

Let $\nabla_i^* = \sup I\{f_i, g_i, \text{Ker}\,(dh_i)\}$, and let $L_i = \{ad_{f_i}^{j-1} g_i, j > 0\}_{\text{L.A.}}$. We can write:

$$L_i = \nabla_i^* \oplus \nabla_i^o$$

As we know that ∇_i^* and L_i are involutive, we can, from lemma B.1, build up a coordinate frame $z_i \rightarrow (\xi_i^1, \xi_i^2)$, so that (III) can be written as:

$$\begin{aligned} \dot{\xi}_i^1 &= f_i^1(\xi_i^1) + g_i^1(\xi_i^1) \cdot v_i \\ \dot{\xi}_i^2 &= f_i^2(\xi_i^1, \xi_i^2) + g_i^2(\xi_i^1, \xi_i^2) \cdot v_i \end{aligned}$$

Moreover, since $[f_i, \nabla_i^*] \subset \nabla_i^* \subset \text{Ker}\,(dh_i)$ and $[g_i, \nabla_i^*] \subset \nabla_i^* \subset \text{Ker}\,(dh_i)$ it is easy to show that $y_i = h_i(\xi_i^1)$.

Note again that the change of coordinates can always be built so that ξ_i^1 fulfils a differential system of type (IV).

Remark that a feedback like $v_i = \phi_i(\xi_i^1, \xi_i^2) + \theta_i(\xi_i^1, \xi_i^2) \cdot w_i$ can make ξ_i^2 observable. On the other hand, the subsystem (f_i, g_i) being strongly accessible, we can expect internal stabilization with the given feedback, like in the well-known linear situation. However, the reader must keep in mind that, in a nonlinear situation, the strong accessibilty does not necessarily imply stabilizability. This analysis, more accurate than the preceding ones already performed, show that, *a priori*, in some situations it will be possible to decouple while preserving "internal stability", but with a loss of linearity.

2.5 Decoupling solution through dynamic compensation

Above, we saw that static state feedback decoupling was achievable when rank $\Delta(x) = p$, on a submanifold M_0 of $\mathbb{R}^n$.

If this rank condition is not fulfilled, we must search for a dynamic solution; in other words, we must search for more general feedbacks than the simple static state feedback.

Before going to the core topic, we shall detail the classification, and even the clarification, concerning the feedbacks used in control problems. This classification, mainly due to the French school, is rather recent and can be found in several publications by Delaleau and Fliess (1992a), Fliess, Lévine, Martin and Rouchon (1992a), Martin (1992). The matter developed hereafter has been communicated by Lévine (1993).

2.5.1 New general concepts

2.5.1.1 The class of feedbacks

Let us introduce the three main types of dynamic feedbacks occurring in control problems.

First, remark that a feedback can be understood simply as a change of coordinates for the input function of the state.

Indeed, let us consider the system $\dot{x} = f(x, u)$. A regular static state feedback is a transformation $u = \alpha(x, v)$ with α invertible with respect to v. This transformation turns the open-loop system into the closed-loop system $\dot{x} = f(x, \alpha(x, v))$. Both systems are then said to be equivalent under static state feedback. This concept can be generalized if we furthermore consider a state change of coordinates $z = \phi(x)$ with ϕ invertible. In other words, both systems are equivalent under regular static state feedback, if there exist invertible transformations ϕ and α like $x = \phi(z)$ and $u = \alpha(x, v)$.

This last point of view allows us to provide definitions, more general, of feedbacks in terms of changes of state coordinates and input coordinates. We shall bring foward three main sets:

1. The exogenous feedbacks are the most general dynamic feedbacks. Within the framework of a state representation, they are written as: $u = \alpha(x, z, v)$ and $\dot{z} = a(x, z, v)$.

2. The endogenous feedbacks take into account the variables of the system as well as their derivatives. In fact, they are **invertible and without integration** transformations of the type:

$$\tilde{x} = a(x, u, \dot{u}, \ldots, u^{(r)}) \quad (2.1)$$
$$\tilde{u} = \alpha(x, u, \dot{u}, \ldots, u^{(s)}) \quad (2.2)$$

and

$$x = \tilde{a}(\tilde{x}, \tilde{u}, \dot{\tilde{u}}, \ldots, \tilde{u}^{(\tilde{r})}) \quad (2.3)$$
$$u = \tilde{\alpha}(\tilde{x}, \tilde{u}, \dot{\tilde{u}}, \ldots, \tilde{u}^{(\tilde{s})}). \quad (2.4)$$

3. The quasi-static state feedbacks are endogenous feedbacks which preserve the state:

$$v = \alpha(x, u, \dot{u}, \ldots, u^{(s)}). \quad (2.5)$$

The quasi-static state feedbacks are basic tools for solving the dynamic decoupling problem, as shown by Delaleau (1993). The endogenous feedbacks have been introduced by Martin (1992). They appear, for the first time, in A. M. Perdon *et al.* (1989), but under no particular name.

Definition 2.4

A feedback is endogenous if it has the form:

$$v = \alpha(x, u, \dot{u}, \ldots, u^{(r)}) \tag{2.6}$$

which is invertible, in other words there exists $z = \kappa(x, u, \dot{u}, \ldots, u^{(r)})$ *(possibly empty) such that* $u = \beta(x, z, v, \dot{v}, \ldots, v^{(s)})$.

Obviously, the feedback is static only in the particular case when it is function of (x, u) *but not of any derivative of the input* u. *The invertibility is equivalent to* rank $(\frac{\partial \alpha}{\partial u}) = m$.

Thus, an endogenous feedback is a dynamic feedback using no variable external to the system. Indeed, for this kind, we can express v in terms of x, u and a finite number of its derivatives and, conversely, we can express u in terms of x, v and a finite number of its derivatives and possibly an additional state z, which can be built up from the variables of the system.

Example 2

Add to the original system the integrator $\dot{u} = v$. It is an endogenous feedback, since $v = \dot{u}$ is actually a function of x, of u and of $\dot{u}$. Conversely, if we add the variable $z = u$, obviously we have $u = z$, which is really a function of x, z, v and of its derivatives, and z is thus an endogenous variable, and therefore we can derive the following definition.

2.5.1.2 The link with the decoupling problem

Definition 2.5

We say that the original system is decouplable, if there exists an endogenous feedback such that the closed-loop system is decoupled.

Like in section 2.3, we shall differentiate the output y but now find an expression of y and its derivative functions of u and its derivatives where the rank, with respect to u and its derivatives, is maximal. The procedure provides, in fact, a correct algorithm to determine an endogenous decoupling feedback. We shall point out, through examples, that the system is decouplable if and only if the rank equals the number of outputs, say p ($p \leq m$). As we shall see below, the latter condition is equivalent to the (right) invertibility of the original system.

Let us start with an example where static state feedback is achievable.

Example 3

Let us consider a three-dimensional system with two inputs and two outputs:

$$\left\{ \begin{array}{rcl} \dot{x}_1 & = & u_1 x_3 \\ \dot{x}_2 & = & u_2 x_1 \\ \dot{x}_3 & = & x_2 + u_1 x_2 \\ y_1 & = & x_1 \\ y_2 & = & x_2 \end{array} \right. \tag{2.7}$$

Calculate the output derivatives.

We have $\dot{y}_1 = u_1 x_3$ and $\dot{y}_2 = u_2 x_1$. We verify immediately that, with the static state feedback $u_1 = \dfrac{v_1}{x_3}$ and $u_2 = \dfrac{v_2}{x_1}$, we get the linear decoupled system:

$$\begin{cases} \dot{y}_1 &= v_1 \\ \dot{y}_2 &= v_2. \end{cases} \tag{2.8}$$

By the way, note that the closed-loop system loses one state through this procedure.

Example 4

Now, we consider again the same example but with a different output.

$$\begin{cases} \dot{x}_1 &= u_1 x_3 \\ \dot{x}_2 &= u_2 x_1 \\ \dot{x}_3 &= x_2 + u_1 x_2 \\ y_1 &= x_1 \\ y_2 &= x_3 \end{cases} \tag{2.9}$$

As previously, calculate the derivatives of the outputs.

We have $\dot{y}_1 = u_1 x_3$ and $\dot{y}_2 = x_2(1 + u_1)$.

The input u_2 does not appear in these first derivatives, which implies that the rank of the Δ matrix of the differentials of $\dot{y}_1$ and $\dot{y}_2$ with respect to (u_1, u_2) is 1. Indeed,

$$\Delta = \begin{pmatrix} x_3 & 0 \\ x_2 & 0 \end{pmatrix}.$$

Now, we must find u_2. Calculate $\ddot{y}_1 = \dot{u}_1 x_3 + u_1 x_2(1 + u_1)$ and $\ddot{y}_2 = \dot{u}_1 x_2 + u_2 x_1(1 + u_1)$. After that, the Δ_1 matrix of differentials of $\ddot{y}_1$ and $\ddot{y}_2$ with respect to $(u_1, u_2, \dot{u}_1, \dot{u}_2)$ has a rank equal to 2:

$$\Delta_1 = \begin{pmatrix} x_2(1 + 2u_1) & 0 & x_3 & 0 \\ u_2 x_1 & x_1(1 + u_1) & x_2 & 0 \end{pmatrix}.$$

The endogenous dynamic compensator is given by $\dot{u}_1 x_3 + u_1 x_2(1 + u_1) = v_1$ and $\dot{u}_1 x_2 + u_2 x_1(1 + u_1) = v_2$, which yields the linear decoupled system:

$$\begin{cases} \ddot{y}_1 &= v_1 \\ \ddot{y}_2 &= v_2. \end{cases} \tag{2.10}$$

To check that the above compensator is endogenous, it is sufficient to express u as a function of v, of x and of an endogenous state z, eventually. Let us pose $z = u_1$. We then have (exercise) $u_2 = \dfrac{v_2 x_3 - v_1 x_2 + z x_2(1 + z)}{x_1 x_3 (1 + z)}$. Thus, the system is decouplable and linearizable using an endogenous feedback. To summarize, the rank of Δ_1 ($= 2$) plays a key role in the existence of an endogenous feedback which decouples and linearizes simultaneously. Note that the dimension of the state x of the original system is 3, while the decoupled system has a state, for instance $(y_1, \dot{y}_1, y_2, \dot{y}_2)$ of dimension 4, which makes it necessary to introduce an additional state z in order to inverse the feedback.

Before giving the characterization of dynamically decouplable systems, let us underscore that the respective ranks of the matrices Δ and Δ_1, in the second example, are the respective dimensions of the vector spaces generated by $(\mathrm{d}\dot{y}_1, \mathrm{d}\dot{y}_2)$ and $(\mathrm{d}\dot{y}_1, \mathrm{d}\dot{y}_2, \mathrm{d}\ddot{y}_1, \mathrm{d}\ddot{y}_2)$, where $dy_i^{(j)}$ denotes the differential of the jth derivative, with respect to the time, of the output y_i. It is the Jacobian matrix of $y_i^{(j)}$ with respect to $x, u, \dot{u}, \ldots, u^{(j-1)}$, where $u^{(r)}$ denotes the derivative of order r, with respect to the time, of the input u.

2.5.1.3 The Hilbert polynomial and related concepts

As sketched just above, a suitable way for calculating those ranks amounts to evaluating the sequel of vector spaces $\mathcal{E}_k$ for $k \geq 0$, called state-input filtration, by means of the formula:

$$\mathcal{E}_0 = \mathrm{sp}\{dx, dy_i,\ i = 1, \ldots, m\} \tag{2.11}$$

and

$$\mathcal{E}_k = \mathrm{sp}\{dx, dy_i, dy_i^{(j)}, i = 1, \ldots, m; 1 \leq j \leq k\} \tag{2.12}$$

We shall admit the following result

Theorem 2.5

There exist three integers k_0, ρ and δ such that

$$\dim \mathcal{E}_k = \rho(k+1) + \delta \quad \forall k \geq k_0 \tag{2.13}$$

with $\rho \leq m$, $\delta \leq n$, $k_0 \leq n-1$ and

$$\dim \mathcal{E}_k \neq \rho(k+1) + \delta \quad \forall k < k_0 \tag{2.14}$$

Definition 2.6

- *The polynomial $P(k) = \rho k + \delta$ is named the* Hilbert polynomial *associated with the filtration $\{\mathcal{E}_k\}_{k \geq 0}$.*
- *k_0 is named the* index.
- *ρ is named the* output differential rank.
- *δ is named the* input–output defect.

The interpretation of these integers can be performed through examples 3 and 4.

In example 3, we had:

- $\mathcal{E}_0 = \mathrm{sp}\{dx_1, dx_2, dx_3\}$ and so $\dim \mathcal{E}_0 = 3$,
- $\mathcal{E}_1 = \mathrm{sp}\{dx_1, dx_2, dx_3, du_1, du_2\}$, or $\dim \mathcal{E}_1 = 5$,

which gives: $\dim \mathcal{E}_1 - \dim \mathcal{E}_0 = 2 = m$. Thus, we have

- $\rho = m = 2$,
- $k_0 = 0$,
- $\delta = 1$ since $\delta = \dim \mathcal{E}_0 - 2(0+1) = 3 - 2$.

The output differential rank equals the number of independent inputs ($m \leq p$) or outputs ($p \leq m$). In some sense, it measures the degree of independence of the outputs. This concept, introduced by Fliess in 1985, is equivalent, in the nonlinear case, to the transfer matrix rank of a linear system. The differential rank can be calculated, in practice, by means of other algorithms not reported here. The interested reader can study the paper by Di Benedetto, Grizzle and Moog (1989). Note that, from an engineer viewpoint, all the systems have a generic output differential rank equal to the output number. Such systems, with (diff. rank $y = p$) are *right invertible*. The index (plus one) measures the number of derivatives after which all the inputs appear in the outputs explicitly, but also, the maximum order of the decoupled system. The input–output defect measures the "maximal number of states" which disappear under decoupling, in other words which become unobservable after feedback.

In example 4, we had:

- $\mathcal{E}_0 = \mathrm{sp}\{\mathrm{d}x_1, \mathrm{d}x_2, \mathrm{d}x_3\}$, or $\dim \mathcal{E}_0 = 3$,
- $\mathcal{E}_1 = \mathrm{sp}\{\mathrm{d}x_1, \mathrm{d}x_2, \mathrm{d}x_3, \mathrm{d}u_1\}$, or $\dim \mathcal{E}_1 = 4$,
- $\mathcal{E}_2 = \mathrm{sp}\{\mathrm{d}x_1, \mathrm{d}x_2, \mathrm{d}x_3, \mathrm{d}u_1, \mathrm{d}\dot{u}_1, \mathrm{d}u_2\}$, or $\dim \mathcal{E}_2 = 6$.

We get

- $\rho = \dim \mathcal{E}_2 - \dim \mathcal{E}_1 = 2 = m$,
- $k_0 = 1$,
- $\delta = \dim \mathcal{E}_1 - \rho(1+1) = 4 - 4 = 0$.

The state does not become unobservable under feedback. The order of the decoupled system equals 2.

Definition 2.7

A compensator, or a dynamic feedback, with the "output" u is said to be regular if the differential rank of u equals m.

Definition 2.8

The nonlinear system under consideration ($p \leq m$) is said to be (right) invertible if

$$\rho = p \leq m$$

where

$$\dim \mathcal{E}_n - \dim \mathcal{E}_{n-1} = \rho$$

When p = m, right invertibility does not differ from invertibility.

2.5.2 More on the decoupling problem

2.5.2.1 The main result

Theorem 2.6

The nonlinear system (h, f, g) is decouplable (by regular endogenous feedback) if and only if it is (right) invertible, on M_0.

The proof of this result can be performed through different ways. The best is, without any doubt, provided by Fliess (1985). Other proofs, as well as detailed constructions of the compensator, can be found in several other works.

Practically speaking, it is enough to follow the calculations performed in the above examples, building up, at each output differentiation step, the corresponding Jacobian.

At the limit step k_0, in other words, when the upper bound of the Hilbert polynomial is reached, two situations must be investigated:

1. either the system is invertible ($\rho = m$) and one can calculate the endogenous dynamic feedback with a proper choice of a basis of $\mathcal{E}_{k_0+1}$ and at last by equalling the highest order output derivatives, in this basis, with the new input v, or
2. the system is not invertible ($\rho < m$), and thus it is not decouplable.

We must notice that, a quasi-static state feedback, which is an elegant theoretical tool, is not physically useable since it involves derivatives of the new input v. For practical reasons, it is then necessary to introduce internal variables, like $z = \dot{v}$, which amounts to introducing integration rather than differentiation.

On the other hand, and always from a practical viewpoint, it can be more powerful to proceed algorithmically. The approach taken below, uses an algorithmic construction of systems Σ_k which "converge" to Σ_K for $K < n$, under the assumption of right invertibility, Σ_K being statically decouplable. It can be found in Descusse and Moog (1985, 1987), who provided the first solution to the dynamic decoupling problem. Other approaches are available. They can be found in the book by Isidori (1989).

The system Σ_K being statically decouplable, it suffices to apply the results of theorem 2.1.

2.5.2.2 A practical solution

Definition 2.9

We say that a row (respectively, a column) of an (analytic) matrix is essential when it is not linearly dependent (on the ring of analytic functions) of other rows (respectively, columns) of this matrix.

Hereafter, ness $\Delta(x)$ denotes the non essential rows of $\Delta(x)$.

Algorithm 1

Step 1 Let us consider the system $\Sigma_0(h, f, g)$.
Calculation of ρ_i and of $\Delta(x)$.
If rank $\Delta(x) = m$, on some submanifold M_0 of $\mathbb{R}^n$: then stop.

Step 2 We define $\text{Ker}\,(\text{ness}\,\Delta(x)) = G_2$, and we choose G_1 such that $[G_1, G_2]$ is regular.
We write, $u = G_1 v_1 + G_2 v_2$

Remark 2.3

ness $\Delta(x) = \Phi \iff \text{rank}\,\Delta(x) = m \iff$ *the system is decouplable statically.*

Step 3 We put integrators in series with v_1. We get the extended system Σ_1:

$$\begin{aligned} \dot{x} &= f + g \cdot G_1 \cdot v_1 + g \cdot G_2 \cdot v_2 \\ \dot{v}_1 &= w_1 \end{aligned}$$

Step 4 Go back to step 1, turning Σ_0 into Σ_1.

Example 5

Let us consider again the system Σ, defined by:

$$\begin{pmatrix} \dot{x}_1 \\ \dot{x}_2 \\ \dot{x}_3 \end{pmatrix} = \begin{pmatrix} 0 \\ x_3 \\ 0 \end{pmatrix} + \begin{pmatrix} x_1 & 0 \\ -x_3 & 0 \\ 0 & x_1 \end{pmatrix} \cdot \begin{pmatrix} u_1 \\ u_2 \end{pmatrix}$$

$$\begin{aligned} y_1 &= x_1 \\ y_2 &= x_2 \end{aligned}$$

A simple calculation gives $\rho_1 = \rho_2 = 1$, and

$$\Delta(x) = \begin{pmatrix} x_1 & 0 \\ -x_3 & 0 \end{pmatrix}$$

It is clear that $\text{rank}\,\Delta(x) = 1$

Conclusion

The system is not decouplable using static state feedback.

Looking at $\Delta(x)$, we can indeed easily see that the input u_2 operates with a "delay" on the outputs contrary to the input u_1. It follows that a static state feedback cannot achieve decoupling. A dynamic compensator is necessary to expect decoupling.

As an exercise, the reader will check that the above algorithm delays the action of u_1, putting one integrator in series with this input, which allows to u_2 to play its role of decoupling. The extended system we get does not differ from example 1.

2.5.3 *More on the index, the output differential rank and the input–output defect*

These integers are closely related to some observability and stabilizability properties of decouplable systems. Their respective interpretations are sketched below.

2.5.3.1 Input–output defect, infinite zero structure

Let us recall the main results from the preceding sections:

Theorem 2.7

If the nonlinear system (h, f, g) is (right) invertible, there exist a local diffeomorphism ϕ and a decoupling endogenous feedback κ which turn the state x and the input u of the system (h, f, g) into the vectors $(z, \xi) = \phi(x, u, \dot{u}, \ldots, u^{(k_0+1)})$ and $(w, v) = \kappa(x, u, \dot{u}, \ldots, u^{(k_0+1)})$, with $\dim z = n - \delta$, $\dim \xi = \delta$, $\dim v = m$ and $\dim \bar{w} \leq m(k_0 + 1)$ and the dynamic of which is described by:

$$\begin{cases} \dot{z}_i & = & A_i z_i + w_i \\ w_i^{(\chi_i)} & = & v_i \ , \quad i = 1, \ldots, m, \\ \dot{\xi} & = & \psi(\xi, z, \bar{w}, v) \\ y_i & = & z_i^1 \ , \quad i = 1, \ldots, m. \end{cases} \tag{2.15}$$

where the size of the matrix A_i is $n_i^{'} \times n_i^{'}$, with $\sum_{i=1}^{m} n_i^{'} = n - \delta$:

$$A_i = \begin{pmatrix} 0 & 1 & 0 & \cdots & 0 \\ 0 & 0 & 1 & \cdots & 0 \\ & & & \ddots & \\ 0 & & \cdots & & 1 \\ 0 & 0 & 0 & 0 & 0 \end{pmatrix},$$

where z_i^1 is the first component of z_i, where the χ_i are integers (possibly zero) less than or equal to $k_0 + 1$ and where $\bar{w} = (w, \dot{w}, \ldots, w^{(\alpha)})$.
Moreover, $(z, \xi) = \phi(x, u, \dot{u}, \ldots, u^{(k_0+1)})$ and $y_i = z_i^1$ imply $x = \tilde{\phi}(y, \dot{y}, \ldots, y^{(k_0+1)}, \xi)$. On the other hand, $u = \tilde{\kappa}(y, \dot{y}, \ldots, y^{(k_0+1)}, \xi)$ where $\tilde{\phi}$ and $\tilde{\kappa}$ represent the inverses of ϕ and κ.

We can emphasis the relationships that exist between, on the one hand, the index, the output differential rank and the input–output defect and, on the other hand, the concept of the infinite zero structure of the nonlinear system (h, f, g). The latter is not new. It has been used commonly in the literature for many years; see, for example, the paper by Di Benedetto, Grizzle and Moog (1989).

Note:

$$\nu_i = \dim \mathcal{E}_i / \mathcal{E}_{i-1} \ , \quad i \geq 1 \ , \quad \nu_O = 0 \ , \tag{2.16}$$

the sentence of integers issued from the dimensions of the state-input filtration. From the Hilbert theorem, we have $\nu_{k_0+1} = \rho$. We also introduce the sentence of integers:

$$n_i^{'} = \#(\{\nu_j \geq i \mid j \leq k_0 + 1\}) \ , \quad i = 1, \ldots, \rho. \tag{2.17}$$

where we write $\#(A)$ the cardinal of the elements of a set A.

Definition 2.10

The infinite zero structure of a nonlinear system (h, f, g) is defined as the sentence of the following integers $\{n_1^{'}, \ldots, n_\rho^{'}\}$.

Theorem 2.8

The infinite zero structure of a nonlinear system (h, f, g) verifies

- $\delta = n - \sum_{i=1}^{m} n_i'$,
- $k_0 + 1 = \max_{i=1,\ldots,m} n_i'$.

Moreover, the dimension $\bar{n}$ of the closed-loop system verifies

$$n \leq \bar{n} \leq m(k_0 + 1) + \delta \tag{2.18}$$

and the dimension $\bar{m}$ of the decoupling compensator verifies

$$0 \leq \bar{m} \leq m(k_0 + 1) - \sum_{i=1}^{m} n_i'. \tag{2.19}$$

When a static state feedback decouples the system, we have $n_i = n_i'$, $i = 1, \ldots, m$ where the n_i denote the row infinite zeros of (h, f, g).

The input–output defect δ measures the dimension of the zero dynamics of (h, f, g).

The infinite zero structure of a nonlinear system (h, f, g), the name of which comes from the frequential theory of linear systems, measures the respective relative orders of the decoupled chains. For example 3, we easily check that only $\nu_1 = \dim \mathcal{E}_1 - \dim \mathcal{E}_0 = 5-3 = 2$ is defined since $k_0 = 0$, which leads to $n_1' = \#(\{\nu_1 \geq 1\}) = 1$ and $n_2' = \#(\{\nu_1 \geq 2\}) = 1$. The infinite zero structure is thus $\{1, 1\}$, which means that the decoupled system has two decoupled chains, each one having a relative order equal to 1, what is in accordance with the preceding solution: $\dot{y}_i = v_i$, $i = 1, 2$. We also verify that $\delta = 1 = n - n_1' - n_2'$ and that the dimension $\bar{m}$ of the compensator is $0 \leq \bar{m} \leq m(k_0 + 1) - n_1' - n_2' = 2 - 2 = 0$ which is consistent with a static state feedback.

We suggest that the reader also performs the above calculations for example 4, as an exercise.

2.5.3.2 Decoupling and stabilizability

Let us consider again example 3.

If we wish both to decouple and stabilize around the origin, we must stabilize each decoupled chain, and therefore we must have, at least:

$$\dot{y}_i = -K_i y_i\,, \quad i = 1, 2\,,$$

with $K_i > 0$. Recall that the feedback is given by $u_1 = \dfrac{v_1}{x_3}$, $u_2 = \dfrac{v_2}{x_1}$. Thus the closed-loop input u_1 is $u_1 = \dfrac{-K_1 y_1}{x_3} = \dfrac{-K_1 x_1}{x_3}$ and the unobservable dynamic is then given by

$$\dot{x}_3 = x_2(1 - \frac{K_1 x_1}{x_3})$$

The point $x_3 = 0$ cannot be an equilibrium point, which finally proves that decoupling and stabilization around the origin are not simultaneously achievable.

2.6 Some practical difficulties

1. **The state has to be measured.** If this is not the case, almost nothing can be done, at the present time.

   - The theory of nonlinear observers has not been brought to enough maturity contrary to the linear situation.
   - A priori, the separation principle is no longer available.

2. **The practical determination of the singular points** of a vector field is not an easy task to perform.

3. As for the **regularity of the decoupling matrix, there are two difficulties:**

   - (a) calculation of the generic rank of $\Delta(x)$
   - (b) when the generic rank of $\Delta(x)$ equals n, determination of the singular points of $\Delta(x)$

4. The **equibrilium points of f are not always kept** in the transformation $f \rightarrow f + g\alpha$. Such a situation may happen, especially when an equilibrium point of f is also a singular point of $\Delta(x)$.

5. **Going through a singularity of** $\Delta(x)$ remains an open problem, because, at the present time, we do not have the right techniques to get over the singularities. The compulsory constraint we have to respect, is to keep the natural directions of the vector fields, at the singular points.

6. Some classes of **systems** can be **mathematically decouplable** but the **feedback laws** are **physically unuseable**. For instance, this is the case when singular perturbations must be considered, situation which generates "high gains" and yields nonrobust solutions.

7. If the **zero dynamics** are **not stable in the sense of Lyapunov**, we do not know how to decouple and stabilize in a general case. This situation widely differs from the linear case, where the problem is solved completely.

8. In section 2.4, we saw that **the change of coordinates Φ and Ψ have antagonist roles.** The first one easily gives a subclass of the decoupling feedbacks, while the second one explicitly gives $f_{m+1}(z)$ and the $g_i^{m+1}(z)$. Conversely, to get the whole class of decoupling feedbacks, we need to calculate Ψ^{-1}, an operation which is not easy to perfom, in general. The calculation of decoupling feedbacks, as a function of the state x, is not straight forward.

   As for the practical determination of the change of coordinates, other difficulties emerge. They will be reported at the end of appendix B, dedicated to mathematical tools necessary for evaluation.

2.7 Conclusion

In this chapter, we pointed out a necessary and sufficient decoupling condition and gave solutions by means of some particular static or dynamic decoupling feedbacks. We saw that the latter should be completed by new feedbacks in order to get suitable dynamics.

The determination of the whole class of decoupling feedbacks as well as the analysis of the stability of the closed-loop system require a decoupled canonical form. When the sum of the row of infinite zeros equals n, the situation is comfortable: no unobservable dynamics and an easy characterization of all the decoupling feedbacks (for the square case). In the contrary situation, things are not so good and the necessity of an attractive decoupled canonical form becomes evident. We already stressed the fact that the determination of the decoupled canonical form as well as the one of the class of decoupling feedbacks could not be, in practice, performed simultaneouly. We use Ψ for the canonical form and Φ for the class of decoupling feedbacks. The determination of Φ and, more particularly the one of Φ_{m+1}, can be improved with the help of the developments used to obtain Ψ.

We have to pay special interest to the change of coordinates which is needed in the static state feedback decoupling theory.

This is the reason why we show in appendix B the fundamental theorems useful to rectify vector fields and coordinate frames and the Frobenius theorem. They are crucial for the calculation of Ψ.

2.8 Bibliography

[1] CARTAN, E. (1945), *Les systèmes différentiels extérieurs et leurs applications géométriques*, Hermann Paris.

[2] GODBILLON, C. (1969), *Géométrie différentielle et mécanique analytique*, Hermann Paris - Collection Méthodes.

[3] SPIVAK, M. (1979), *Differential geometry*, Publish or Perish, Washington, USA.

[4] BOOTHBY, WILLIAM M. (1975), *An introduction to differentiable manifolds and Riemannian geometry*, Academic Press, New-York, USA.

[5] BOURNONVILLE, F. (1984), *Quelques éléments de géométrie différentielle à l'usage de l'automatique non-linéaire*, E.N.S.M. Nantes, France. "*Découplage des systèmes non linéaires affines*", Thesis, E.N.S.M. Nantes, France.

[6] FLIESS, M. (1985), A new approach to the non-interacting control problem in nonlinear system theory, *Proc. 23rd Allerton Conf.,* 123–129.

[7] DESCUSSE, J. and MOOG, C.H. (1985), Decoupling with dynamic compensation for strong invertible affine nonlinear systems, *Internat. J. Control,* **42**, 1387–1398.

[8] NIJMEIJER, H. and SCHUMACHER, J.M. (1986), The regular local noninteracting control problem for nonlinear control systems, *SIAM J. Control & Optimization,* **24**, 1232–1245.

[9] DESCUSSE J. and MOOG, C. (1987), Dynamic decoupling for right-invertible nonlinear system, *System and control Letters*, **8**, 345–349.

[10] DI BENEDETTO, M.D., GRIZZLE, J.W. and MOOG, C.H. (1989), Rank invariants of nonlinear systems, *SIAM J. Control & Optimization*, **27**, 658–672.

[11] ISIDORI, A. (1989), *Nonlinear Control System*, 2nd Ed. Springer Verlag. Communications and Control Engineering.

[12] WONHAM, MURRAY W. (1989), *Linear Multivariable Control a Geometric Approach*, 2nd Ed. Springer Verlag, Applications of Mathematics, **10**.

[13] PERDON, A.M., CONTE, G. and MOOG, C.H. (1989), *Some canonical properties of nonlinear systems*, Proceedings of the International Symposium MTNS–89, Vol. 1; Birkhäuser; Boston–Basel–Berlin, 1990.

[14] DELALEAU, E. and FLIESS, M. (1992a), Algorithme de structure, filtrations et découplage, *C.R. Acad. Sci. Paris*, **315**, I, 101–106.

[15] DELALEAU, E. and FLIESS, M. (1992b), *An algebraic interpretation of the structure algorithm with an application to feedback decoupling*, Proceedings of IFAC-NOLCOS'92 Conference, 489–494, Bordeaux, France.

[16] DELALEAU, E. (1993), Sur les dérives de l'entrée en représentation et commande des systèmes non linéaires, Thesis, University of Paris-sud (Orsay), 2645.

[17] FLIESS, M., LÉVINE, J., MARTIN, P. and ROUCHON, P. (1992), Sur les systèmes non linéaires différentiellement plats, *C.R. Acad. Sci. Paris*, I, 619–624.

[18] FLIESS, M., LÉVINE, J., MARTIN, P. and ROUCHON, P. (1992), *On differentially flat nonlinear systems*, Proceedings of IFAC NOLCOS'92 Conference, 408–412, Bordeaux, France.

[19] MARTIN, PH. (1992), Contribution à l'étude des systèmes différentiellement plats, Thesis ENSMP, Paris, France.

[20] LÉVINE, J. (1993), Commande non linéaire : aspects qualitatifs, Lecture note given at the École Nationale Supérieure des Mines de Paris (available on demand from the author).

CHAPTER 3

A graph-theoretic approach to input–output decoupling and linearization

J. LÉVINE

3.1 Introduction

In this chapter, we develop an algorithmic point of view of input–output decoupling in terms of the *system graph*. This approach was first introduced in the nonlinear context and for the decoupling problem by *static* state feedback, (Kasinski and Lévine, 1984), and then improved De Luca *et al.* (1985) and D'Andrea-Novel and Lévine (1986) (for linear systems, the reader may refer to the books of Murota (1980) and Reinschke (1988) and the references therein). The extension of graph techniques to dynamic feedback decoupling has, until now, only received preliminary answers (Svaricek, 1992; Svaricek and Schwarz, 1993). We here present new results on the subject, completing the ones described in Chapter 2. They will be detailed in section 3 of this chapter, after having recalled the former static state feedback method of Kasinski and Lévine (1984).

We consider an input–output representation of the form:

$$\begin{cases} \dot{x} &= f(x,u) \\ y &= h(x,u) \end{cases} \tag{3.1}$$

where the state x lives in an n-dimensional analytic manifold X and the input u and the output y are m-dimensional (square system). f and h are analytic.

Recall that the input–output decoupling and linearization problem by static feedback consists in finding a feedback law of the type $u = \alpha(x, v)$, defined in an open subset V of X and invertible with respect to v for every x, such that the ith output of the closed-loop system is the output of a linear decoupled system, depending only on v_i, $i = 1, \ldots, m$, *i.e.*, for suitable integers k_i

$$\frac{d^{k_i} y_i}{dt^{k_i}} = v_i\,, \quad i = 1, \ldots, m\,. \tag{3.2}$$

The input–output decoupling and linearization problem by dynamic feedback consists in finding a dynamic compensator $\dot{w} = a(x, w, v)$, $u = \alpha(x, w, v)$, defined in an open subset

$V \times W$ of $X \times \mathbb{R}^q$ where q is the dimension of the vector w, such that, as before, the ith output of the closed-loop system is the output of a linear decoupled system, depending only on v_i, $i = 1, \ldots, m$, of the form (3.2).

Remark 3.1

The input–output approach, very popular among system theorists and control engineers, has, in the nonlinear case, a different status. In what follows, the outputs are used as a means to summarize, more or less artificially, the control objectives, and should not be confused with the measurements, provided by sensors, that describe the only information available on the state of the system to synthesize the feedback. The considered feedback laws, static or dynamic, will indeed need an exact knowledge of the state at every time, or, at least, a state reconstructor such as an observer or a filter.

However, when the outputs are suitably chosen (for example by application of input-to-state methods or by physical considerations), the input–output decoupling and linearizing approach may sometimes lead to a surprisingly simple design.

Remark 3.2

Though unnecessary, the linearity of the vector field f with respect to u is usually assumed. In fact, the system (3.1) naturally satisfies this property up to the addition of an integrator of the form $\dot{u} = v$, since u becomes an extended state and the linearity with respect to v is obvious. Nevertheless, in practice, such an extension can sometimes reveal unacceptable due, for example, to the increase of parameters to be tuned, and a direct design using the form (3.1) may be preferable.

3.2 Decoupling and linearization by static feedback and the system graph

Let us start this section with some recalls about input–output decoupling and linearization by static feedback for systems affine in the control. The graph theoretic algorithm will then be presented in this special case.

3.2.1 Some recalls on static feedback decoupling

Consider the input–output representation:

$$\begin{cases} \dot{x} &= f(x) + \sum_{i=1}^{m} u_i g_i(x) \\ y &= h(x) \end{cases} \tag{3.3}$$

where x evolves in an analytic manifold X of dimension n. For simplicity's sake, we assume that the number of inputs equals the number of outputs, or $y \in \mathbb{R}^m$ and $u \in \mathbb{R}^m$ (square system).

Recall the following notations:

Definition 3.1

The ith characteristic index, *relative to an open set* V, *denoted by* ρ_i, $i = 1, \ldots, m$, *is the unique integer defined by*

$$\rho_i = \inf\{k \in \mathbb{N} | \exists j \in \{1, \ldots, m\} : L_{g_j} L_f^k h_i(x) \neq 0\,,\ \forall x \in V\} \tag{3.4}$$

with the convention $\rho_i = +\infty$ *if* $L_{g_j} L_f^k h_i \equiv 0\ \forall j = 1, \ldots, m$ *and* $\forall k \geq 0$, *and with the notation* $L_{g_j} L_f^0 h_i = L_{g_j} h_i$.

The ith characteristic index ρ_i may be interpreted as the minimum number of integrations after which at least one input starts modifying the ith output.

The *decoupling matrix* Δ is the functional matrix whose entry corresponding to the ith line and jth column, denoted by $\Delta_{i,j}$, is given by:

$$\Delta_{i,j}(x) = L_{g_j} L_f^{\rho_i} h_i(x)\ ,\ i, j = 1, \ldots, m\,. \tag{3.5}$$

We report the following classical result Isidori, 1989; Nijmeijer and Der Schaft, 1990.

Theorem 3.1

The necessary and sufficient condition for the system (3.3) to be decoupled and linearized by static feedback in an open subset V *of* X *is that the decoupling matrix is invertible in* V.

The feedback is then given by

$$u = \Delta^{-1}(-\Delta_0 + v) \stackrel{\text{def}}{=} \alpha + \beta v \tag{3.6}$$

where $\Delta_0(x) = (L_f^{\rho_1+1} h_1(x), \ldots, L_f^{\rho_m+1} h_m(x))^T)$.

Moreover, if Δ *is invertible, we have* $m+\rho_1+\cdots+\rho_m \stackrel{\text{def}}{=} p \leq n$ *and if we set* $\xi_{j+1}^i = L_f^j h_i$, $0 \leq j \leq \rho_i$, $i = 1, \ldots, m$, *there exists* $n-p$ *independent functions of the* $L_f^j h_i$, *denoted by* η_i, $i = 1, \ldots, n-p$, *such that if we note* $\zeta_i = \eta_i(x)$, $i = 1, \ldots, n-p$, *and* $(\xi, \zeta)^T = \Phi(x)$, Φ *is a diffeomorphism of* V *and the system (3.3) with the feedback (3.6) is equivalent to*

$$\left\{\begin{array}{lcl} \left\{\begin{array}{lcl} \dot{\xi}_1^i & = & \xi_2^i \\ & \vdots & \\ \dot{\xi}_{\rho_i+1}^i & = & v_i \end{array}\right. \quad i = 1, \ldots, m & & \\ \dot{\zeta}_1 & = & L_{(f+\sum_{i=1}^m \alpha_i g_i)} \eta_1 \circ \Phi^{-1} + \sum_{i=1}^m v_j \beta_{i,j} L_{g_i} \eta_1 \circ \Phi^{-1} \\ & \vdots & \\ \dot{\zeta}_{n-p} & = & L_{(f+\sum_{i=1}^m \alpha_i g_i)} \eta_{n-p} \circ \Phi^{-1} + \sum_{i=1}^m v_j \beta_{i,j} L_{g_i} \eta_{n-p} \circ \Phi^{-1} \\ y_1 & = & \xi_1^1 \\ & \vdots & \\ y_m & = & \xi_1^m\,. \end{array}\right. \tag{3.7}$$

This result may be interpreted as follows. Use the notation $y_i^{(\nu)} = \dfrac{d^\nu y_i}{dt^\nu}$ for $\nu \geq 0$ and $i = 1, \ldots, m$, with the convention $y_i^{(0)} = y_i$. A direct computation shows that the decoupling matrix is obtained by the formula:

$$\Delta_{i,j}(x) = \frac{\partial y_i^{(\rho_i+1)}}{\partial u_j}_{\mid t=0} \qquad \forall i, j = 1, \ldots, m$$

Indeed, using the Lie derivative formula of $y = h(x)$ along the vector field * $f + \sum_{i=1}^{m} u_j g_j$,

we have $\dot{y}_i = L_f h_i(x(t)) + \sum_{j=1}^{m} u_j L_{g_j} h_i(x(t))$.

By definition of the index ρ_i, $\rho_i = 0$ means that $L_{g_j} h_i \neq 0$ for at least one j. Otherwise, $\rho_i > 0$, $L_{g_j} h_i \equiv 0$. In the first case, we have

$$\frac{\partial y_i^{(\rho_i+1)}}{\partial u_j}_{\mid t=0} = L_{g_j} h_i(x) = L_{g_j} L_f^0 h_i(x) = \Delta_{i,j}(x)$$

for $j = 1, \ldots, m$ with $x(0) = x$. If, on the contrary, $\rho_i > 0$, we have $\dot{y}_i = L_f h_i(x(t))$ and it suffices to rederive this last expression:

$$\ddot{y}_i = L_f^2 h_i(x(t)) + \sum_{j=1}^{m} u_j L_{g_j} L_f h_i(x(t)) .$$

If ρ_i is equal to 1, the announced formula is immediately obtained by differentiating with respect to the inputs and evaluating the formula at time $t = 0$ with $x(0) = x$. Otherwise, we go on differentiating the output y_i up to the order $\rho_i + 1$, the first-order at which the inputs explicitly appear. More precisely, we have $y_i^{(k)} = L_f^k h_i(x(t))$ for all $k \leq \rho_i$ and

$$y_i^{(\rho_i+1)} = L_f^{\rho_i+1} h_i(x(t)) + \sum_{j=1}^{m} u_j L_{g_j} L_f^{\rho_i} h_i(x(t))$$

and the result follows.

The decoupling matrix Δ thus describes the respective contributions of the inputs in the output derivatives. In particular, this matrix is full rank if all the outputs and their derivatives

$$y_1, y_1^{(1)}, \ldots, y_1^{(\rho_1)}, \ldots, y_m, y_m^{(1)}, \ldots, y_m^{(\rho_m)}$$

are linearly independent and if $y_1^{(\rho_1)}, \ldots, y_m^{(\rho_m)}$ can be assigned to arbitrary functions by the inputs.

We now define the sequence of vector spaces, for $k \geq 0$:

$$\mathcal{E}_k = sp\left\{\mathrm{d}x, \mathrm{d}y_i^{(j)} | i = 1, \ldots, m; 0 \leq j \leq k\right\}$$

where $\mathrm{d}y_i^{(j)}$ denotes the differential of the function $y_i^{(j)}$ with respect to $x, u, \dot{u}, \ldots$

According to the above results, if the matrix Δ is invertible and if we note

$$\rho = \max\{\rho_i | i = 1, \ldots, m\}, .$$

then $\dim \mathcal{E}_{k+1} - \dim \mathcal{E}_k = m$ for every $k \geq \rho$. This remark will be useful in the dynamic feedback case.

3.2.2 *The system graph* (Kasinski and Lévine, 1984)

Definition 3.2

The system graph associated with (3.3), denoted by Γ, defined in an open subset V of the manifold X is the oriented and weighted graph comprising

(i) **nodes.** *They are made up of the input nodes $(u_1, \ldots, u_m)$, the state nodes $(x_1, \ldots, x_n)$ and the output nodes $(y_1, \ldots, y_m)$*

(ii) **oriented arcs and weights.**

- from an input node to a state node *there exists an oriented arc from u_j to x_k, of weight $\varpi_{j,k}$, if, and only if, the kth component of the vector field g_j, noted $g_{j,k}$, is nonzero in V. Its weight is then defined by $\varpi_{j,k} = g_{j,k}$, $j = 1, \ldots, m$, $k = 1, \ldots, n$.*
- from one state node to another state node *there exists an oriented arc from x_k to x_l, of weight $\omega_{k,l}$, if, and only if, the lth component of the vector field f, noted f_l, depends on x_k, i.e., if $\dfrac{\partial f_l}{\partial x_k}$ is nonzero in V; its weight is then defined by $\omega_{k,l} = \dfrac{\partial f_l}{\partial x_k}$, $k, l = 1, \ldots, n$, $k \neq l$.*
- from a state node to an output node *there exists an oriented arc from x_l to y_i, of weight ω_l^i, if, and only if, the ith output h_i depends on x_l, i.e. $\dfrac{\partial h_i}{\partial x_l}$ is nonzero in V; its weight is then defined by $\omega_l^i = \dfrac{\partial h_i}{\partial x_l}$, $l = 1, \ldots, n$, $i = 1, \ldots, m$.*

Example 1

Consider the system

$$\left\{ \begin{array}{rcl} \dot{x}_1 & = & f_1(x_1, x_2, x_3) + u_1 g_{1,1}(x_1, x_2, x_3) \\ \dot{x}_2 & = & f_2(x_1, x_2) \\ \dot{x}_3 & = & f_3(x_1, x_2, x_3) + u_2 g_{2,3}(x_1, x_2, x_3) \\ y_1 & = & h_1(x_2) \\ y_2 & = & h_2(x_3) \end{array} \right. \tag{3.8}$$

We have:

$$\omega_2^1 = \frac{\partial h_1}{\partial x_2},\ \omega_3^2 = \frac{\partial h_2}{\partial x_3},$$
$$\omega_{1,2} = \frac{\partial f_2}{\partial x_1},\ \omega_{2,1} = \frac{\partial f_1}{\partial x_2},\ \omega_{1,3} = \frac{\partial f_3}{\partial x_1},\ \omega_{3,1} = \frac{\partial f_1}{\partial x_3}, \omega_{2,3} = \frac{\partial f_3}{\partial x_2},$$
$$\varpi_{1,1} = g_{1,1},\ \varpi_{2,3} = g_{2,3}.$$

One can easily check that

$$\frac{\partial \dot{y}_1}{\partial u_1} = L_{g_1} h_1 = \frac{\partial \dot{y}_1}{\partial u_2} = L_{g_2} h_1 = 0,\ \frac{\partial \dot{y}_2}{\partial u_1} = L_{g_1} h_2 = 0,\ \frac{\partial \dot{y}_2}{\partial u_2} = L_{g_2} h_2 = \varpi_{2,3}\omega_3^2,$$
$$\frac{\partial \ddot{y}_1}{\partial u_1} = L_{g_1} L_f h_1 = \varpi_{1,1}\omega_{1,2}\omega_2^1,\ \frac{\partial \ddot{y}_1}{\partial u_2} = L_{g_2} L_f h_1 = 0.$$

Thus, the characteristic index ρ_1 associated to y_1 is 0 and ρ_2, associated with y_2, is 1. Moreover, the decoupling matrix Δ is given by

$$\Delta = \begin{pmatrix} \varpi_{1,1}\omega_{1,2}\omega_2^1 & 0 \\ 0 & \varpi_{2,3}\omega_3^2 \end{pmatrix}$$

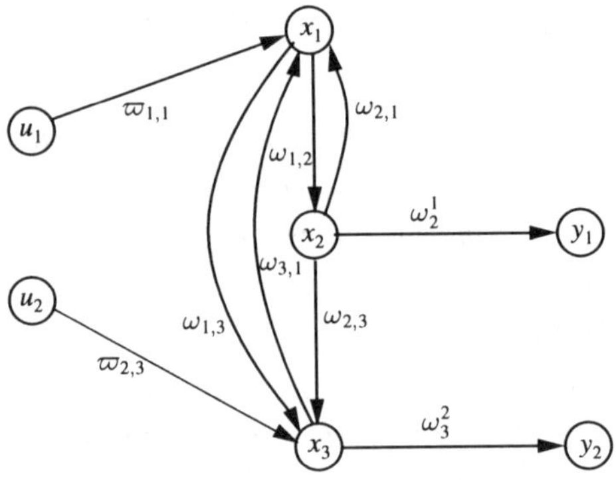

Figure 3.1: The graph of system (3.8).

Now, if we look at the graph, the path containing the minimum number of arcs joining any input to y_1, called *minimal path*, is $u_1 \to x_1 \to x_2 \to y_1$ and, if we define its weight as the product of the weights of each arc, it is precisely equal to $\varpi_{1,1}\omega_{1,2}\omega_2^1$. Following the same lines, the minimal path joining any input to y_2 is $u_2 \to x_3 \to y_2$, and its weight is precisely $\varpi_{2,3}\omega_3^2$.
To summarize, in this example, all the information needed to compute the static state feedback that decouples and linearizes the system can be obtained from the minimum subgraph defined by the above minimal paths. The next result shows that this property, though not general, is *generic* in some sense.

Definition 3.3

We define the distance from u_j to y_i, noted $d(u_j, y_i)$, as the minimum number of arcs of an oriented path from u_j to y_i. We denote by $\Gamma_{i,j}$ the subgraph made of all the minimal oriented paths from u_j to y_i, and thus of length $d(u_j, y_i)$.

The weight of a path from u_j to y_i is the product of the weights of the arcs that compose it.

The weight of the subgraph $\Gamma_{i,j}$, denoted $\Pi_{i,j}$, is the sum of the weights of the paths that compose it.

The minimum distance from the inputs to the ith output y_i, denoted σ_i, is given by:

$$\sigma_i = \min\{d(u_j, y_i) | j = 1, \ldots, m\} \; i = 1, \ldots, m$$

Remark that the definition of Γ depends on the choice of coordinates associated to the system realization (3.3). Consequently, different lengths may correspond to different choices of coordinates. Nevertheless, theorem 3.2 shows that pathologic systems of coordinates are not generic in a certain sense.

Theorem 3.2 (Kasinski and Lévine, 1984; D'Andréa-Novel and Lévine, 1986)
We have:

(i) $\sigma_i - 2 \le \rho_i$ for every $i = 1, \ldots, m$.

(ii) If $\Pi_{i,j} \ne 0$ for at least one j, then $\sigma_i - 2 = \rho_i$ and $\Delta_{i,j} = \Pi_{i,j} \; \forall j = 1, \ldots, m$.

(iii) If $\sigma_i > n+1$ then $\sigma_i = \rho_i = +\infty$.

The proof can be found in Kasinski and Lévine (1984). A proof in a more general context will be given in the next section.

In addition to the interpretation of the characteristic indexes in terms of minimal paths, this result shows that one can easily compute the elements of the decoupling matrix Δ thanks to the graph Γ, at least when the system is expressed in a generic system of coordinates, corresponding to the open relation "$\Pi_{i,j} \neq 0$ for at least one j".

The feedback is then obtained by theorem 3.1.

Remark 3.3

In non generic cases, it might happen that the decoupling matrix is not computable from the graph. The reader will find examples of this kind in Geromel et al. (1984).

3.3 The extended graph for dynamic feedback decoupling

3.3.1 *More on dynamic feedback decoupling*

In Chapter 2, several examples were presented, showing that dynamic feedback is necessary to decouple and linearize the system when the decoupling matrix (3.5) is not full rank.

Let us recall the following simple example.

Example 2

We consider the three-dimensional system with two inputs and two outputs:

$$\begin{aligned} \dot{x}_1 &= u_1 x_3 \\ \dot{x}_2 &= u_2 x_1 \\ \dot{x}_3 &= x_2 + u_1 x_2 \\ y_1 &= x_1 \\ y_2 &= x_3 \end{aligned}$$

The derivatives of the outputs are $\dot{y}_1 = u_1 x_3$ (thus $\rho_1 = 0$) and $\dot{y}_2 = x_2(1+u_1)$ (thus $\rho_2 = 0$ too). Since u_2 is not present in these two expressions, the decoupling matrix Δ, obtained by differentiating $\dot{y}_1$ and $\dot{y}_2$ with respect to (u_1, u_2), is singular:

$$\Delta = \begin{pmatrix} x_3 & 0 \\ x_2 & 0 \end{pmatrix}.$$

Therefore the system is not decouplable by static feedback.
However, if we compute $\ddot{y}_1$ and $\ddot{y}_2$, we find: $\ddot{y}_1 = \dot{u}_1 x_3 + u_1 x_2(1+u_1) = \dot{u}_1 x_3 + u_1 x_2 + u_1^2 x_2$

and $\ddot{y}_2 = \dot{u}_1 x_2 + u_2 x_1(1+u_1) = \dot{u}_1 x_2 + u_2 x_1 + u_1 u_2 x_1$, where u_2 is now present. Indeed, computing Δ_1 the Jacobian of the pair $(\ddot{y}_1, \ddot{y}_2)$ with respect to $(u_1, u_2, \dot{u}_1, \dot{u}_2)$, we have:

$$\Delta_1 = \begin{pmatrix} x_2(1+2u_1) & 0 & x_3 & 0 \\ u_2 x_1 & x_1(1+u_1) & x_2 & 0 \end{pmatrix}, \quad \text{rank } \Delta_1 = 2 .$$

The dynamic compensator is thus given by $\dot{u}_1 = v_1$, $v_1 x_3 + u_1 x_2(1+u_1) = v_1'$ and $v_1 x_2 + u_2 x_1(1+u_1) = v_2$, producing the linear decoupled closed-loop system: $\ddot{y}_1 = v_1'$, $\ddot{y}_2 = v_2$. Note that in the (endogeneous) dynamic compensator $\dot{u}_1 = v_1$, $v_1 = \dfrac{v_1' - u_1 x_2(1+u_1)}{x_3}$ and $u_2 = \dfrac{v_2 - v_1 x_2}{x_1(1+u_1)}$, u_1 may be considered as an extended state variable. The distinction between state and input variables thus becomes artificial and is abolished hereafter.

We now return to system (3.1).

As announced in the above example, we do not distinguish between input and state variables. Moreover, since the number of differentiations required for the computation of the dynamic feedback is not *a priori* fixed, we do not want to impose restrictions on it. We introduce the new coordinate system (in infinite dimension):

$$z \triangleq (z_1, \ldots, z_n, z_{n+1}, \ldots, z_{n+m}, \ldots, z_{n+jm+1}, \ldots, z_{n+(j+1)m}, \ldots) \tag{3.9}$$

with $z_i = x_i$ for $i = 1, \ldots, n$, $z_{n+j} = u_j$ for $j = 1, \ldots, m$, and $z_{n+jm+k} = u_k^{(j)}$ for $k = 1, \ldots, m$ and $j \geq 1$, where $u_k^{(j)} = \dfrac{d^j u_k}{dt^j}$. We adopt the convention $u_j^{(0)} = u_j$ for $j = 1, \ldots, m$.

In these extended coordinates, the image F of the vector field f is written:

$$F_i(z) \triangleq \begin{cases} f_i(x,u) & \text{if } i = 1, \ldots, n \\ z_{i+m} & \text{if } i \geq n \end{cases} \tag{3.10}$$

Note that the components of F for $i \geq n$ simply express the fact that

$$\dot{z}_{n+jm+k} = \frac{du_k^{(j)}}{dt} = u_k^{(j+1)} = z_{n+(j+1)m+k}$$

Accordingly, we define

$$H_i(z) = h_i(x,u) \,, \quad i = 1, \ldots, m$$

The interested reader may find a complete presentation of the underlying infinite dimensional geometric concepts in Krasil'shchick *et al.* (1986) and Zharinov (1992) and some of their applications to control problems in Fliess *et al.* (1993a; 1993b; 1994; 1995).

We assume from now on that z belongs to an open subset V of the above infinite dimensional manifold Z (endowed with the Fréchet topology, see Krasil'shchick *et al.*, 1986; Zharinov, 1992). The Lie derivative of h_i along the vector field f in the extended coordinates rewrites:

$$L_F H_i = \sum_{j \geq 1} F_j \frac{\partial H_i}{\partial z_j} \tag{3.11}$$

Recall from Chapter 2 that the system (3.1) is decouplable by dynamic feedback if and only if it is invertible. Its invertibility is, in turn, characterized by the properties of the following filtration:

$$\mathcal{E}_{-1} = \mathrm{sp}\{\mathrm{d}z_1, \ldots, \mathrm{d}z_n\}\ , \quad \mathcal{E}_0 = \mathrm{sp}\{\mathrm{d}z_1, \ldots, \mathrm{d}z_n, \mathrm{d}y_i, i = 1, \ldots, m\}\ ,$$
$$\mathcal{E}_r = \mathrm{sp}\left\{\mathrm{d}z_1, \ldots, \mathrm{d}z_n, \mathrm{d}y_i^{(k)}, i = 1, \ldots, m; 1 \le k \le r\right\}\ , \quad r \ge 1\ .$$

Here, $\mathcal{E}_r$ is a vector space for each r and the symbol "sp" means that linear combinations which are analytic functions of x, u and a finite number of derivatives of u are considered.

Proposition 3.1

There exist 3 integers ρ, δ and k_0, called respectively the differential output rank, the input–output defect and the index, such that $0 \le \rho \le m$, $0 \le \delta \le n$, $0 \le k_0 \le n-1$, and

$$\dim \mathcal{E}_k = \rho(k+1) + \delta\ , \quad \forall k \ge k_0\ .$$

The polynomial $P_{\mathcal{E}}(k) = \rho(k+1) + \delta$ is called the Hilbert polynomial associated to the filtration $\{\mathcal{E}_k\}$ (see El Asmi and Fliess, 1991).

Definition 3.4

The system is invertible if, and only if, the differential output rank is equal to the number of inputs, i.e. $\rho = m$.

It is computed, at a generic (regular) point (see Martin, 1993; Di Benedetto *et al.*, 1989; Nijmeijer and Van Der Schaft, 1990), by evaluating the ranks of the sequence of Jacobian matrices $J_r(\mathcal{E})$ of size $(r+1)m \times (r+1)m$

$$J_0(\mathcal{E}) = \begin{pmatrix} \dfrac{\partial y_1}{\partial z_{n+1}} & \cdots & \dfrac{\partial y_1}{\partial z_{n+m}} \\ \vdots & & \vdots \\ \dfrac{\partial y_m}{\partial z_{n+1}} & \cdots & \dfrac{\partial y_m}{\partial z_{n+m}} \end{pmatrix} \tag{3.12}$$

and, for every $r \ge 1$:

$$J_r(\mathcal{E}) = \begin{pmatrix} \dfrac{\partial y_1^{(0)}}{\partial z_{n+1}} & \cdots & \dfrac{\partial y_1^{(0)}}{\partial z_{n+(r+1)m}} \\ \vdots & & \vdots \\ \dfrac{\partial y_m^{(0)}}{\partial z_{n+1}} & \cdots & \dfrac{\partial y_m^{(0)}}{\partial z_{n+(r+1)m}} \\ \vdots & & \vdots \\ \dfrac{\partial y_1^{(r)}}{\partial z_{n+1}} & \cdots & \dfrac{\partial y_1^{(r)}}{\partial z_{n+(r+1)m}} \\ \vdots & & \vdots \\ \dfrac{\partial y_m^{(r)}}{\partial z_{n+1}} & \cdots & \dfrac{\partial y_m^{(r)}}{\partial z_{n+(r+1)m}} \end{pmatrix} . \tag{3.13}$$

Proposition 3.2

We have, at every generic point, rank $J_r(\mathcal{E}) = \dim \mathcal{E}_r$ *for every* $r \ge 0$.

3.3.2 More on regular invertibility

Let us now compare the Hilbert polynomial of the system to the one associated to the tangent linear system at an equilibrium point. Given an equilibrium point $\overline{z} = (\overline{x}, \overline{u}, 0, \ldots, 0, \ldots)$, the tangent linear system is defined by

$$\left\{ \begin{array}{lcl} \dfrac{d\delta z}{dt} & = & \dfrac{\partial F}{\partial z}(\overline{z})\delta z \\ \delta y & = & \dfrac{\partial H}{\partial z}(\overline{z})\delta z \end{array} \right. \tag{3.14}$$

The filtration attached to (3.14), denoted $\{\mathrm{T}\mathcal{E}_r\, ,\ r \geq -1\}$, is defined as above by replacing the system (3.1) by the linear system (3.14). The associated Hilbert polynomial is denoted $P_{\mathrm{T}\mathcal{E}}(k)$.

We also need to introduce the restriction $\{\mathcal{E}_r(\overline{z})\ ,\ r \geq -1\}$ at the equilibrium point $\overline{z}$ of the filtration $\{\mathcal{E}_r\ ,\ r \geq -1\}$. It is obtained by expressing $\mathrm{d}z_1, \ldots, \mathrm{d}z_n, \mathrm{d}y_i^{(k)}, i = 1, \ldots, m; 1 \leq k \leq r$ in the basis of the cotangent space to Z at $\overline{z}$, namely $\mathrm{d}z_1(\overline{z}), \ldots$, $\mathrm{d}z_n(\overline{z})$, $\mathrm{d}y_i^{(k)}(\overline{z}), i = 1, \ldots, m; 1 \leq k \leq r$. The associated Hilbert polynomial is noted $P_{\mathcal{E}(\overline{z})}(k)$.

Proposition 3.3 (Martin, 1993)

$$\forall r \in \mathbb{N}\, ,\ \dim \mathrm{T}\mathcal{E}_r = \dim \mathcal{E}_r(\overline{z})\ .$$

Consequently, $P_{\mathrm{T}\mathcal{E}}(k) = P_{\mathcal{E}(\overline{z})}(k)$ *for every* $k \in \mathbb{N}$.

The following definitions are adapted from Martin, 1993:

Definition 3.5

(i) *(3.1) is called* **regular** *at* $\overline{z}$ *if and only if there exists* $\kappa \in \mathbb{N}$ *such that* $\dim \mathrm{T}\mathcal{E}_k = \dim \mathcal{E}_k$ *for every* $k \geq \kappa$.

(ii) *(3.1) is called* **strongly regular** *at* $\overline{z}$ *if and only if* $\dim \mathrm{T}\mathcal{E}_k = \dim \mathcal{E}_k$ *for every* $k \geq 0$.

(iii) *(3.1) is called* **regularly invertible** *(resp. strongly regularly invertible) at* $\overline{z}$ *if and only if it is regular (resp. strongly regular) at* $\overline{z}$ *and invertible.*

Theorem 3.3 (Martin, 1993)

If the system (3.1) is regular, its differential output rank and defect are equal to those of (3.14). In addition, if (3.1) is strongly regular, its differential output rank, defect and index coincide with those of (3.14).
Moreover, if (3.1) is regularly invertible, it is regularly decouplable.

3.3.3 *The extended system graph*

We now introduce the following definition of the *extended system graph*:

Definition 3.6

The extended system graph associated with (3.3), denoted by Γ_{ext}, is the oriented and weighted graph comprising

(i) **nodes.** *They are made up of the extended state nodes $(z_1, \dots, z_n, z_{n+1}, \dots, z_{n+m}, \dots)$ and the output nodes $(y_1, \dots, y_m)$;*

(ii) **oriented arcs and weights:**

- from one extended state node to another extended state node *there exists an oriented arc from z_k to z_j, of weight $\omega_{k,j}$, if and only if the jth component of the vector field F, noted F_j, depends on z_k, i.e. if $\dfrac{\partial F_j}{\partial z_k}$ is nonzero in V; its weight is then defined by $\omega_{k,j} = \dfrac{\partial F_j}{\partial z_k}$, $k, j \geq 1$.*
- from an extended state node to an output node *there exists an oriented arc from z_j to y_i, of weight ω_j^i, if and only if the ith output H_i depends on z_j, i.e. $\dfrac{\partial H_i}{\partial z_j}$ is nonzero in V; its weight is then defined by $\omega_j^i = \dfrac{\partial H_i}{\partial z_j}$, $j \geq 1$, $i = 1, \dots, m$.*

To each output y_i, $i = 1, \dots, m$, we associate a sequence of oriented subgraphs $\{\Gamma_i^r | r \geq 0\}$ constructed by induction on the length (number of arcs). We generate backward, from the node $\{y_i\}$, the set of nodes that are related to it in the graph Γ_{ext}. We distinguish between the set Ω_i^r of nodes related to y_i by a path of length r and appearing for the first time at length r, and the set $\tilde{\Omega}_i^r$ of nodes related to y_i by a path of length r and which appeared for the first time at a length $k < r$.

We set:

$$\Omega_i^0 = \{j \in \mathbb{N} | \omega_j^i \neq 0\} = \tilde{\Omega}_i^0 \tag{3.15}$$

Then, for every r:

$$\Omega_i^r = \{j \in \mathbb{N} - \tilde{\Omega}_i^{r-1} | \exists j_{r-1} \in \tilde{\Omega}_i^{r-1}, \dots, \exists j_0 \in \tilde{\Omega}_i^0 \ s.t. \ \omega_{j,j_{r-1}} \omega_{j_{r-1},j_{r-2}} \cdots \omega_{j_0}^i \neq 0\}\,, \tag{3.16}$$

$$\tilde{\Omega}_i^r = \Omega_i^r \cup \tilde{\Omega}_i^{r-1}\,. \tag{3.17}$$

We define the subgraph Γ_i^O as the graph made of the single output node $\{y_i\}$ and its weight as

$$W_i^0 = H_i\,, \quad i = 1, \dots, m \tag{3.18}$$

Further, the subgraph Γ_i^r of Γ_{ext}, $r \geq 1$, is defined as the set of oriented paths $(j_{r-1}, \dots, j_0)$ of length r such that $j_\alpha \in \tilde{\Omega}_i^\alpha$ for every $\alpha = 0, \dots r-1$, and

$$\omega_{j_{r-1},j_{r-2}} \cdots \omega_{j_1,j_0} \omega_{j_0}^i \neq 0 \tag{3.19}$$

Its associated weight is, by definition:

$$W_i^r = \sum_{(j_{r-1},\ldots,j_0)\in\Gamma_i^r} F_{j_{r-1}}\omega_{j_{r-1},j_{r-2}}\cdots\omega_{j_1,j_0}\omega_{j_0}^i \,. \tag{3.20}$$

The same construction can be carried out with the tangent linear system (3.14). The associated graph is denoted $\mathrm{T}\Gamma_{ext}$, the subgraph sequence $\{\mathrm{T}\Gamma_i^r\ ,\ i=1,\ldots,m, r\geq 0\}$, and the tangent graph Hilbert polynomial $P_{\mathrm{T}\mathcal{G}}(k)$.

The aim of the graph method is precisely to compute the coefficients of the Hilbert polynomial $P_{\mathcal{E}}(k)$, in a simplified way.

We thus mimic the filtration $\{\mathcal{E}_k\}$ by another filtration $\{\mathcal{G}_k\}$ associated with the graph Γ_{ext}:

$$\mathcal{G}_{-1} = \mathrm{sp}\{\mathrm{d}z_1,\ldots,\mathrm{d}z_n\}\,,$$

$$\mathcal{G}_0 = \mathrm{sp}\{\mathrm{d}z_1,\ldots,\mathrm{d}z_n,\mathrm{d}W_i^0\}\,,$$

$$\mathcal{G}_r = \mathrm{sp}\left\{\mathrm{d}z_1,\ldots,\mathrm{d}z_n,\mathrm{d}W_i^{(k)}, i=1,\ldots,m; 1\leq k\leq r\right\}\,,\quad r\geq 1\,.$$

Let us denote by $\rho_{\mathcal{G}}, \delta_{\mathcal{G}}$ and $k_0^{\mathcal{G}}$ the associated differential output rank, defect and index respectively. The graph Hilbert polynomial is thus $P_{\mathcal{G}}(k)=\rho_{\mathcal{G}}(k+1)+\delta_{\mathcal{G}}$ and we have:

$$\dim\mathcal{G}_k = P_{\mathcal{G}}(k)\,,\quad \forall k\geq k_0^{\mathcal{G}}\,.$$

The tangent graph filtration $\{\mathrm{T}\mathcal{G}_k\}$ and its Hilbert polynomial $P_{\mathrm{T}\mathcal{G}}(k)$ can be defined in the same way.

At a generic point, $P_{\mathcal{G}}(k)$ can be practically computed, as before, by evaluating the ranks of the sequence of Jacobian matrices $J_r(\mathcal{G})$ of size $(r+1)m\times(r+1)m$

$$J_0(\mathcal{G}) = \begin{pmatrix} \dfrac{\partial W_1^0}{\partial z_{n+1}} & \cdots & \dfrac{\partial W_1^0}{\partial z_{n+m}} \\ \vdots & & \vdots \\ \dfrac{\partial W_m^0}{\partial z_{n+1}} & \cdots & \dfrac{\partial W_m^0}{\partial z_{n+m}} \end{pmatrix} \tag{3.21}$$

and, for every $r\geq 1$

$$J_r(\mathcal{G}) = \begin{pmatrix} \dfrac{\partial W_1^0}{\partial z_{n+1}} & \cdots & \dfrac{\partial W_1^0}{\partial z_{n+(r+1)m}} \\ \vdots & & \vdots \\ \dfrac{\partial W_m^0}{\partial z_{n+1}} & \cdots & \dfrac{\partial W_m^0}{\partial z_{n+(r+1)m}} \\ \vdots & & \vdots \\ \dfrac{\partial W_1^r}{\partial z_{n+1}} & \cdots & \dfrac{\partial W_1^r}{\partial z_{n+(r+1)m}} \\ \vdots & & \vdots \\ \dfrac{\partial W_m^r}{\partial z_{n+1}} & \cdots & \dfrac{\partial W_m^r}{\partial z_{n+(r+1)m}} \end{pmatrix} \tag{3.22}$$

We have, at every generic point, rank $J_r(\mathcal{G}) = \dim\mathcal{G}_r$ for every $r\geq 0$.

Before stating the results on the computation of the Hilbert polynomial, we relate the weights W_i^r of the subgraphs Γ_i^r to the rth iteration of the Lie derivative of H_i.

Proposition 3.4

We have:

$$\frac{d^r y_i}{dt^r} = L_F^r H_i = W_i^r + \lambda_{i,r} \tag{3.23}$$

where the functions $\lambda_{i,r}$ satisfy:

$$\frac{\partial \lambda_{i,r}}{\partial z_j} = 0\,, \quad \forall j \notin \tilde{\Omega}_i^{r-1}\,. \tag{3.24}$$

For linear systems, this formula holds with $\lambda_{i,r} \equiv 0$ for every i and r.

Proof

(By induction.) The result is obvious for $r = 0$ with $\lambda_{i,0} = 0$. Assume that it holds up to r. For $r + 1$, we have:

$$L_F^{r+1} H_i = L_F(L_F^r H_i) = L_F(W_i^r + \lambda_{i,r})\,.$$

Clearly,

$$\begin{aligned} L_F W_i^r &= \sum_{j\geq 1} F_j \frac{\partial W_i^r}{\partial z_j} = \sum_{j\in\Omega_i^r\cup\tilde{\Omega}_i^{r-1}} F_j \frac{\partial}{\partial z_j}\left(\sum_{(j_{r-1},\ldots,j_0)\in\Gamma_i^r} F_{j_{r-1}}\omega_{j_{r-1},j_{r-2}}\cdots\omega_{j_1,j_0}\omega_{j_0}^i\right) \\ &= W_i^{r+1} + \sum_{j\in\tilde{\Omega}_i^{r-1}} \sum_{(j_{r-1},\ldots,j_0)\in\Gamma_i^r} F_j F_{j_{r-1}} \frac{\partial}{\partial z_j}\left(\omega_{j_{r-1},j_{r-2}}\cdots\omega_{j_1,j_0}\omega_{j_0}^i\right)\,. \end{aligned}$$

Also, according to (3.24)

$$L_F \lambda_{i,r} = \sum_{j\in\tilde{\Omega}_i^{r-1}} F_j \frac{\partial \lambda_{i,r}}{\partial z_j}\,.$$

Thus, $L_F^{r+1} H_i = W_i^{r+1} + \lambda_{i,r+1}$ with

$$\lambda_{i,r+1} = \sum_{j\in\tilde{\Omega}_i^{r-1}} \sum_{(j_{r-1},\ldots,j_0)\in\Gamma_i^r} F_j F_{j_{r-1}} \frac{\partial}{\partial z_j}\left(\omega_{j_{r-1},j_{r-2}}\cdots\omega_{j_1,j_0}\omega_{j_0}^i\right) + \sum_{j\in\tilde{\Omega}_i^{r-1}} F_j \frac{\partial \lambda_{i,r}}{\partial z_j}\,.$$

The property that $\dfrac{\partial \lambda_{i,r+1}}{\partial z_j} = 0\,, \quad \forall j \notin \tilde{\Omega}_i^r$ follows from the definition of the Ω_i^α. Moreover, for a linear system, the weights $\omega_{j,k}$ and ω_j^k are constant and

$$\frac{\partial}{\partial z_j}\left(\omega_{j_{r-1},j_{r-2}}\cdots\omega_{j_1,j_0}\omega_{j_0}^i\right) \equiv 0\,.$$

The result readily follows.

We can now state our main result

Corollary

Let $\overline{z}$ be an equilibrium point. We have $P_{\mathrm{T}\mathcal{E}}(k) = P_{\mathrm{T}\mathcal{G}}(k)$ for every $k \in \mathbb{N}$.
Moreover, if the system (3.1) is regularly invertible at $\overline{z}$, there exists an integer $\kappa \in \mathbb{N}$ such that $P_{\mathrm{T}\mathcal{G}}(k) = P_{\mathcal{G}}(k) = P_{\mathcal{E}}(k)$ for every $k \geq \kappa$.
If the system (3.1) is strongly regularly invertible at $\overline{z}$, $P_{\mathrm{T}\mathcal{G}}(k) = P_{\mathcal{G}}(k) = P_{\mathcal{E}}(k)$ for every $k \geq k_0^{\mathcal{G}}$.

Proof

This is a simple consequence of the fact that $\lambda_{i,r}$ is null for linear systems, combined with Theorem 3.3 and Proposition 3.4.

Remark 3.4

In the strongly regular case, the dynamic feedback can be directly obtained from the graph. This is no longer true, even in the regular case. Note that in any case, the matrices $J_r(W)$ and $J_r(\mathcal{E})$ can have quite different terms, due to the presence of nonzero $\lambda_{i,r}$ expressions, but in the strongly regularly invertible case, these terms do not play any role in computing the structure of the dynamic feedback. Finally, in the irregular case, it can be shown that if the tangent graph is invertible, then the system is too.

3.4 Bibliography

[1] D'ANDRÉA-NOVEL, B. and LÉVINE, J. (1986), CAD for nonlinear systems decoupling, perturbations rejection and feedback linearization with applications to the dynamic control of a robot arm. in *Algebraic and Geometric Methods in Nonlinear Control Theory*, M. Fliess and M. Haazewinkel (eds), Reidel Pub. Co., 545–572.

[2] DE LUCA, A., ISIDORI, A. and NICOLO, F. (1985), *An application of nonlinear model matching to the dynamic control of robot arm with elastic joints.* Report 04.85, Univ. "La Sapienza", Roma.

[3] DI BENEDETTO, M.D., GRIZZLE, J.W. and MOOG, C.H. (1989), Rank invariants of nonlinear systems, *SIAM J. Control & Optimization*, **27**, 658–672.

[4] EL ASMI, S. and FLIESS, M. (1991), Formules d'inversion. in *Analysis of Controlled Dynamical Systems*, B. Bonnard, B. Bride, J.P. Gauthier, I. Kupka, eds., Birkhäuser, Boston, 201–210.

[5] FLIESS, M., LÉVINE, J., MARTIN, P. and ROUCHON, P. (1993), Linéarisation par bouclage dynamique et transformations de Lie-Bäcklund, *C. R. Acad. Sci. Paris*, **317** I, 981–986.

[6] FLIESS, M., LÉVINE, J., MARTIN, P. and ROUCHON, P. (1996), A Lie-Bäcklund approach to dynamic feedback equivalence and flatness, in *Robust control via Viarable Structure and Lyapunov techniques*, F. Garofalo and L. Glielmo (eds), Lect. Notes in Control and info-Sciences, Springer-Verlag, London, 245–268.

[7] FLIESS, M., LÉVINE, J., MARTIN, P. and ROUCHON, P. (1994), *Nonlinear control and Lie-Bäcklund transformations: towards a new differential geometric standpoint,* Proc. 33rd IEEE CDC, Lake Buena Vista, FL.

[8] FLIESS, M., LÉVINE, J., MARTIN, P. and ROUCHON, P. (1995), *Design of trajectory stabilization for flat driftless systems,* Proc. ECC 95, Rome, Italy.

[9] GEROMEL, F., LÉVINE, J. and WILLIS, P. (1984), A fast algorithm for systems decoupling using formal calculus, in *Analysis and Optimization of Systems,* A. Bensoussan, J.L. Lions (eds), Lect. Notes in Control and Information Sciences, **63**, II, 378–390, Springer Verlag.

[10] ISIDORI, A. (1989), *Nonlinear Control Systems.* 2nd ed., Springer-Verlag, New York.

[11] KAILATH, T. (1980), *Linear Systems,* Information and Systems Sciences Series, Prentice Hall.

[12] KASINSKI, A. and LÉVINE, J. (1984), *A fast graph theoretic algorithm for the feedback decoupling problem of nonlinear systems,* in Lecture Notes in Control and Information Sciences, 58, 550–562, P.A. Fuhrmann (ed.), Springer-Verlag, New York.

[13] KRASIL'SHCHICK, I.S., LYCHAGIN, V.V. and VINOGRADOV, A.M. (1986), *Geometry of Jet Spaces and Nonlinear Partial Differential Equations,* Gordon and Breach, New York.

[14] MARTIN, P. (1993), An intrinsic sufficient condition for regular decoupling, *Systems & Control Letters*, **20**, 383–391.

[15] MUROTA, K. (1980), *Systems Analysis by Graphs and Matroids,* Springer-Verlag, New York.

[16] NIJMEIJER, H. and VAN DER SCHAFT, A. (1990), *Nonlinear Dynamical Control Systems,* Springer-Verlag, New York.

[17] REINSCHKE, K.J. (1988), *Multivariable Control – A Graph-theoretic Approach,* Springer Verlag, Berlin.

[18] SINGH, S.N. (1981), A modified algorithm for invertibility in nonlinear systems *IEEE Trans. Automatic Control,* **26**, 595–598.

[19] SVARICEK, F. (1992), *A graph-theoretic approach for the determination of the structure at infinity of nonlinear systems.* Proc. of the IFAC NOLCOS, Bordeaux, 124–129.

[20] SVARICEK, F. and SCHWARZ, H. (1993), *Graph-theoretic determination of the nonlinear zeros at infinity. Computational results,* Proc. ECC 93, Groningen.

[21] WONHAM, W.M. (1974), *Linear Multivariable Control: a Geometric Approach,* Springer-Verlag, New York.

[22] ZHARINOV, V.V. (1992), *Geometrical Aspects of Partial Differential Equations,* World Scientific, Singapore.

CHAPTER 4

Static and dynamic state feedback linearization

J. LÉVINE

4.1 Linearization by diffeomorphism and static state feedback

4.1.1 Problem formulation

The problem of feedback linearization consists in transforming (in a given open domain) a nonlinear system into a linear one by state feedback and state space diffeomorphism. Indeed, such transformations do not exist in general (or rather rarely) but, if they exist, they allow the use of classic linear stabilizing techniques such as pole placement, quadratic synthesis, robust synthesis and so on. The price of such a simplification is paid, on the one hand, by the fact that such transformations are not everywhere defined and may generate a blow-up of controls. On the other hand, the computations of the feedback may be very heavy and digital synthesis may decrease the feedback performances. Finally, the feedback efficiency dramatically depends on the accuracy with which the model represents the real plant.

To evaluate the degree of rarity of feedback linearizable systems, the concept of largest feedback linearizable subsystem is introduced. One can also try to find larger classes of feedback laws to ensure the linearization. In this spirit, we propose at the end of this chapter an extension to dynamic feedback linearization, namely using nonlinear dynamic compensators and extended state transformations to linearize the system. Sufficient conditions for the existence of such linearizing compensators are presented as well as examples of applications.

Apart from the dynamic feedback aspects, this presentation is widely inspired by recent books (D'Andréa-Novel, 1988; Nijmeijer and Van Der Schaft, 1990; Slotine and Li, 1991) where the reader may find other fruitful insights.

During the last 3 years, important developments in the field of dynamic feedback linearization, and more precisely on the concept of *differential flatness*, have given birth to a rapidly growing literature. It turns out that many systems encountered in practice are flat (dynamic feedback linearizable by a special type of dynamic feedback called endogeneous) and particularly in the field of holonomic and non holonomic mechanical systems. We encourage the reader to complete his knowledge on this question by referring to [13–24, 36–38, 46–47]. Other interesting approaches can also be found in [29, 30, 42, 44, 48, 52]. The geometric view point presented here, though limited in many aspects such as the fact that the space dimension is fixed, is a first step towards a deeper understanding of dynamic feedback and the structure of linearizable systems.

Let us now give a precise definition of a linearizable system by diffeomorphism and feedback. Let us consider the following nonlinear state space representation:

$$(\Sigma) \quad \dot{x} = f(x) + \sum_{i=1}^{m} u_i g_i(x)$$

where f and $g_1, \ldots, g_m$ are C^∞ vector fields on a C^∞ manifold X of dimension n, satisfying $f(0) = 0$ and rank $\{g_1(0), \ldots, g_m(0)\} = m$.

We are looking for two things

1. *a regular change of coordinates*

$$\xi = \phi(x) \tag{4.1}$$

   where ϕ is a C^∞ diffeomorphism (a C^∞ invertible mapping with a C^∞ inverse – see Appendix A) from a neighborhood V of the origin in X, satisfying $\phi(0) = 0$; and

2. *a static state feedback*

$$u = \alpha(x) + \beta(x)v \tag{4.2}$$

   with $\alpha(0) = 0$, $\beta(x)$ invertible for all x in V and α and β of class C^∞

such that, once transformed, the system (Σ) rewrites

$$(\Lambda) \quad \dot{\xi} = A\xi + Bv$$

the pair (A, B) being controllable.

Remark 4.1

The assumption that the origin coincides with a regular equilibrium point of (Σ) for $u = 0$, ($f(0) = 0$ and rank $\{g_1(0), \ldots, g_m(0)\} = m$*), as well as $\phi(0) = 0$ and $\alpha(0) = 0$, does not alter the generality of the above definitions. Indeed, if $(\bar{x}, \bar{u})$ is an equilibrium point of $f + gu$ ($f(\bar{x}) + \sum_{i=1}^{m} \bar{u}_i g_i(\bar{x}) = 0$), it suffices to consider the following affine change of coordinates: $\tilde{x} = x - \bar{x}$ and $\tilde{u} = u - \bar{u}$. Setting $\tilde{f}(\tilde{x}) = f(\tilde{x} + \bar{x}) + \sum_{i=1}^{m} \bar{u}_i g_i(\tilde{x} + \bar{x})$ and $\tilde{g}_i(\tilde{x}) = g_i(\tilde{x} + \bar{x})$, $i = 1, \ldots, m$, we get the desired properties for $\dot{\tilde{x}} = \tilde{f}(\tilde{x}) + \sum_{i=1}^{m} \tilde{u}_i \tilde{g}_i(\tilde{x})$, namely $\tilde{f}(0) = 0$ and* rank $\{\tilde{g}_1(0), \ldots, \tilde{g}_m(0)\} = m$.

Finally, since the equilibrium point of (Σ) must be transformed by ϕ into the one of (Λ) corresponding to $v = 0$ (otherwise ϕ could not be a diffeomorphism – this is left as an exercise), the conditions $\phi(0) = 0$ and $\alpha(0) = 0$ immediately follow.

Remark 4.2

Subsequently, we shall consider the more general feedback linearization problem for the system $\dot{x} = f(x, u)$ *and show that it can be solved by the same kind of methods (see section 4.1.3.3). The assumption that the control appears linearly is only introduced for simplicity's sake.*

Remark 4.3

The controllability of the pair (A, B) *is useful if one wants to follow arbitrary trajectories (remaining in a prescribed neighborhood* V*) with stability since, in the linear case, controllability implies stabilizability.*

Remark 4.4

If we restrict ourselves to the single input case and if we choose $\xi_1 = \phi_1(x)$ *(first component of the diffeomorphism) as the output of the system* (Σ)*, we recover the input–output linearization results, the only difference being that the zero dynamics (the dynamics made unobservable by feedback, nonlinear extension of the notion of zero-pole cancelation of the closed-loop transfer matrix) is reduced to 0. The internal stability is therefore automatically guaranteed by suitably placing the poles of the closed-loop system.*

Remark 4.5

Feedback linearization can also be viewed as a tool to classify those nonlinear systems which are equivalent to a linear system up to a change of coordinates on the state and input space. The remaining systems are thus true nonlinear ones. We will see hereafter that this classification may be refined thanks to the concept of largest linearizable subsystem and linearization by dynamic feedback.

Example 1

Let us consider a rigid robot arm with n degrees of freedom and n actuators, each one acting on a different link. The mechanical model can be written:

$$\Gamma_0(z)\ddot{z} + \Gamma_1(z, \dot{z}) = Q(z, \dot{z})u \tag{4.3}$$

where z is the vector of relative angles between two consecutive links, $\dot{z}$ is the vector of the corresponding angular velocities, u is the vector of applied torques by the actuators (with $Q(z, \dot{z})$ invertible for all $(z, \dot{z})$), $\Gamma_0(z)$, invertible for all z, is the inertia matrix, $\Gamma_1(z, \dot{z})$ is the vector of resisting torques, centrifugal, Coriolis, *etc.* (D'Andréa-Novel, 1988)

Setting $x_1 = z$ and $x_2 = \dot{z}$, (4.3) becomes

$$\left\{ \begin{array}{rcl} \dot{x}_1 & = & x_2 \\ \dot{x}_2 & = & -(\Gamma_0(x_1))^{-1}(\Gamma_1(x_1, x_2) - Q(x_1, x_2)u) \end{array} \right. \tag{4.4}$$

and it suffices, to linearize, to choose the identity as diffeomorphism and the right-hand side of the equation (4.4) as feedback, namely

$$\xi = (x_1, x_2)^T \,, \quad v = -(\Gamma_0(x_1))^{-1}(\Gamma_1(x_1, x_2) - Q(x_1, x_2)u)\,. \tag{4.5}$$

For a long time this method has been well-known in robotics, under the name of *computed torque*.

Let us note that the present theory is superfluous in such a simple case. This example is nevertheless interesting to illustrate the fact that the dynamical complexity of a rigid robot with n degrees of freedom and n actuators is equivalent to n decoupled second-order linear systems. Furthermore, it suffices to assign $z = x_1$ to an arbitrary reference trajectory to guarantee the closed-loop stability by pole placement.

However, if the dimension of the vector z differs from the number of actuators, the computed torque method no longer applies (since then $Q(z, \dot{z})$ is not invertible) and the results presented in the next sections will be helpful to prove the impossibility of static feedback linearization.

4.1.2 *Explicit computation in the single-input case*

We restrict ourselves for the moment to the single input case ($m = 1$) to highlight the main ideas. Their extension to the multi-input case follows the same lines, up to computational difficulties that will be detailed in the next section.

4.1.2.1 Reduction to the canonical form

To dramatically simplify the analysis, one may remark that, according to the classical results on canonical forms of controllable linear systems (Kailath, 1980), the linear controllable system (Λ) can be transformed, by an invertible change of basis $\zeta = T\xi$ and a linear invertible feedback $v = K_1\xi + K_2\tilde{v}$ with K_2 invertible, into the controllable canonical form

$$\dot{\zeta} = \begin{pmatrix} 0 & 1 & & \mathbf{0} \\ & \ddots & \ddots & \\ & & \ddots & 1 \\ & \mathbf{0} & & 0 \end{pmatrix} \zeta + \begin{pmatrix} 0 \\ \vdots \\ 0 \\ 1 \end{pmatrix} \tilde{v} \tag{4.6}$$

or

$$\left\{ \begin{array}{lcl} \dot{\zeta}_1 & = & \zeta_2 \\ \vdots & & \\ \dot{\zeta}_{n-1} & = & \zeta_n \\ \dot{\zeta}_n & = & \tilde{v}\,. \end{array} \right. \tag{4.7}$$

One can easily check that the invertible transformation $\zeta = T\phi(x)$ is a diffeomorphism in the form (4.1) and that the feedback

$$u = \alpha(x) + \beta(x)(K_1\phi(x) + K_2\tilde{v}) \stackrel{\text{def}}{=} \tilde{\alpha}(x) + \tilde{\beta}(x)\tilde{v}$$

satisfies the required invertibility assumption with $\tilde{\alpha}(0) = 0$. It suffices then to find transformations for which the linear controllable system (Λ) is obtained in the canonical form (4.7).

The problem may be thus reformulated as follows: one looks for a diffeomorphism (4.1) and a feedback (4.2), under the same assumptions as before, such that (Σ) is transformed into

$$(\Lambda_c)\begin{cases} \dot{\xi}_1 &= \xi_2 \\ \vdots & \\ \dot{\xi}_{n-1} &= \xi_n \\ \dot{\xi}_n &= v\,. \end{cases}$$

4.1.2.2 Characterization of the diffeomorphism and feedback

Let us recall that if $x(t)$ is an integral curve of (Σ), generated by the feedback (4.2), then (see Apendix A, definition A.8) $\xi_i(t) = \phi_i(x(t))$ satisfies, for all $i = 1, \ldots, n$

$$\begin{aligned}\dot{\xi}_i &= L_f\phi_i(x(t)) + uL_g\phi_i(x(t)) = (L_f\phi_i(x(t)) + \alpha(x(t))L_g\phi_i(x(t))) \\ &\qquad + v\beta(x(t))L_g\phi_i(x(t))\,. \end{aligned} \tag{4.8}$$

This remark is essential to prove the lemma 4.1:

Lemma 4.1

A necessary condition for (Σ) to be linearizable by diffeomorphism and feedback is that there exists a solution, in a neighborhood V of the origin, to the system of partial differential equations

$$\begin{cases} L_g\phi_1 &= 0 \\ L_gL_f\phi_1 &= 0 \\ \vdots & \\ L_gL_f^{n-2}\phi_1 &= 0 \end{cases} \tag{4.9}$$

with $L_gL_f^{n-1}\phi_1 \neq 0$ in V.
Furthermore

$$\phi_i = L_f^{i-1}\phi_1 \;\forall\, i = 2, \ldots, n \tag{4.10}$$

and

$$\alpha = -\frac{L_f^n\phi_1}{L_gL_f^{n-1}\phi_1}\,, \qquad \beta = \frac{1}{L_gL_f^{n-1}\phi_1}\,. \tag{4.11}$$

Proof

Applying formula (4.8) for $i = 1$, since $\dot{\xi}_1 = \xi_2 = \phi_2$, we obtain:

$$(L_f\phi_1(x(t)) + \alpha(x(t))L_g\phi_1(x(t))) + v\beta(x(t))L_g\phi_1(x(t)) = \phi_2(x(t)) \quad \forall v\,. \tag{4.12}$$

Since the right-hand side is independent of the input variable v, one immediately deduces $L_g\phi_1 \equiv 0$ and $\phi_2 = L_f\phi_1$. One shows, in the same way (computing by induction $\dot{\xi}_i = \xi_{i+1}$ and using (4.8)), that $L_gL_f^k\phi_1 \equiv 0$ and $\phi_{k+1} = L_f^k\phi_1$ for all $k = 0, \ldots, n-2$. Finally, computing $\dot{\xi}_n$ by the same technique, we get: $L_f^n\phi_1 + \alpha L_gL_f^{n-1}\phi_1 = 0$ and $\beta L_gL_f^{n-1}\phi_1 = 1$, which proves (4.9), (4.10) and (4.11).

We now prove that this condition is also sufficient. To this end, it suffices to show that the mapping

$$\xi = (\phi_1(x), L_f\phi_1(x), \ldots, L_f^{n-1}\phi_1(x))^T$$

properly defines a local diffeomorphism. We begin by reducing the system of PDEs of order $n-1$ (4.9) to a first-order system by using the Lie brackets of f and g. Recall the notation $ad_f^k g = [f, ad_f^{k-1}g]$ for $k \geq 1$, with the convention $ad_f^0 g = g$.

Lemma 4.2

A necessary and sufficient condition for (Σ) *to be linearizable by diffeomorphism and feedback is that there exists a solution, in a neighborhood* V *of the origin, to the system of first-order partial differential equations:*

$$\left\{\begin{array}{lll} L_g\phi_1 & = & 0 \\ L_{ad_f g}\phi_1 & = & 0 \\ \vdots & & \\ L_{ad_f^{n-2}g}\phi_1 & = & 0 \end{array}\right. \tag{4.13}$$

with $L_{ad_f^{n-1}g}\phi_1 \neq 0$ *in* V.

Proof

The relations $L_g\phi_1 = 0$ and $L_gL_f\phi_1 = 0$ imply that $L_{ad_f g}\phi_1 = L_fL_g\phi_1 - L_gL_f\phi_1 = 0$. They are thus equivalent to $L_g\phi_1 = 0$ and $L_{ad_f g}\phi_1 = 0$. It is then easily shown by induction that (4.9) is equivalent to (4.13), and that $L_gL_f^{n-1}\phi_1 \neq 0$ in V is equivalent to $L_{ad_f^{n-1}g}\phi_1 \neq 0$ in V.

Conversely, if ϕ_1 is a solution to the system (4.13), according to proposition A.3, we get, for all i and j such that $i + j \leq n-2$: $< ad_f^j g, dL_f^i\phi_1 >= 0$. Let us construct the matrix M whose entry corresponding to line i and column j is $M_{i,j} =< ad_f^{n-j}g, dL_f^{i-1}\phi_1 >$. it is easily shown that M is a lower triangular matrix with non zero diagonal elements due to $L_{ad_f^{n-1}g}\phi_1 \neq 0$. Thus, M is invertible. According to (4.10), we have $d\phi_i = dL_f^{i-1}\phi_1$ for all $i = 1, \ldots, n$ and we can check that

$$M = \begin{pmatrix} d\phi_1 \\ \vdots \\ d\phi_n \end{pmatrix} (ad_f^{n-1}g \cdots g)\,. \tag{4.14}$$

The invertibility of M immediately implies (this is an elementary result of linear algebra) that

$$d\phi = \begin{pmatrix} d\phi_1 \\ \vdots \\ d\phi_n \end{pmatrix}$$

is invertible and that the vectors $ad_f^{n-1}g, \ldots, g$ are linearly independent. However, the invertibility of $d\phi$ is equivalent to the fact that ϕ is a local diffeomorphism (proposition A.1). (Σ)'s linearizability follows from (4.11) by inverting the order of the computations in lemma 4.1.

Let us now consider the distributions of vector fields for $i = 1, \ldots, n$:

$$G_i = \text{sp}\{g, ad_f g, \ldots, ad_f^i g\} \tag{4.15}$$

for $i = 1, \ldots, n$.

We are now ready to state and prove the main result of this section:

Theorem 4.1

A necessary and sufficient condition for (Σ) to be linearizable by diffeomorphism and feedback is that:

(i) *G_{n-2} is involutive with constant rank in V*

(ii) *G_{n-1} has rank n.*

Proof

Let us assume that (Σ) is linearizable by diffeomorphism and feedback. Then there exists a diffeomorphism ϕ such that ϕ_1 is solution of (4.13) and satisfies (4.10). By Frobenius' theorem, the existence of a solution to (4.13) is equivalent to (i). Furthermore, the condition $L_{ad_f^{n-1}g}\phi_1 \neq 0$ in V, with $d\phi_1 \neq 0$ in V, implies that $ad_f^{n-1}g \notin G_{n-2}$. Furthermore, following the same lines as in lemma 4.2, the fact that M is invertible and that ϕ is a diffeomorphism, implies that the n vectors $g, ad_f g, \ldots, ad_f^{n-1}g$ are independent in V, which proves (ii).

Conversely, if (i) is satisfied, by Frobenius' theorem, the system (4.13) admits a solution satisfying $L_{ad_f^{n-1}g}\phi_1 \neq 0$ in V. Also, rank $G_{n-2} \leq n-1$, which, combined with (ii), gives $ad_f^{n-1}g \notin G_{n-2}$ (otherwise G_{n-1} would not have rank n). Thus there exists ϕ_1 such that $L_{ad_f^{n-1}g}\phi_1 \neq 0$ in V. The linearizability follows from lemma 4.2.

4.1.3 The multi-input case

We now turn to the general case with m inputs. As before, transforming a nonlinear system (Σ) into a linear controllable one amounts to transforming it into a linear system in the controllability canonical form. In this case, such a canonical form is given by:

$$\dot{\xi} = \begin{pmatrix} A_1 & & & \\ & A_2 & 0 & \\ & & \ddots & \\ & 0 & & \ddots & \\ & & & & A_m \end{pmatrix} \xi + \begin{pmatrix} B_1 & & & \\ & B_2 & 0 & \\ & & \ddots & \\ & 0 & & \ddots & \\ & & & & B_m \end{pmatrix} v \tag{4.16}$$

with A_i of size $k_i \times k_i$, B_i of size $k_i \times 1$ and k_i being the ith controllability index, $i = 1, \ldots, m$, with $k_1 + \cdots + k_m = n$ and

$$A_i = \begin{pmatrix} 0 & 1 & 0 & \\ & \ddots & \ddots & \\ & & \ddots & 1 \\ & 0 & & 0 \end{pmatrix} \quad \text{and} \quad B_i = \begin{pmatrix} 0 \\ \\ 1 \end{pmatrix}. \tag{4.17}$$

Without loss of generality, up to an input reordering, one can assume that the sequence of cotrollability indices is non increasing: $k_1 \geq \ldots \geq k_m$.

4.1.3.1 An intrinsic characterization of feedback equivalence to a linear system

Before stating the main result, we introduce the following sequence of distributions of vector fields:

$$\begin{aligned} G_0 &= \mathrm{sp}\{g_1, \ldots, g_m\} \\ G_i &= G_{i-1} + ad_f G_{i-1} \quad , \quad i \geq 1 \end{aligned} \tag{4.18}$$

Theorem 4.2 *(Jakubczyk and Respondek, 1980; Hunt et al., 1983)*
A necessary and sufficient condition for (Σ) to be linearizable by diffeomorphism and feedback is that:

(i) G_i is involutive with constant rank in V for $i = 0, \ldots, n-2$

(ii) G_{n-1} has rank n.

The idea of the proof is the same as in the single-input case and we just sketch the main steps. The following notations will be useful: the state vector ξ of the linear system (4.16) is noted by blocks (we recall that $k_1 + \cdots + k_m = n$)

$$\xi = (\xi_1^1, \ldots, \xi_{k_1}^1, \ldots, \xi_1^m, \ldots, \xi_{k_m}^m)^T = (\xi^1, \ldots, \xi^m)^T . \tag{4.19}$$

Similarly, the components of the diffeomorphism ϕ are noted

$$\phi = (\phi_1^1, \ldots, \phi_{k_1}^1, \ldots, \phi_1^m, \ldots, \phi_{k_m}^m)^T = (\phi^1, \ldots, \phi^m)^T . \tag{4.20}$$

Lemma 4.3
A necessary and sufficient condition for (Σ) to be linearizable by diffeomorphism and feedback is that there exists a solution $(\phi_1^1, \ldots, \phi_1^m)$, in a neighborhood V of the origin, to the system of first-order partial differential equations:

$$L_{ad_f^k g_j} \phi_1^i = 0 \quad \forall j = 1, \ldots, m, \quad \forall k = 0, \ldots, k_i - 2, \quad \forall i = 1, \ldots m \tag{4.21}$$

such that the matrix Δ defined by:

$$\Delta_{i,j} = L_{ad_f^{k_i - 1} g_j} \phi_1^i \quad i, j = 1, \ldots, m. \tag{4.22}$$

is invertible in V.
The other components of the diffeomorphism ϕ are then deduced by the relations

$$\phi_{k+1}^i = L_f^k \phi_1^i \;\; k = 1, \ldots, k_i - 1 \; , \; i = 1, \ldots, m \tag{4.23}$$

and the feedback $u = \alpha + \beta v$ is given by:

$$\alpha = -\Delta^{-1} K \begin{pmatrix} L_f^{k_1} \phi_1^1 \\ \vdots \\ L_f^{k_m} \phi_1^m \end{pmatrix} \text{ and } \beta = -\Delta^{-1} K , \tag{4.24}$$

with

$$K = \begin{pmatrix} (-1)^{k_1 - 1} & & \\ 0 & \ddots & 0 \\ & & (-1)^{k_m - 1} \end{pmatrix} . \tag{4.25}$$

Proof

We proceed by induction on each block corresponding to the decomposition (4.19), (4.20), exactly as in the single-input case. We get

$$\dot{\xi}_k^i = L_f^k\phi_1^i + \sum_{j=1}^m(\alpha_j + \sum_{l=1}^m v_l\beta_{l,j})L_{g_j}L_f^k\phi_1^i = \xi_{k+1}^i = \phi_{k+1}^i$$

for $k = 1,\dots,k_i-2, i = 1,\dots,m$. Thus $L_{g_j}L_f^k\phi_1^i = 0$ for $j = 1,\dots,m, k = 0,\dots,k_i-2$ and $i = 1,\dots m$. The same relation for $k = k_i$ allows us to establish that the matrix Δ' defined by $\Delta'_{i,j} = L_{g_j}L_f^{k_i-1}\phi_1^i$ for $i,j = 1,\dots,m$ is invertible.

We then proceed as in lemma 4.2 to obtain the system of first-order PDEs (4.21) and one easily checks that $L_{g_j}L_f^{k_i-1}\phi_1^i = (-1)^{k_i-1}L_{ad_f^{k_i-1}g_j}\phi_1^i$, and thus $\Delta' = K\Delta$ with K defined by (4.25), which proves the result.

Conversely, we will show that a solution $(\phi_1^1,\dots,\phi_1^m)$ of (4.21) such that the matrix Δ defined by (4.22) is invertible, is such that the mapping ϕ obtained by (4.20) and (4.23) is a diffeomorphism in V. The remainder of the proof is then straightforward since the formulas (4.24) give an explicit construction of the linear system (4.16). It suffices then to prove that $d\phi$ is invertible in V. For this purpose, let us introduce the matrix $M = \Phi\,.\,\Gamma$, of size $n \times n$

$$\begin{aligned}\Phi^T = \; & (d\phi_1^1,\cdots,d\phi_1^m,\cdots,dL_f^{k_m-1}\phi_1^1,\cdots,dL_f^{k_m-1}\phi_1^m,dL_f^{k_m}\phi_1^1,\cdots,dL_f^{k_m}\phi_1^{m-1},\\ & \cdots,dL_f^{k_2-1}\phi_1^1,dL_f^{k_2-1}\phi_1^2,dL_f^{k_2}\phi_1^1,\cdots,dL_f^{k_1-1}\phi_1^1)\end{aligned} \tag{4.26}$$

and

$$\begin{aligned}\Gamma = \; & (ad_f^{k_1-1}g_1,\cdots,ad_f^{k_m-1}g_m,\cdots,ad_f^{k_1-k_m}g_1,\cdots,g_m,ad_f^{k_1-k_m-1}g_1,\cdots,\\ & ad_f^{k_{m-1}-k_m-1}g_{m-1},\cdots,ad_f^{k_1-k_{m-1}}g_1,\cdots,g_{m-1},\\ & \cdots,ad_f^{k_1-k_2}g_1,g_2,ad_f^{k_1-k_2-1}g_1,\cdots,g_1)\,.\end{aligned} \tag{4.27}$$

The reader can check that M, Φ and Γ are of size $n \times n$ and that M is lower triangular by blocks, each block of the diagonal being invertible thanks to the invertibility of Δ. It results that M is invertible and Φ and Γ too. Now, completing the definition of ϕ by (4.23), the matrix Φ is, up to a matrix of permutation, the linear tangent mapping of ϕ, which means, as claimed, that ϕ is a local diffeomorphism. The proof is easily completed as indicated above.

Proof of the theorem 4.2:

It remains to check that the conditions (i) and (ii) are equivalent to the existence of a solution to (4.21) such that Δ is invertible in V.

Let us assume that (i) and (ii) are true. It is easily proven that one can select, up to a reordering of the indices of the input vector fields, a basis of each G_i as follows:

- if $i \leq k_m$, then
$$G_i = \text{sp}\{g_1,\dots,g_m,\dots,ad_f^i g_1,\dots,ad_f^i g_m\}$$
- if $k_m \leq i \leq k_{m-1}$, then
$$G_i = \text{sp}\{g_1,\dots,g_m,\dots,ad_f^{k_m-1}g_1,\dots,ad_f^{k_m-1}g_m,\dots,ad_f^i g_1,\dots,ad_f^i g_{m-1}\}$$

and so on. We deduce that the distribution generated by the set of vector fields appearing in the system of PDEs (4.21) is involutive and that the columns of Γ, defined by (4.27), form a basis of G_{n-1} of dimension n. The existence of a solution to (4.21) such that Δ is invertible follows from Frobenius theorem and from the fact that the vectors $ad_f^{k_1-1}g_1, \ldots, ad_f^{k_m-1}g_m$ and $ad_f^k g_j$ are independent for $j = 1, \ldots, m$ and $k \leq k_i - 2$. Conversely, let us assume that the system (4.21) has a solution such that Δ is invertible in V. According to lemma 4.3, we know that this solution defines a diffeomorphism in V by (4.23). By the same lemma, we have:

$$\begin{aligned} \mathrm{sp}\{dL_f^j\phi_1^1, \ldots, dL_f^j\phi_1^m\}^\perp &= \mathrm{sp}\{ad_f^k g_i, k \leq k_m - j - 2, i = 1, \ldots, m\} \\ &\qquad \forall j = 0, \ldots, k_m - 2 \end{aligned} \tag{4.28}$$

which proves that $L_f^j\phi_1^1, \ldots, L_f^j\phi_1^m$, $j = 0, \ldots, k_m - 2$, are $m \times (k_m - 1)$ independent first integrals for the distributions of vector fields defined by the right-hand side of (4.28). The involutivity and non degeneracy in V of these distributions, which can be identified, respectively, to G_{k_m-2}, ..., G_0, follow from Frobenius theorem. Proceeding in a similar way, we see that, for $j = 0, \ldots, k_{m-1} - k_m - 1$

$$\begin{aligned} \mathrm{sp}\{dL_f^j\phi_1^1, \ldots, dL_f^j\phi_1^{m-1}\}^\perp &= \mathrm{sp}\{ad_f^k g_i, k_m - 1 \leq k \leq k_{m-1} - j - 2, i = 1, \ldots, m\} \\ &\qquad + G_{k_m-1} \end{aligned} \tag{4.29}$$

and that the distributions G_{k_m-1}, ..., $G_{k_{m-1}-2}$ are involutive with constant rank. Similarly, we show step by step that the distributions up to G_{k_1-2} are involutive with constant rank in V and finally that $G_{k_1-1} = G_{k_1} = \cdots = G_{n-1}$, with rank $G_{k_1-1} = k_1 + \cdots + k_m = n$ by assumption, which achieves the proofof the theorem.

4.1.3.2 Computation of the diffeomorphism and feedback

It results from lemma 4.3 that it suffices to compute the m functions $(\phi_1^1, \ldots, \phi_1^m)$ to obtain at the same time the diffeomorphism and the feedback by the formulaes (4.23) and (4.24). The only remaining difficulty concerns the determination of the *controllability indices* $(k_1, \ldots, k_m)$ which will be shown to be deduced from the construction of the distributions G_i defined by (4.18).

We consider the following integers:

$$r_i = \dim G_i - \dim G_{i-1}\ ,\ i \geq 1 \text{ and } r_0 = \dim G_0\ . \tag{4.30}$$

Proposition 4.1

The controllability indices $(k_1, \ldots, k_m)$ of lemma 4.3 are given by:

$$k_i = \mathrm{Card}\,\{j \geq 0 \mid r_j \geq i\}\ ,\ i = 1, \ldots, m\ . \tag{4.31}$$

These integers are a set of invariants of the system (Σ), i.e. they are independent of the choice of coordinates.

Proof

From the proof of the theorem 4.2, it results, by suitably choosing a basis of the G_i as indicated in the proof, that $r_i = (m - i)(k_{m-i} - k_{m-i-1})$. The expression of the k_i as functions of the r_j follows immediately. In addition, one easily checks that the k_i only depend on the dimensions of the distributions G_j, invariant by change of coordinates, which proves the result.

Therefore, the construction is done as follows:

1. Compute the $ad_f^i g_j$, and then the G_i, test the involutivity by computing the brackets $[G_i, G_i]$. If $[G_i, G_i]$ is not contained in G_i, for at least one i, the feedback linearization is impossible.
2. Compute the rank of the G_i, the r_i (formula (4.30)) and then the k_i (formula (4.31)). If rank $G_{n-1} < n$, and consequently $k_1 + \cdots + k_m < n$, the system is not controllable and thus not feedback linearizable though its controllable part is.
3. Write the system of partial differential equations (4.21) (whose solution exists by construction) and solve it (explicitly if possible!).
4. Compute the diffeomorphism and feedback by the formulae (4.23) and (4.24).

Remark 4.6

Every system of dimension of at most 2 whose tangent approximation is controllable is feedback linearizable. The case of dimension 1 is trivial. In 2 dimensions, two cases may happen: either the rank of G_0 is 2 ($m = 2$), or G_0 has rank 1 and G_1 has rank 2. In both cases, the assumptions of theorem 4.2 are clearly satisfied.

Remark 4.7

In the single-input case, the necessary and sufficient conditions are apparently simpler since it suffices to check the involutivity and non degeneracy of G_{n-2} and G_{n-1} only. One can easily check that, since there is only one input and therefore only one controllability index $k_1 = n$ (as in the converse part of the proof of theorem 4.2), that the involutivity and non degeneracy of all the G_i readily follow.

Remark 4.8

Let us evaluate the conditions of theorem 4.2 in the case of a linear system of the form (Λ) (which amounts to verifying that every linear controllable system is feedback linearizable!).

We have $G_0 = \text{sp}\{b_1, \ldots, b_m\} = ImB$ where b_i is the ith column of the matrix B. G_0 is involutive since, the vector fields b_i being constant, every bracket $[b_i, b_j]$ vanishes. One easily checks that since $ad_{Ax}^k b_i = (-1)^k A^k b_i$, we have $G_k = Im(B\ AB \cdots A^k B)$, which is involutive for every k for the same reason. Feedback linearization conditions can thus be interpreted as the necessary and sufficient conditions for the distributions G_i of (Σ) to be transformable (by a single diffeomorphism) into the sequence of reachable subspaces of a linear system. Let us stress that in the linear case, all the reachable subspaces are trivial distributions (affine spaces) which are naturally involutive, and the conditions on the distributions G_i associated with (Σ) mean that, though distorted by a diffeomorphism, they are the image of a sequence of affine spaces.

The comparison between feedback linearization and tangent linearization methods is indeed a natural question. The next result shows that feedback linearization is only of interest if one leaves the domain where the tangent approximation is acceptable.

Proposition 4.2

If the system (Σ) is feedback linearizable in a neighborhood of an equilibrium point, then its tangent approximation at this point is controllable.

Proof

It suffices to compute the brackets $ad_f^k g_j(0)$ at the origin by making use of the fact that $f(0) = 0$. We have $ad_f^k g_j(0) = (-1)^k F^k \gamma_j$, where we have denoted $F = \frac{\partial f}{\partial x}(0)$ and $\gamma_j = g_j(0)$. One readily checks that the condition (ii) of theorem 4.2 evaluated at the origin gives exactly Kalman's controllability criterion for the tangent approximation.

4.1.3.3 Extension to the " nonlinear in the input" case

The fact that the controls appear linearly in the system dynamics is by no means necessary. In fact, the previous definition may be adapted as follows: the system

$$\dot{x} = f(x, u) \tag{4.32}$$

with $f(0,0) = 0$ and rank $\frac{\partial f}{\partial u}(0,0) = m$, is feedback linearizable if there exists a diffeomorphism $\xi = \phi(x)$ in a neighborhood V of the origin, satisfying $\phi(0) = 0$ and a static state feedback $u = \alpha(x, v)$ satisfying $\alpha(0,0) = 0$ and rank $\frac{\partial \alpha}{\partial v}(x, v) = m$ for $(x, v) \in V \times W$, W neighborhood of 0 in $\mathbb{R}^p$, such that the transformed system is of the form (Λ), linear and controllable in the new input v.

We now consider these distributions, defined in $V \times W \subset \mathbb{R}^n \times \mathbb{R}^m$:

$$\begin{aligned} G_0 &= \text{sp}\left\{\frac{\partial f}{\partial u_1}, \ldots, \frac{\partial f}{\partial u_m}, \frac{\partial}{\partial u_1}, \ldots, \frac{\partial}{\partial u_m}\right\} \\ G_i &= G_{i-1} + ad_f G_{i-1} \qquad\qquad i \geq 1 \end{aligned} \tag{4.33}$$

Theorem 4.3

The system (4.32) is feedback linearizable if and only if the distributions (4.33) satisfy the conditions of theorem 4.2.

Proof

We only outline the adaptations needed with respect to the proof of theorem 4.2 in the linear-in-the-input case. As in the previous case, we compute $\dot{\xi}_1^i = \frac{d}{dt}\phi_1^i = L_f \phi_1^i = \xi_2^i$, which provide $\frac{\partial}{\partial u_j} L_f \phi_1^i = L_{\frac{\partial f}{\partial u_j}} \phi_1^i = 0$ for every j. By induction, one shows that $\dot{\xi}_k^i = L_f^k \phi_1^i$ and that $L_{\frac{\partial f}{\partial u_j}} L_f^k \phi_1^i = 0$ for every i, j and $k \leq k_i - 2$. The remainder of the proof remains unchanged up to considering a neighborhood $V \times W$ of $(0,0)$, formula (4.24) being replaced by an implicit equation whose solution exists if $\frac{\partial \Delta}{\partial v}$ is invertible. The reader may easily check that the construction of the diffeomorphism and feedback is obtained by following the same steps 1 to 4 as before.

4.1.4 Examples

4.1.4.1 Second-order nonlinear systems

Let us consider the following system

$$\begin{cases} \dot{x}_1 &= x_2 \\ \dot{x}_2 &= f(x_1, x_2, u) \end{cases} \tag{4.34}$$

where x_1 and x_2 have dimension $\frac{n}{2}$ (the state $x = (x_1, x_2)^T$ being thus n-dimensional), where the input vector u has dimension $m = \frac{n}{2}$ and where

$$\text{rank}\left\{\frac{\partial f}{\partial u_1}(0,0,0), \ldots, \frac{\partial f}{\partial u_m}(0,0,0)\right\} = m,$$

with f an analytic vector field satisfying $f(0,0,0) = 0$.

One sees without computation that (4.34) is (globally) feedback linearizable, the diffeomorphism being the identity of $\mathbb{R}^n$ and the feedback verifying the implicit equation $v = f(x_1, x_2, u)$, which has a local solution thanks to the assumption on $\frac{\partial f}{\partial u}(0,0,0)$.

The closed-loop system is thus given by blocks:

$$\begin{cases} \dot{x}_1^i &= x_2^i \\ \dot{x}_2^i &= v^i \end{cases}, \quad i = 1, \ldots, m\,. \tag{4.35}$$

One easily checks that the same result may be obtained by applying theorem 4.3.

Let the neighborhood V of $(0,0,0)$ be defined as the largest open set containing $(0,0,0)$ such that the rank of $\frac{\partial f}{\partial u}(x_1, x_2, u)$ remains constant. If $V = \mathbb{R}^n \times \mathbb{R}^p$, the feedback linearization is globally defined. Otherwise, the matrix $\frac{\partial f}{\partial u}(x_1, x_2, u)$ is singular on V's boundary and the feedback is no longer defined at such points. In this case, the feedback linearization is local.

Note that this example contains the aforementioned class of mechanical systems.

4.1.4.2 Application to the control of a robot arm with pneumatic actuators

Let us consider (Lévine, 1988) a robot arm made of an arbitrary number, n_1, of segments. The actuators (servo-valves) inject air under pressure at the top and the bottom of each balloon with flow rate u_1 (resp. u_2) (Figure 4.1). We denote by p_1 the n_1-dimensional vector of the pressures in the upper part of each balloon and p_2 the n_1-dimensional vector of the pressures in the lower part of each balloon, the air injection in the upper or lower part of each balloon inducing a rotation of the associated segments in the positive or negative sense. x_1 is the n_1-dimensional vector of relative angular positions, x_2 is the n_1-dimensional vector of relative angular velocities. The dynamical model of the robot is

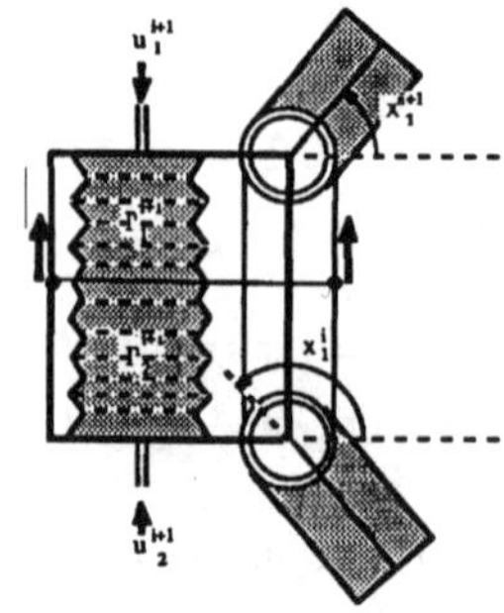

Figure 4.1: two segments with their pneumatic actuator.

thus given by:

$$\left\{\begin{array}{lcl}\dot{x}_1 & = & x_2 \\ \dot{x}_2 & = & \Gamma_0^{-1}(x_1)(\Gamma(x_1,x_2)+Q(x_1,x_2,p_1,p_2)) \\ \dot{p}_1 & = & a_1(x_1,x_2,p_1,p_2)+b_1(x_1,x_2,p_1,p_2)u_1 \\ \dot{p}_2 & = & a_2(x_1,x_2,p_1,p_2)+b_2(x_1,x_2,p_1,p_2)u_2\end{array}\right. \tag{4.36}$$

where the (x_1, x_2) part corresponds to the mechanical part

$$\Gamma_0(x_1)\ddot{x}_1 = \Gamma(x_1,\dot{x}_1) + Q(x_1,\dot{x}_1,p_1,p_2)$$

and where the (p_1, p_2) part corresponds to the model of pressure induced torques, which does not need to be precised since the linearizability property of the system is independent of the exact form of a and b. We only assume that one can find two real numbers λ_1 and λ_2, one of them, at least, being nonzero, such that

$$\text{rank }(\lambda_2\frac{\partial Q}{\partial p_1} - \lambda_1\frac{\partial Q}{\partial p_2}) = n_1 \ \forall(x,p) \tag{4.37}$$

Remark that the p part of the model may be considered as a dynamic feedback where p_1 and p_2 play the role of control variables for the mechanical part in x. The presence of such a nonlinear integrator is generally the source of persistent state dependent oscillations. It results that the tangent linearization methods are difficult to apply to follow reference trajectories. The next proposition gives a way to circumvent this problem.

Proposition 4.3

The change of coordinates $\xi = \varphi(x_1,x_2,p_1,p_2)$ and the static state feedback $u = \alpha(x_1,x_2,p_1,p_2)+\beta(x_1,x_2,p_1,p_2)v$, with φ, α, β given by (4.39), (4.40), (4.41) and (4.42) below, are such that $\xi(t) = \varphi(x(t),p(t))$ satisfies

$$\left\{\begin{array}{lcl}\dot{\xi}_1 & = & \xi_2 \\ \dot{\xi}_2 & = & \xi_3 \\ \dot{\xi}_3 & = & v_1 \\ \dot{\xi}_4 & = & v_2\end{array}\right. \tag{4.38}$$

with

$$\varphi(x_1, x_2, p_1, p_2) = \begin{pmatrix} x_1 \\ x_2 \\ \Gamma_0^{-1}(x_1)(\Gamma(x_1, x_2) + Q(x_1, x_2, p_1, p_2)) \\ \lambda_1 p_1 + \lambda_2 p_2 \end{pmatrix} \tag{4.39}$$

$$\begin{aligned} \alpha(x,p) &= M^{-1}(x,p) \begin{pmatrix} \gamma(x,p) \\ -\lambda_1 a_1(x,p) - \lambda_2 a_2(x,p) \end{pmatrix} \\ \beta(x,p) &= M^{-1}(x,p) \begin{pmatrix} \Gamma_0(x) & 0 \\ 0 & I \end{pmatrix} \end{aligned} \tag{4.40}$$

$$M(x,p) = \begin{pmatrix} \frac{\partial Q}{\partial p_1} b_1(x,p) & \frac{\partial Q}{\partial p_2} b_2(x,p) \\ \lambda_1 b_1(x,p) & \lambda_2 b_2(x,p) \end{pmatrix} \tag{4.41}$$

and

$$\gamma(x,p) = -\Gamma_0(\frac{\partial}{\partial x_1}(\Gamma_0^{-1}(\Gamma+Q))x_2 + (\frac{\partial}{\partial x_2}(\Gamma_0^{-1}(\Gamma+Q)))(\Gamma_0^{-1}(\Gamma+Q))) - \frac{\partial Q}{\partial p_1} a_1 - \frac{\partial Q}{\partial p_2} a_2 \tag{4.42}$$

Consequently, this control law can easily be implemented by classical linear observer–controller method, even if the state is only partially measured. It suffices for example to measure x_1 and p. Furthermore, this control law is robust in the sense that small modeling errors or small perturbations will not affect the stability of the linear closed-loop system, even in the case where fast reference trajectories are followed, until the constraints on the actuators are saturated. However, when errors are too large, the linear structure of the closed-loop system may be destroyed by reinjecting nonlinearities in (4.38) by φ and the feedback. In this case, the model must be refined. However, if exogeneous perturbations or noises influence the system, their characteristics can be taken into account in the gain design of the observer.

4.1.4.3 A five-dimensional example

Let us consider the system

$$\left\{ \begin{array}{rcl} \dot{x}_1 &=& x_2 \\ \dot{x}_2 &=& x_5 \\ \dot{x}_3 &=& u_2 \\ \dot{x}_4 &=& x_3(1 - x_5) \\ \dot{x}_5 &=& u_1 \; . \end{array} \right. \tag{4.43}$$

We have $f = x_2 \frac{\partial}{\partial x_1} + x_5 \frac{\partial}{\partial x_2} + x_3(1 - x_5)\frac{\partial}{\partial x_4}$, $g_1 = \frac{\partial}{\partial x_5}$ and $g_2 = \frac{\partial}{\partial x_3}$.
One verifies that

$$G_0 = \mathrm{sp}\left\{ \frac{\partial}{\partial x_3}, \frac{\partial}{\partial x_5} \right\} = \overline{G}_0 \; .$$

In order to check that G_1 is involutive, one has to compute the brackets $\mathrm{ad}_f g_1$ and $\mathrm{ad}_f g_2$:

$$\mathrm{ad}_f g_1 = [f, g_1] = -\frac{\partial f}{\partial x_5} = -\frac{\partial}{\partial x_2} + x_3 \frac{\partial}{\partial x_4}$$

$$\mathrm{ad}_f g_2 = [f, g_2] = -\frac{\partial f}{\partial x_3} = -(1 - x_5)\frac{\partial}{\partial x_4}$$

which provide

$$G_1 = G_0 + \mathrm{sp}\left\{\mathrm{ad}_f g_1, \mathrm{ad}_f g_2\right\} = \mathrm{sp}\left\{\frac{\partial}{\partial x_2}, \frac{\partial}{\partial x_3}, \frac{\partial}{\partial x_4}, \frac{\partial}{\partial x_5}\right\} = \overline{G}_1 \, .$$

One finally checks that, since $\mathrm{ad}_f^2 g_1 = \frac{\partial}{\partial x_1}$, $G_2 = T\mathbb{R}^5$. Thus $r_0 = \dim G_0 = 2$, $\dim G_1 = 4$, $\dim G_2 = 5$, $r_1 = \dim G_1 - \dim G_0 = 2$ and $r_2 = \dim G_2 - \dim G_1 = 1$.
The controllability indices k_1 and k_2 are given by the formulaes

$$k_1 = \mathrm{Card}\,\{j \geq 0 | r_j \geq 1\} = 3 \qquad \text{and} \qquad k_2 = \mathrm{Card}\,\{j \geq 0 | r_j \geq 2\} = 2.$$

The two chains correspond to the choice of the basis

$$\{g_1, \mathrm{ad}_f g_1, \mathrm{ad}_f^2 g_1, g_2, \mathrm{ad}_f g_2\} \, .$$

The diffeomorphism is thus computed by the relations

$$\mathrm{d}\phi_1 \perp \mathrm{sp}\{g_1, \mathrm{ad}_f g_1, g_2, \mathrm{ad}_f g_2\} \, , \; < \mathrm{d}\phi_1, \mathrm{ad}_f^2 g_1 > \neq 0 \, ,$$

$$\mathrm{d}\phi_2 \perp \mathrm{sp}\{g_1, g_2\} \, , \; < \mathrm{d}\phi_2, \mathrm{ad}_f g_2 > \neq 0$$

or

$$\frac{\partial \phi_1}{\partial x_5} = 0 \, , \; \frac{\partial \phi_1}{\partial x_2} + x_3 \frac{\partial \phi_1}{\partial x_4} = 0 \, , \; \frac{\partial \phi_1}{\partial x_3} = 0 \, , \; (1 - x_5)\frac{\partial \phi_1}{\partial x_4} = 0 \, , \; \frac{\partial \phi_1}{\partial x_1} \neq 0$$

$$\frac{\partial \phi_2}{\partial x_5} = 0 \, , \; \frac{\partial \phi_2}{\partial x_3} = 0 \, , \; (1 - x_5)\frac{\partial \phi_2}{\partial x_4} \neq 0 \, .$$

A possible choice is $\phi_1 = x_1$ and $\phi_2 = x_4$. The diffeomorphism is thus given by $\xi_1^1 = x_1$, $\xi_2^1 = L_f \phi_1 = x_2$, $\xi_3 = L_f^2 \phi_1 = x_5$ and $\xi_1^2 = x_4$, $\xi_2^2 = L_f \phi_2 = x_3(1 - x_5)$, and the corresponding closed-loop system

$$\dot{\xi}_1^1 = \xi_2^1 \, , \; \dot{\xi}_2^1 = \xi_3^1 \, , \; \dot{\xi}_3^1 = u_1 = v_1$$

$$\dot{\xi}_1^2 = \xi_2^2 \, , \; \dot{\xi}_2^2 = u_2(1 - x_5) - u_1 x_3 = v_2 \, .$$

Note that the diffeomorphism is not globally defined since its Jacobian matrix is singular on the hyperplane $\{x_5 = 1\}$ (recall that the equilibrium point is such that $x_5 = 0$).

4.1.4.4 A counterexample

We know by remark 4.6 that every system of dimension less than or equal to 2 whose tangent approximation is controllable, is feedback linearizable. We now present a simple three-dimensional system which is not feedback linearizable. It is often referred to as a system in "chain form".

$$\dot{x} = \begin{pmatrix} 0 \\ 1 \\ 0 \end{pmatrix} u_1 + \begin{pmatrix} 1 \\ 0 \\ x_2 \end{pmatrix} u_2 \tag{4.44}$$

or

$$\begin{cases} \dot{x}_1 & = & u_2 \\ \dot{x}_2 & = & u_1 \\ \dot{x}_3 & = & u_2 x_2 \, . \end{cases} \tag{4.45}$$

In the notations of theorem 4.2, we have $x = (x_1, x_2, x_3)^T$, $f(x) = (0,0,0)^T$, $g_1(x) = (0,1,0)^T$ and $g_2(x) = (1,0,x_2)^T$. However, $G_0 = \text{sp}\{g_1, g_2\}$. But G_0 is not involutive since the bracket $[g_1, g_2] = (0,0,1)^T$ does not belong to G_0. Thus the system is not feedback linearizable.

In fact, the reason of this obstruction may be understood without computations: the system is made of two coupled chains of integrators $\dot{x}_2 = u_1$ on the one hand and $\dot{x}_1 = u_2$, $\dot{x}_3 = u_2 x_2$ on the other hand. Therefore, one cannot linearize x_3 without destroying the linearity on the x_1 dynamics. Note that x_3 corresponds to the non involutive direction $[g_1, g_2]$. Hereafter, we shall see that this can be fully justified and that the number of directions which produce non involutivity corresponds to the minimal number of nonlinearizable variables. We shall also see that this obstruction to linearizability disappears if we use dynamic rather than static feedback.

4.1.5 Comments on the implementation

4.1.5.1 Difficulties related to the feedback computation

We have shown that the computations of the diffeomorphism and feedback are done by computing Lie brackets, ranks of families of vector fields and the solution of a system of first-order partial differential equations. In practice, those calculations are subject to the following severe limitations.

Even if the system has the form (Σ), the vector fields f, g_1, ..., g_m may be given as a table of numbers resulting, for example, from an identification algorithm or from the numerical solution of an implicit equation (see Lévine and Rouchon, 1991; Rouchon, 1990). This requires computing the successive derivatives of the components of the vector fields which appear in the Lie brackets by numerical analytic methods. Therefore, after a given, generally small, number of numerical derivatives, the result might lose its meaning due to numerical errors. Modeling thus appears very important in this context, especially for what concerns preserving the underlying structure of the system (e.g. controllability indices) by numerical treatment.

More precisely, we aim at checking rank conditions on the one hand and obtaining explicit formulaes for the diffeomorphism and feedback on the other hand. As far as rank computations are concerned, it appears in some cases (see Bastin and Lévine, 1993;

El Moubaraki *et al.*, 1993), that physical constraints imply a given system structure independently of the precision to compute the Lie brackets. Otherwise stated, the ranks of the distributions G_i are equal to their "generic rank" and a solution to the feedback linearization problem might also exist independently of the numerical values of the vector fields. However, the diffeomorphism and feedback accuracy generally directly depend on the numerical precision of the data and the applicability of the results may be examined case by case.

When the system vector fields are given analytically, another source of difficulties might arise when using symbolic computation programs for iterated Lie brackets. Dimensions of the system state and input thus impose serious constraints. For example, computing iterated Lie brackets at order 3, or more, with 2 inputs and in state dimension 12, or more, without simplifications due to the specific structure of the vector fields, requires an enormous amount of memory, even with the current computers and languages. Moreover, even when such symbolic computation can be done in a reasonable time, the result generally consists of a list of several pages. Therefore, detecting simplifications, if any, in such a jungle requires a huge effort. However, if those results are to be used for simulation purposes, tools for automatic generation of Fortran programs, or such, might be helpful.

When one can overcome these difficulties, one is left with the resolution of a system of partial differential equations. An explicit solution can sometimes be obtained thanks to the method of characteristics presented in the Appendix B. But in the general case, the characteristics may be obtained by numerical integration and one must carefully check that the numerical errors do not too much corrupt the closed-loop system.

Finally, it is worth noting that implementing the obtained control law requires the use of discretization methods and the choice of a suitable sampling period. Approximation errors may then be asymptotically compensated by adequately placing the poles of the closed-loop discrete-time system. Remark that, since the closed-loop continuous-time one is linear and admits an exact discretization, the control law constitutes the main source of errors, assuming that the state is perfectly measured. Indeed, errors mainly result from the difference between the control value at each time and its sampled approximation. For a small enough sampling period, asymptotic compensation of small modeling and observation errors and perturbations is easily proven. If, moreover, fast stable dynamics have been neglected, the sampling period may be chosen larger than the time constants of the neglected parasitic dynamics.

4.1.5.2 The problems caused by singularities

In the above examples, the diffeomorphism and feedback are generally only locally defined. The impossibility of defining them globally comes from the existence of singularities. They are in fact one of the characteristics of the nonlinear nature of a system. Their cause may be twofold: the topology of the manifold (*e.g.* the non existence of a global smooth diffeomorphism from the sphere to a plane), and the geometry of the distribution of vector fields (*e.g.* loss of controllability rank). Some of these causes are analyzed in Dayawansa *et al.* (1985). One should note however that singularities cannot be detected on the closed-loop system which remains linear with respect to the auxiliary input v with bounded state on any finite interval if v is bounded. However, the diffeomorphism stops existing and the feedback generally blows up. The more one approaches a singularity, the greater the control sensitivity to errors becomes, since its gradient with respect to state, which plays the role of a nonlinear counterpart of the control gain of linear control, increases in norm and thus amplifies the effect of perturbations.

In specific and, of course, non generic cases, a singular point of the diffeomorphism may coincide with a singular point of the controlled vector fields. Then one can pass through this singularity if a simplification of the type $\frac{0}{0}$ appears in the feedback law.

However, in the general case, singularities will impose limits on the choice of reference trajectories that one wishes to follow. Reference trajectories must indeed avoid approaching too closely such singularities to avoid saturating the control constraints, which are not taken into account in this approach, and degrading the overall performances of the closed-loop system. If, according to the given practical objectives, the reference trajectory does cross a singularity, the control may switch, in a neighborhood of this point, to a different feedback law during the crossing. This technique, however, may not be useful for a too large number of crossings per time unit.

4.1.6 The largest feedback linearizable subsystem

In regard to the small number of feedback linearizable systems, one may want to know if there always exists a largest feedback linearizable subsystem and to understand its structure. The interest of this question is threefold:

1. The coordinates in which the largest feedback linearizable subsystem can be expressed may be useful to determine the output functions that can be assigned linearly and such that the number of state components that remain nonlinear and unobservable by feedback are minimal. This property may be related to the concept of *zero dynamics*, nonlinear generalization of the notion of finite zeros for linear systems.

2. From a more structural viewpoint, we have seen that the controllability indices of feedback linearizable systems exist and are invariant by static state feedback and diffeomorphism. An analog property holds for the largest feedback linearizable subsystem.

3. The minimum number of nonlinear components of a system may give an idea of its complexity and can be used as a classification tool for nonlinear systems.

4.1.6.1 Marino's theorem

Definition 4.1

The system (Σ) *is said to be partially linearizable with controllability indices* $k_1, \ldots, k_m$, *if there exists a diffeomorphism* $\xi = \phi(x)$ *and a static state feedback* $u = \alpha(x) + \beta(x)v$, *satisfying the same assumptions as before, such that the closed-loop system rewrites, in the new coordinates* $\xi = (\xi_1, \xi_2)^T$,

$$\left\{ \begin{array}{rcl} \dot{\xi}_1 & = & A\xi_1 + Bv \\ \dot{\xi}_2 & = & a(\xi_1, \xi_2) + b(\xi_1, \xi_2)v \end{array} \right. \tag{4.46}$$

with the pair (A, B) *in controllability canonical form with indices* $k_1, \ldots, k_m$.

We now extend the definition of controllability indices that have been defined for feedback linearizable systems. To this aim, we need to introduce the distributions

$$\begin{aligned} Q_0 &= G_0 \\ Q_i &= \overline{Q}_{i-1} + ad_f^i Q_0 \ , \ i \geq 1 \end{aligned} \tag{4.47}$$

where $\overline{Q}$ denotes the involutive closure of Q (see the definition A.14).

These new distributions coincide with the G_i's when they are involutive:

Proposition 4.4

If Q_i is involutive for every $i \leq I$, with $I \leq \infty$, then $Q_i = G_i$ for every $i \leq I$.

Proof

If $Q_0 = G_0$ is involutive, we have $Q_1 = G_0 + ad_f G_0 = G_1$. By induction, if $Q_j = G_j$ is involutive for all $j \leq i$, one has $Q_{i+1} = G_i + ad_f^{i+1} G_0 = \sum_{j=0}^{i+1} ad_f^j G_0$. Since G_{i+1} is also involutive, one easily checks that $G_{i+1} = \sum_{j=0}^{i+1} ad_f^j G_0$ and thus $Q_{i+1} = G_{i+1}$.

Let us now define the sequence of integers s_i:

$$\begin{aligned} s_0 &= \dim Q_0 \\ s_i &= \dim Q_i - \dim \overline{Q}_{i-1} \ , \ i \geq 1 . \end{aligned} \tag{4.48}$$

One can show (exercise) that the sequence of s_i is non increasing. It immediately results from proposition 4.4 that if the Q_i are all involutive, we have $s_i = r_i$ for every i, the r_i being defined in (4.30).

Definition 4.2

We call controllability indices associated to (Σ) the sequence of integers $k_1^*, \ldots, k_m^*$ given by:

$$k_i^* = \text{Card}\,\{j \geq 0 \mid s_j \geq i\} \ i = 1, \ldots, m\,. \tag{4.49}$$

Clearly, if (Σ) is feedback linearizable, one has $s_i = r_i$ for every i and $k_i^* = k_i$ for every $i = 1, \ldots, m$. This definition thus generalizes the notion introduced for feedback linearizable systems.

The largest feedback linearizable subsystem can be defined by the following result:

Theorem 4.4 *(Marino, 1986)*

To every system (Σ) one can associate in a unique way (up to input indices permutation) the list of controllability indices $k_1^*, \ldots, k_m^*$. Moreover, the largest feedback linearizable subsystem of (Σ) always exists and has dimension $n_\lambda = k_1^* + \cdots + k_m^*$.

The proof, long and difficult, is established in Marino (1986). Since it is not necessary to understand the rest of this chapter, we have not reproduced it here.

Let us remark that the dimension $q = n - n_\lambda$ plays a crucial role since it represents the minimum number of components of the state that cannot be linearized. This integer is invariant by change of coordinates and an immediate consequence of theorem 4.4 is that for an arbitrary choice of static state feedback, one would obtain a linearizable system of dimension lower than or equal to n_λ, with moreover controllability indices $\kappa_i \leq k_i^*$ for every $i = 1, \ldots, m$. The integers k_i^* are thus the largest input by input indices. Finally, the number q of nonlinear coordinates corresponds by construction to the number of directions of $\overline{Q}_i$ that do not belong to Q_i. One can summarize this remark by saying that the minimum number of nonlinear coordinates corresponds to the total number of non involutive directions.

We now give a construction of the largest feedback linearizable subsystem. It follows the same lines as the one for feedback linearizable systems.

1. Compute the $ad_f^i g_j$, the Q_i, check the involutivity by the computing the brackets $[Q_i, Q_i]$. If $[Q_i, Q_i]$ is not contained in Q_i, one computes its involutivity closure (by computing $[Q_i, [Q_i, Q_i]]$, $[Q_i, [Q_i, [Q_i, Q_i]]]$, *etc.*).

2. Compute the rank of the Q_i, the s_i (formula (4.48)) and finally the k_i^* (formula (4.49)). We then compute $n_\lambda = k_1^* + \cdots + k_m^*$.

3. Write the system of partial differential equations (4.21) with the k_i^* in place of the k_i. A solution of this system gives the linearizable part of the diffeomorphism

$$\xi_1 = (\phi_1^1(x), \ldots, L_f^{k_1^*-1}\phi_1^1(x), \ldots, \phi_1^m(x), \ldots, L_f^{k_m^*-1}\phi_1^m(x))^T .$$

4. Compute the diffeomorphism by adding to ξ_1 $q = n - n_\lambda$ arbitrary independent functions and compute the feedback by (4.24) where Δ is computed with the indices k_i^* in place of the k_i. Note that if Q_0 is involutive, one can always choose coordinates and a feedback such that $b(\xi_1, \xi_2) \equiv 0$.

4.1.6.2 Links with input–output approaches, zero dynamics

The above maximally linearizing transformation $\xi_1 = \phi_1(x)$ has the following property: the outputs $y_i = \phi_1^i(x) = \xi_1^i$ for $i = 1, \ldots, m$, may be seen as the outputs of the transformed linear system, of dimension n_λ that takes the form $\dot{\xi}_1 = A\xi_1 + Bv, y = \xi_1$. The corresponding feedback may be interpreted as a decoupling and linearizing feedback for the original system with outputs $y_1, \ldots, y_m$. It is indeed easily checked that the decoupling matrix associated to the outputs $y_1, \ldots, y_m$ is equal to $K\Delta$, obtained by (4.22) and (4.25) after changing the k_i into the k_i^* in K and Δ. It results that the (minimally) nonlinearizable part ξ_2 is made unobservable by feedback.

Let us note that, contrary to the linearizable part, only the dimension of the unobservable part is intrinsic: its control structure depends on the choice of the $n - n_\lambda$ independent functions that define ξ_2. In a neighborhood of an equilibrium point, its closed-loop structure is analyzed, as in the input–output approach (see chapter 3, devoted to input–ouput decoupling), by separating the decoupled and nonlinearly controllable part (controllable but uncontrollable at the first-order), the controllable and coupled part and the non controllable one.

If we want to stabilize an equilibrium point, or equivalently, up to a change of coordinates, the origin, one can compute a feedback such that y_i exponentially tends to 0, $i = 1, \ldots, m$. However, the internal stability will only be guaranteed if ξ_2 remains bounded for every initial condition in a suitable open and bounded domain. The center manifold techniques may be useful in studying the behavior around the equilibrium point of the unobservable part independently of the observable one, namely the one corresponding to $\dot{\xi}_2 = a(0, \xi_2)$, which can be interpreted as a slow dynamics, the linearized dynamics being chosen much faster than the latter. This residual dynamics is generally called the *zero dynamics.* The following local result can be found in Isidori (1989)

Theorem 4.5 *(Byrnes and Isidori, 1989)*

If the zero dynamics are locally asymptotically stable, there exists a feedback (of the above type) which exponentially stabilizes the outputs with local internal asymptotic stability.

It may be noted that the largest linearizable subsystem is not necessarily well suited to stabilization. One might sometimes prefer different output functions for which the linearizable subsystem is smaller but with a naturally locally asymptotically stable zero dynamics (D'Andréa-Novel and Praly, 1988).

4.2 Dynamic feedback linearization

Most of the proofs of the results of this section have been withdrawn since they are very technical and do not reveal new insight. The interested reader will find the necessary details in the references given throughout the text, the main ones being Charlet *et al.*, (1989; 1991).

The notion of dynamic feedback has long since been introduced in many aspects of linear system theory such as the problem of decoupling with stability (Wonham, 1974). In the nonlinear input–output context, they were introduced by S.N. Singh (1981) to solve the inversion problem; for a complete solution see Fliess (1989). From then on, it appeared that the status of dynamic feedback was quite different in the nonlinear case. The idea of trying to enlarge the applicability of feedback linearization by introducing dynamic feedback is thus quite natural. Precisely, we call *dynamic compensator* a system

$$\begin{aligned} \dot{w} &= a(x,w) + B(x,w)v \qquad w \in \mathbb{R}^q \\ u &= \alpha(x,w) + \beta(x,w)v \qquad v \in \mathbb{R}^{m'}, \quad m' \geq m. \end{aligned} \tag{4.50}$$

with $\alpha(0,0) = 0$, $a(0,0) = 0$. Clearly, the static state feedbacks are contained in this class. Note however that, so defined, this class is too large since in particular nothing guarantees that the number of independent inputs remains unchanged after closing the loop. This is why we shall only deal with *nondegenerated* or *regular compensators.* Before describing them, let us introduce an important subclass of non trivial dynamic compensators made of chains of pure integrators. We note

$$u^{(\mu)} = \frac{d^\mu u}{dt^\mu}, \qquad w = \left(u_1, \ldots, u_1^{(\mu_1)}, \ldots, u_m, \ldots, u_m^{(\mu_m)}\right)$$

and

$$\begin{pmatrix} u_1^{(\mu_1)} \\ \vdots \\ u_m^{(\mu_m)} \end{pmatrix} = \alpha(x,w) + \beta(x,w) \begin{pmatrix} v_1 \\ \vdots \\ v_{m'} \end{pmatrix} \tag{4.51}$$

with $\mu_i \geq 0$, $1 \leq i \leq m$, $\mu = \sum_{i=1}^{m} \mu_i$, $\alpha(0,0) = 0$, $\beta(x, w)$ of full rank m ($m' \geq m$) in V, a neighborhood of the origin in $\mathbb{R}^{n+\mu}$. This compensator can be realized as follows

$$\begin{aligned}
\dot{w}_i^j &= w_{i+1}^j && 1 \leq i \leq \mu_j - 1,\ 1 \leq j \leq m,\ \mu_j > 0 \\
\dot{w}_{\mu_j}^j &= \alpha_j(x, w) + \sum_{l=1}^{m'} \beta_{j,l}(x, w) v_l(t) \ = \ v_j'(t) && 1 \leq j \leq m,\ \mu_j > 0 \\
u_j &= w_1^j && 1 \leq j \leq m,\ \mu_j > 0 \\
u_j &= \alpha_j(x, w) + \sum_{l=1}^{m'} \beta_{j,l}(x, w) v_l(t) \ = \ v_j'(t) && 1 \leq j \leq m,\ \mu_j = 0
\end{aligned} \tag{4.52}$$

Let us now return to the general compensators of the type (4.50).

The system (Σ) put together with the compensator (4.50) is called the *extended system.*

$$\begin{aligned}
\dot{x} &= f(x) + G(x)\alpha(x, w) + G(x)\beta(x, w)v \\
\dot{w} &= a(x, w) + B(x, w)v
\end{aligned} \tag{4.53}$$

or

$$\dot{\tilde{x}} = \tilde{f}(\tilde{x}) + \tilde{G}(\tilde{x})v = \tilde{f}(\tilde{x}) + \sum_{i=1}^{m'} \tilde{g}_i(\tilde{x})v_i \tag{4.54}$$

where $\tilde{x} = (x, w)$ is an *extended state* and

$$\tilde{f} = \begin{pmatrix} f(x) + G(x)\alpha(x, w) \\ a(x, w) \end{pmatrix}, \quad \tilde{G} = \begin{pmatrix} G(x)\beta(x, w) \\ B(x, w) \end{pmatrix}. \tag{4.55}$$

4.2.1 Regular compensators

We are now able to give a precise definition of a regular dynamic compensator that extends the fact that, in the static case, $\beta(x)$ is invertible for all x. The idea is that the closed-loop system may not lose input channels. This property may be rigorously stated in terms of the output differential rank

If $u_j = \alpha_j(\tilde{x}) + \sum_{i=1}^{m'} \beta_{j,i}(\tilde{x})v_i$, $1 \leq j \leq m$, are considered as m outputs for the system (4.54), the differential output rank (Di Benedetto *et al.*, 1989; Fliess, 1989) is computed as follows. Denote by $u^{(k)}$ the time derivative of u of order k, u being considered as an output for (4.54). We define the sequence of vector spaces $E_0 \subset E_1 \subset \ldots \subset E_{n+q}$ by:

$$E_k = \text{sp}\{d\tilde{x}, d\dot{u}, \ldots, du^{(k)}\}.$$

The differential output rank $d^\circ(u)$ is then:

$$d^\circ(u) = \dim E_{n+q} - \dim E_{n+q-1}.$$

In the single input case ($m = 1$), one shows in Di Benedetto *et al.* (1989) that the computation of the differential output rank amounts to the classical computation of the rank of the decoupling matrix; (see for example Isidori (1989)).

In this case, the characteristic index ν is defined by

$$\begin{aligned} \nu &= 0 \text{ if } \beta_i(\tilde{x}) \neq 0 \text{ for at least one } i,\ 1 \leq i \leq m' \\ \nu &= \min\{r \mid L_{\tilde{g}_i} L_{\tilde{f}}^{r-1} \alpha(\tilde{x}) \neq 0 \text{ for at least one } i,\ 1 \leq i \leq m'\} \end{aligned} \tag{4.56}$$

Let

$$\delta_i(\tilde{x}) = \begin{cases} \beta_i(\tilde{x}) & \text{if } \nu = 0 \\ L_{\tilde{g}_i} L_{\tilde{f}}^{\nu-1} \alpha(\tilde{x}) & \text{if } \nu > 0 \end{cases} \tag{4.57}$$

When ν is finite, the matrix of size $1 \times m'$

$$D(\tilde{x}) = (\delta_1(\tilde{x}), \ldots, \delta_{m'}(\tilde{x})) \tag{4.58}$$

is called the decoupling matrix of the compensator (4.50) for the system (Σ) with $m = 1$. In such a case, $d^\circ(u) = \text{rank } D$.

Definition 4.3

The dynamic compensator (4.50) is called regular for the system (Σ) if the corresponding differential output rank $d^\circ(u)$ *is equal to m in V, a neighborhood of the origin in $\mathbb{R}^{n+q}$.*

Definition 4.4

The system (Σ) is said to be locally dynamic feedback linearizable if there exists a regular dynamic compensator (4.50) and a diffeomorphism of the extended state space $\xi = \phi(\tilde{x})$ defined in a neighborhood of $(x, w) = (0, 0)$, that transform (Σ) into the linear controllable system (Λ) of dimension $n' = n + q$, and such that $\phi(0, 0) = (0, 0)$.

Remark 4.9

M. Fliess (1990) introduced a different notion of dynamic feedback linearization by means of the generalized controller canonical form. It consists in finding a diffeomorphism of the original state space (non extended) and a system of differential equations on the inputs (dynamic compensator) such that the closed-loop system (without the compensator) is equivalent to a linear controllable system (thus of the same dimension as the original, non extended, system). For such a notion, every nonlinear system is dynamic feedback linearizable. However, since the dynamic compensator itself is not linearized, contrary to the above approach, its stability is not guaranteed. In fact, one can check on many examples that this dynamic compensator may blow up in finite time, which significantly decreases its practical interest.

4.2.2 Comparison between static and dynamic feedbacks in the single-input case

Let us first try to describe the potential improvements of regular dynamic feedback with respect to static feedback. We start with a result stating that in the single input case, there is no hope of improving the static feedback linearization results (Charlet *et al.*, 1989; 1991). In other words, dynamic feedback may only be useful for linearization in the multi-input case.

Theorem 4.6 *(Charlet et al., 1989; 1991)*
The following assertions are equivalent:

(i) *The system* (Σ) *with* $m = 1$ *is locally static feedback linearizable.*

(ii) *The system* (Σ) *with* $m = 1$ *is locally dynamic feedback linearizable.*

On the other hand, we have seen in the proposition 4.2 that a system whose tangent approximation is not controllable cannot be static feedback linearizable. The same property holds for dynamic feedback (see Charlet *et al.*, 1991).

Theorem 4.7 *(Charlet et al., 1991)*
If the system (Σ) *is locally dynamic feedback linearizable, then its linear tangent approximation* $\dot{z} = \dfrac{\partial f}{\partial x}(0)z + G(0)u \stackrel{\text{def}}{=} Fz + Gu$ *is controllable, i.e.* rank $(G\, FG \cdots F^{n-1}G) = n$.

4.2.3 Sufficient conditions

We now turn to sufficient conditions for dynamic linearizability in the particular case of the dynamic compensators described by (4.51) or (4.52), which we call *simple chains*. Note that these compensators are regular if β has rank m in the given neighborhood. This assumption is supposed to hold from now on. The system (Σ) in closed-loop can be written as

$$\begin{cases} \dot{x} &= f(x) + \sum\limits_{j,\mu_j \geq 1} w_1^j g_j(x) + \sum\limits_{j,\mu_j = 0} v_j'(t) g_j(x) \\ \dot{w}_i^j &= w_{i+1}^j \qquad 1 \leq j \leq m,\ \mu_j \geq 1,\ 1 \leq i \leq \mu_j - 1 \\ \dot{w}_{\mu_j}^j &= v_j' \qquad 1 \leq j \leq m,\ \mu_j \geq 1, \end{cases} \tag{4.59}$$

or

$$\dot{\tilde{x}} = \tilde{f}(\tilde{x}) + \sum_{j=1}^{m} \tilde{g}_j(\tilde{x}) v_j'(t) \tag{4.60}$$

where

$$\begin{aligned} \tilde{x} &= (x, w_1^j, \ldots, w_{\mu_j}^j ; 1 \leq j \leq m, \mu_j \geq 1) = (x, w) \\ \tilde{f} &= f + \sum_{j,\mu_j \geq 1} w_1^j g_j + \sum_{j,\mu_j \geq 2} \sum_{i=1}^{\mu_j - 1} w_{i+1}^j \frac{\partial}{\partial w_i^j} \\ \tilde{g}_j &= \begin{cases} g_j & \text{if } \ \mu_j = 0 \\ \dfrac{\partial}{\partial w_{\mu_j}^j} & \text{if } \ \mu_j \geq 1 \end{cases} \end{aligned} \tag{4.61}$$

The distributions $\tilde{G}_i$ associated with the system (4.60) are defined in the extended space (x, w) as

$$\tilde{G}_i = \text{sp}\{\text{ad}_{\tilde{f}}^k \tilde{g}_j \,,\ 0 \leq k \leq i \,,\ 1 \leq j \leq m\}. \tag{4.62}$$

We now define another family of distributions on the original space, distributions depending on the sequence of indices $\{\mu_1, \ldots, \mu_m\}$ and that play a major role in the linearization problem:

$$\begin{aligned} \Delta_0 &= \mathrm{sp}\{g_k\,,\ 1 \le k \le m\,,\ \mu_k = 0\} \\ \Delta_{i+1} &= \Delta_i + \mathrm{ad}_f \Delta_i + \mathrm{sp}\{g_k\,,\ 1 \le k \le m\,,\ \mu_k = i+1\}. \end{aligned} \tag{4.63}$$

By definition, $\Delta_0 \subset \Delta_1 \subset \cdots \subset \Delta_i \subset \cdots$. The following lemma gives sufficient conditions under which the sequence of distributions $\tilde{G}_i$ in the extended state space is related to the sequence of distributions Δ_i in the original (non extended) state space.

Lemma 4.4 *(Charlet et al., 1991)*
If for every i, $i \ge 0$ and every j, $1 \le j \le m$, such that $\mu_j \ge 1$

$$[g_j, \Delta_i] \subset \Delta_{i+1}$$

then, for $i \ge 0$

$$\tilde{G}_i = \Delta_i + \mathrm{sp}\left\{ \frac{\partial}{\partial w^j_{\mu_j}} \mid \mu_j \ge 1 \right\} + \cdots + \mathrm{sp}\left\{ \frac{\partial}{\partial w^j_{\mu_j - i}} \mid \mu_j \ge i+1 \right\} \tag{4.64}$$

Under the assumptions of lemma 4.4, the involutivity of the $\tilde{G}_i$'s (in the extended space) can easily be translated in terms of involutivity of the Δ_i's (in the original state space):

Theorem 4.8 *(Charlet et al., 1991)*
If for a sequence of integers $\{\mu_1, \ldots, \mu_m\}$, $0 \le \mu_1 \le \cdots \le \mu_m$, $\mu = \sum_{i=1}^{m} \mu_i$, the distributions defined by (4.63), up to a permutation on the input vector fields, are such that in U, a neighborhood of the origin,

(i) *Δ_i is involutive with constant rank, $0 \le i \le n + \mu_m - 1$*

(ii) rank $\Delta_{n+\mu_m-1} = n$

(iii) $[g_j, \Delta_i] \subset \Delta_{i+1}$ *for every j, $1 \le j \le m$, such that $\mu_j \ge 1$ and i, $0 \le i \le n + \mu_m - 1$*

then (Σ) is locally dynamic feedback linearizable by a simple dynamic compensator (4.51) having $\mu_1, \ldots, \mu_m$ for indices, and a local diffeomorphism in a neighborhood V of the origin in the extended state space $\mathbb{R}^{n+\mu}$.

A complete construction (analogous to the static feedback case) of the diffeomorphism and the dynamic compensator, as well as practical criteria to choose the integers μ_i, can be found in Charlet *et al.* (1991).

When the system has $m = n - 1$ inputs (number of inputs equal to the number of states minus 1), the theorem 4.8 becomes remarkably simple, converse to theorem 4.7.

Corollary *(Charlet et al., 1991).*
Consider the system (Σ) with $m = n - 1$. The following assertions are equivalent:

(i) *(Σ) is locally dynamic feedback linearizable.*

(ii) *The tangent approximation of* (Σ) *at the origin is controllable.*

Proof

$(i) \Rightarrow (ii)$ is just a reformulation of theorem 4.7.
$(ii) \Rightarrow (i)$. Since the tangent approximation of (Σ) at the origin is controllable, there exists a vector field g_i such that $\mathrm{ad}_f g_i$ does not belong to G_0. Moreover, $G_1 = T\mathbb{R}^n$. Two cases are thus possible. Either G_0 is involutive which implies, with $G_1 = T\mathbb{R}^n$, that theorem 4.2 applies and thus that (Σ) is static feedback linearizable. Or in the opposite case, we set $\mu_k = 1$ if $k \neq i$ and $\mu_i = 0$, and we have

- $\Delta_0 = \mathrm{sp}\{g_i\}$ which has rank 1 and is involutive.
- $\Delta_1 = G_0 + \mathrm{sp}\{\mathrm{ad}_f g_i\} = T\mathbb{R}^n$ since $n = m + 1$.

The assumptions of theorem 4.8 are thus satisfied and (Σ) is locally dynamic feedback linearizable.

Remark 4.10

This result applies to the counterexample (4.44) or (4.45) of section 4.1.4 which was proven non linearizable by static feedback. Precisely, recall that the system is

$$\left\{\begin{array}{rcl} \dot{x}_1 & = & u_2 \\ \dot{x}_2 & = & u_1 \\ \dot{x}_3 & = & u_2 x_2 \end{array}\right.$$

It is easily checked that the compensator $\dot{u}_2 = v_2$ *satisfies the conditions of theorem 4.8 and its corollary and that, for* $u_2 \neq 0$, *the extended state diffeomorphism is given by*

$$\xi_1^1 = x_1 \ , \ \xi_1^2 = u_2 \ , \ \xi_2^1 = x_3 \ , \ \xi_2^2 = u_2 x_2$$

with the feedback $v_1 = v_2 x_2 + u_2 u_1$.
This example is a particular case of the so-called chained systems (see for instance Murray and Sastry, 1993) for which the result of theorem 4.8 also apply.

Remark 4.11

The conditions of theorem 4.8 are invariant by diffeomorphism of the original state space, but are not invariant by feedback. This means, in particular, that the above considered class of simple dynamic compensators (4.51) or (4.52), is still too poor and that the conditions of dynamic feedback linearizability may hopefully be weakened.

4.2.4 Example: rigid body dynamics

We consider a general model of rigid body dynamics that includes the case of aircraft dynamics. Our purpose here is to show that theorem 4.8 applies to important classes of mechanical systems rather than deal with specific applications of aircraft control. Other applications of dynamic feedback linearization can be found in [6, 13, 18, 21, 22, 23].

Let (x, y, z) be the coordinates of the center of mass in an absolute frame with the vertical z-axis oriented downward, (u, v, w) the velocity components in a relative frame linked to the rigid body, (p, q, r) the components of the kinetic moment in the relative frame, and (Φ, Θ, Ψ) the roll, pitch, and yaw angles, respectively. Let $\xi = (x, y, z, u, v, w, \Phi, \Theta, \Psi)^T$ be the state vector.

We consider p, q, r, and ρ (ρ is the thrust) as control variables. The equations of motion are

$$\begin{aligned}
\dot{x} &= u \cos\Psi \cos\Theta \\
&\quad +v(\cos\Psi \sin\Theta \sin\Phi - \sin\Psi \cos\Phi) \\
&\quad +w(\cos\Psi \sin\Theta \cos\Phi + \sin\Psi \sin\Phi) \\
\dot{y} &= u \sin\Psi \cos\Theta \\
&\quad +v(\sin\Psi \sin\Theta \sin\Phi + \cos\Psi \cos\Phi) \\
&\quad +w(\sin\Psi \sin\Theta \cos\Phi - \cos\Psi \sin\Phi) \\
\dot{z} &= -u \sin\Theta + v \cos\Theta \sin\Phi + w \cos\Theta \cos\Phi \\
\dot{u} &= -g \sin\Theta + rv - qw + \frac{X_0(\xi)}{m} + \frac{J\rho}{m} \\
\dot{v} &= g \cos\Theta \sin\Phi - ru + pw + \frac{Y(\xi)}{m} \\
\dot{w} &= g \cos\Theta \cos\Phi + qu - pv + \frac{Z(\xi)}{m} \\
\dot{\Phi} &= p + \mathrm{tg}\Theta(q \sin\Phi + r \cos\Phi) \\
\dot{\Theta} &= q \cos\Phi - r \sin\Phi \\
\dot{\Psi} &= \frac{q \sin\Phi + r \cos\Phi}{\cos\Theta}
\end{aligned} \tag{4.65}$$

where $(X_0(\xi) + J\rho, Y(\xi), Z(\xi))$ are the components of the resultant of the exterior forces, excepting gravity. The model is of the type

$$\dot{\xi} = f(\xi) + pg_1(\xi) + qg_2(\xi) + rg_3(\xi) + \rho g_4(\xi) \tag{4.66}$$

with f, g_1, g_2, g_3, and g_4 obtained from (4.65).

The reader can compute the Lie brackets $\mathrm{ad}_{g_i} g_j$ with $1 \leq i, j \leq 4$ and independently of the type of functions that $X_0(\xi)$, $Y(\xi)$, and $Z(\xi)$ may be in specific cases, check that,

- $\mathrm{sp}\{g_1, g_2, g_3\}$ is involutive
- $G_0 = \mathrm{sp}\,\{g_1, g_2, g_3, g_4\}$ is not involutive, rank $G_0 = 4$
- $\overline{G_0} = G_0 + \mathrm{sp}\,\{\mathrm{ad}_{g_4} g_2, \mathrm{ad}_{g_4} g_3\} = \mathrm{sp}\left\{\frac{\partial}{\partial u}, \frac{\partial}{\partial v}, \frac{\partial}{\partial w}, \frac{\partial}{\partial \Phi}, \frac{\partial}{\partial \Theta}, \frac{\partial}{\partial \Psi}\right\}$, rank $\overline{G_0} = 6$
- $G_1 = G_0 + \mathrm{ad}_f G_0 \neq \overline{G_1} = T\mathbb{R}^9$, rank $G_1 = 7$ and rank $\overline{G_1} = 9$

Since G_0 is not involutive, the system is not linearizable by static feedback. However, it is clear from the Lie bracket configuration that all the non involutivities are caused by g_4

and if we choose $\mu_1 = \mu_2 = \mu_3 = 0$ and $\mu_4 = 1$, we have

$$
\begin{aligned}
\Delta_0 &= \mathrm{sp}\{g_1, g_2, g_3\} \\
\Delta_1 &= \Delta_0 + \mathrm{ad}_f \Delta_0 + \mathrm{sp}\{g_4\} &= \overline{G_0} \\
\Delta_2 &= \Delta_1 + \mathrm{ad}_f \Delta_1 &= \mathrm{T}\mathbb{R}^9.
\end{aligned}
$$

Therefore, the sufficient conditions of theorem 4.8 are satisfied, which leads to the compensator

$$
\begin{aligned}
\dot{\rho} &= v_4 \\
p &= v_1 \\
q &= v_2 \\
r &= v_3
\end{aligned} \tag{4.67}
$$

Define the functions

$$
\begin{aligned}
y_1 &= x \\
y_2 &= y \\
y_3 &= z \\
y_4 &= v
\end{aligned}
$$

and the extended space diffeomorphism

$$(\xi, v) \longmapsto \tilde{\xi}$$

where

$$\tilde{\xi} = \left(y_1, L_{\tilde{f}} y_1, L_{\tilde{f}}^2 y_1, y_2, L_{\tilde{f}} y_2, L_{\tilde{f}}^2 y_2, y_3, L_{\tilde{f}} y_3, L_{\tilde{f}}^2 y_3, y_4\right)^T .$$

The dynamic compensator (4.67) together with the extended state feedback

$$\nu = D^{-1}(\nu' - D_0)$$

where

$$
\begin{aligned}
D_0^1 &= \frac{L_f X_0}{m} \cos\Psi \cos\Theta + \frac{L_f Y}{m}(\cos\Psi \sin\Theta \sin\Phi - \sin\Psi \cos\Phi) \\
&\quad + \frac{L_f Z}{m}(\cos\Psi \sin\Theta \cos\Phi + \sin\Psi \sin\Phi) \\
D_0^2 &= \frac{L_f X_0}{m} \sin\Psi \cos\Theta + \frac{L_f Y}{m}(\sin\Psi \sin\Theta \sin\Phi + \cos\Psi \cos\Phi) \\
&\quad + \frac{L_f Z}{m}(\sin\Psi \sin\Theta \cos\Phi - \cos\Psi \sin\Phi) \\
D_0^3 &= \frac{L_f X_0}{m} \sin\Theta + \frac{L_f Y}{m} \cos\Theta \sin\Phi + \frac{L_f Z}{m} \cos\Theta \cos\Phi \\
D_0^4 &= g \cos\Theta \sin\Phi + \frac{Y}{m}
\end{aligned}
$$

and

$$\begin{aligned}
D_1^1 &= \frac{Y}{m}(\cos\Psi\sin\Theta\cos\Phi) - \frac{Z}{m}(\cos\Psi\sin\Theta\sin\Phi) + \varepsilon_1^1 \\
D_1^2 &= -\frac{X_0 + J\rho}{m}(\cos\Psi\sin\Theta\cos\Phi + \sin\Psi\sin\Phi) + \frac{Z}{m}\cos\Psi\cos\Theta + \varepsilon_1^2 \\
D_1^3 &= \frac{X_0 + J\rho}{m}(\cos\Psi\sin\Theta\sin\Phi - \sin\Psi\cos\Phi) - \frac{Y}{m}\cos\Psi\cos\Theta + \varepsilon_1^3 \\
D_1^4 &= \frac{J}{m}\cos\Psi\cos\Theta \\
D_2^1 &= \frac{Y}{m}(\sin\Psi\sin\Theta\cos\Phi - \cos\Psi\sin\Phi) \\
&\qquad -\frac{Z}{m}(\sin\Psi\sin\Theta\sin\Phi + \cos\Psi\cos\Phi) + \varepsilon_2^1 \\
D_2^2 &= -\frac{X_0 + J\rho}{m}(\sin\Psi\sin\Theta\cos\Phi - \cos\Psi\sin\Phi) + \frac{Z}{m}\sin\Psi\cos\Theta + \varepsilon_2^2 \\
D_2^3 &= \frac{X_0 + J\rho}{m}(\sin\Psi\sin\Theta\sin\Phi + \cos\Psi\cos\Phi) - \frac{Y}{m}\sin\Psi\cos\Theta + \varepsilon_2^3 \\
D_2^4 &= \frac{J}{m}\sin\Psi\cos\Theta \\
D_3^1 &= \frac{\cos\Theta}{m}(-Y\cos\Phi + Z\sin\Phi) + \varepsilon_3^1 \\
D_3^2 &= \frac{X_0 + J\rho}{m}\cos\Theta\cos\Phi + \frac{Z}{m}\sin\Theta + \varepsilon_3^2
\end{aligned}$$

$$\begin{aligned}
D_3^3 &= -(\frac{X_0 + J\rho}{m}\cos\Theta\sin\Phi + \frac{Y}{m}\sin\Theta) + \varepsilon_3^3 \\
D_3^4 &= \frac{J}{m}\sin\Theta
\end{aligned}$$

and $D_4^1 = w$, $D_4^2 = 0$, $D_4^3 = -u$ and $D_4^4 = 0$, with

$$\begin{aligned}
\varepsilon_1^i &= \cos\Psi\cos\Theta\frac{L_{g_i}X_0}{m} + (\cos\Psi\sin\Theta\sin\Phi - \sin\Psi\cos\Phi)\frac{L_{g_i}Y}{m} \\
&\qquad +(\cos\Psi\sin\Theta\cos\Phi + \sin\Psi\sin\Phi)\frac{L_{g_i}Z}{m} \qquad 1 \le i \le 3 \\
\varepsilon_2^i &= \sin\Psi\cos\Theta\frac{L_{g_i}X_0}{m} + (\sin\Psi\sin\Theta\sin\Phi + \cos\Psi\cos\Phi)\frac{L_{g_i}Y}{m} \\
&\qquad +(\sin\Psi\sin\Theta\cos\Phi - \cos\Psi\sin\Phi)\frac{L_{g_i}Z}{m} \qquad 1 \le i \le 3 \\
\varepsilon_3^i &= \sin\Theta\frac{L_{g_i}X_0}{m} + \cos\Theta\sin\Phi\frac{L_{g_i}Y}{m} + \cos\Theta\cos\Phi\frac{L_{g_i}Z}{m} \qquad 1 \le i \le 3
\end{aligned}$$

makes the closed-loop system linear and controllable:

$$\frac{d^3 y_i}{dt^3} = \nu_i' \qquad 1 \leq i \leq 3$$

$$\dot{y}_4 = \nu_4'$$

The choice of the function y_4 (obtained as the solution of a system of PDEs which is not detailed here) is not unique but corresponds, in practice, to the fact that $y_4 = 0$ implies skidless trajectories.

4.3 Bibliography

[1] BASTIN, G. and LÉVINE, J. (1993), Structure of reaction networks and controllability of open isothermal reactors, *IEEE Trans. Automatic Control*, **38**, 5, 733–742.

[2] CHARLET, B., LÉVINE, J. and MARINO, R. (1989), On dynamic feedback linearization, *Systems & Control Letters*, **13**, 143–151.

[3] CHARLET, B., LÉVINE, J. and MARINO, R. (1991), Sufficient conditions for dynamic state feedback linearization, *SIAM J. Control and Optimiz.*, **29**, 1, 38–57.

[4] CIBRARIO, M. and LÉVINE, J. (1991), *Saddle–node bifurcation control with application to thermal runaway of continuous stirred tank reactors,* Proc. 30th CDC, Brighton.

[5] D'ANDRÉA-NOVEL, B. (1988), *Commande Non Linéaire des Robots,* Traité des Nouvelles Technologies, Série Automatique, Hermès, Paris.

[6] D'ANDRÉA-NOVEL, B. and LÉVINE, J. (1990), Modelling and nonlinear control of an overhead crane, in *Robust Control of Linear Systems and Nonlinear Control*, M.A. Kaashoek, J.H. van Schuppen and A.C.M. (Ran eds), Vol. II, Birkhaüser Boston, 523–529.

[7] D'ANDRÉA-NOVEL, B. and PRALY, L. (1988), About finite nonlinear zeros for decouplable systems, *Systems & Control Letters*, **10**, 103–109.

[8] DAYAWANSA, W., BOOTHBY, W. and ELLIOTT, D.L. (1985), Global state and feedback equivalence of nonlinear systems, *Systems & Control Letters*, **6**, 4, 229–234.

[9] DI BENEDETTO, M.D., GRIZZLE, J.W. and MOOG, C.H. (1989), Rank invariants of nonlinear systems, *SIAM J. Control & Optimization*, **27**, 658–672.

[10] EL MOUBARAKI, J., BASTIN, G. and LÉVINE, J. (1993), Nonlinear control of biotechnological processes with growth/production decoupling, *Mathematical Biosciences*, **116**, 21–44.

[11] FLIESS, M. (1989), Automatique et corps différentiels, *Forum Math.*, **1**, 227–238.

[12] FLIESS, M. (1990), Generalized controller canonical forms for linear and nonlinear dynamics, *IEEE Trans. Automatic Control*, **35**, 9, 994–1001.

[13] FLIESS, M., LÉVINE, J. and ROUCHON, P. (1991), *A simplified approach of crane control via a generalized state-space model,* Proc. 30th IEEE CDC, Brighton, U.K., 736–741.

[14] FLIESS, M., LÉVINE, J. and ROUCHON, P. (1993), Generalised state variable representation for a simplified crane description, *Int. J. Control*, **58**, 2, 277–283.

[15] FLIESS, M., LÉVINE, J., MARTIN, P. and ROUCHON, P. (1992), Sur les systèmes non linéaires différentiellement plats, *C.R. Acad. Sci. Paris*, **I-315**, 619–624.

[16] FLIESS, M., LÉVINE, J., MARTIN, P. and ROUCHON, P. (1993), On differentially flat nonlinear systems, in *Nonlinear Control Systems Design 1992*, M. Fliess (ed.), Pergamon, Oxford, 159–163.

[17] FLIESS, M., LÉVINE, J., MARTIN, P. and ROUCHON, P. (1993), Défaut d'un système non linéaire et commande haute fréquence, *C.R. Acad. Sci. Paris*, **I-316**, 513–518.

[18] FLIESS, M., LÉVINE, J., MARTIN, P. and ROUCHON, P. (1993), *Commande non linéaire pour l'ingénieur,* Cours INRIA, Rocquencourt, 18–20, INRIA Press.

[19] FLIESS, M., LÉVINE, J., MARTIN, P. and ROUCHON, P. (1993), Linéarisation par bouclage dynamique et transformations de Lie-Bäcklund, *C.R. Acad. Sci. Paris*, **I-317**, 981–986.

[20] FLIESS, M., LÉVINE, J., MARTIN, P. and ROUCHON, P. (1996), A Lie-Bäcklund approach to dynamic feedback equivalence and flatness, in *Robust control via viarable structure and Lyapunov techniques*, F. Garofalo and L. Glielmo (eds), Springer-verlag, London, 245–268.

[21] FLIESS, M., LÉVINE, J., MARTIN, P. and ROUCHON, P. (1994), *Nonlinear control and Lie-Bäcklund transformations: towards a new differential geometric standpoint,* Proc. 33rd IEEE CDC, Lake Buena Vista, FL.

[22] FLIESS, M., LÉVINE, J., MARTIN, P. and ROUCHON, P. (1995), Flatness and defect of nonlinear systems: introductory theory and applications, *Int. J. Control*, **61**, 1327–1361.

[23] FLIESS, M., LÉVINE, J., MARTIN, P. and ROUCHON, P. (1995), *Design of trajectory stabilization for flat driftless systems,* Proc. ECC 95, Roma.

[24] FLIESS, M., LÉVINE, J., MARTIN, P. and ROUCHON, P. (1995), *Flatness and dynamic feedback linearizability.* Proc. ECC 95, Roma.

[25] HUNT, L.R., SU, R. and MEYER, G. (1983), Design for multi-input nonlinear systems, in *Differential Geometric Control Theory*, R. Brockett, R. Millman, H. Sussmann (eds.), Birkhäuser, 268–298.

[26] HAUSER, J., SASTRY, S. and MEYER, G. (1989), *Nonlinear controller design for flight control systems,* Proc. 1st IFAC NOLCOS'89 Conference, Capri.

[27] ISIDORI, A. (1989), *Nonlinear Control Systems.* 2nd ed., Springer-Verlag, New York.

[28] JAKUBCZYK, B. and RESPONDEK, W. (1980), On linearization of control systems, *Bull. Acad. Pol. Sci., Ser. Sci. Math.*, **28**, 9-10, 517–522.

[29] JAKUBCZYK, B. (1993), Remarks on equivalence and linearization of nonlinear systems, in *Nonlinear Control Systems Design 1992*, M. Fliess (ed.), Pergamon, Oxford, 393–397.

[30] JAKUBCZYK, B. (1993), *Invariants of dynamic feedbacks and free systems,* Proc. ECC 93, Groningen, 1510–1513.

[31] KAILATH, T. (1980), *Linear Systems,* Information and Systems Sciences Series, Prentice Hall.

[32] KRASIL'SHCHICK, I.S., LYCHAGIN, V.V. and VINOGRADOV, A.M. (1986), *Geometry of Jet Spaces and Nonlinear Partial Differential Equations,* Gordon and Breach, New York.

[33] LÉVINE, J. (1988), Remarks on some worked out applications of nonlinear control theory, in *New Trends in Nonlinear Control Theory*, J. Descusse, M. Fliess, A. Isidori et D. Leborgne (ed.), Lecture Notes in Control and Info. Sci., Vol. 102, Springer-Verlag, New York, 446–458.

[34] LÉVINE, J. and ROUCHON, P. (1991), Quality control of binary distillation columns via nonlinear aggregated models, *Automatica*, **27**, 3, 463–480.

[35] MARINO, R. (1986), On the largest feedback linearizable subsystem, *Systems & Control Letters*, **6**, 345–351.

[36] MARTIN, P. (1992), Contribution à l'étude des systèmes différentiellement plats, PhD Thesis, ENSMP, Paris.

[37] MARTIN, P. (1993), *Endogeneous feedback and equivalence,* Proc. MTNS 93, Regensburg.

[38] MARTIN, P. and ROUCHON, P. (1995), Any (controllable) driftless system with three inputs and five states is flat, Systems Control Letters, **25**, 167–173.

[39] MEYER, G. and CICOLANI, L. (1981), Application of nonlinear system inverses to automatic flight control designs–system concepts and flight evaluations, in *Theory and Application of Optimal Control in Aerospace Systems*, AGARD–AG251, 10.1–10.29.

[40] MEYER, G., SU, R. and HUNT, L.R., Application of nonlinear transformations to automatic flight control, *Automatica*, **20**, 103–107.

[41] MURRAY, R.M. and SASTRY, S. (1993), Nonholonomic motion planning: steering using sinusoids, *IEEE Trans. Automatic Control*, **38**, 5, 700–716.

[42] VAN NIEUWSTADT, M., RATHINAM, M. and MURRAY, R.M. (1994), Differential flatness and equivalence, Techn. Rep. CIT-CDS 94–006, Caltech, Pasadena.

[43] NIJMEIJER, H. and VAN DER SCHAFT, A. (1990), *Nonlinear Dynamical Control Systems*, Springer-Verlag, New York.

[44] POMET, J.B. (1993), *A differential geometric setting for dynamic equivalence and feedback linearization*, Proc. Workshop Geometry in Nonlinear Control, Banach Center, Warsaw.

[45] ROUCHON, P. (1990), Simulation dynamique et commande non linéaire des colonnes à distiller, PhD Thesis, ENSMP, Paris.

[46] ROUCHON, P., FLIESS, M., LÉVINE, J. and MARTIN, P. (1993), *Flatness and motion planning: the car with n trailers*, Proc. ECC 93, Groningen, 1518–1522.

[47] ROUCHON, P., FLIESS, M., LÉVINE, J. and MARTIN, P. (1993), *Flatness, motion planning and trailer systems*, Proc. 32nd IEEE CDC, San Antonio, TX, 2700–2705.

[48] SHADWICK, W.F. (1990), Absolute equivalence and dynamic feedback linearization, *Systems & Control Letters*, **15**, 35–39.

[49] SINGH, S.N. (1981), A modified algorithm for invertibility in nonlinear systems, *IEEE Trans. Automatic Control*, **26**, 595–598.

[50] SINGH, S.N. and SCHY, A. (1980), Output feedback nonlinear decoupled control synthesis and observer design for maneuvering aircraft, *Int. J. Control*, **31**, 781–806.

[51] SLOTINE, J.J.E. and LI, J.W. (1991), *Applied Nonlinear Control*, Prentice Hall, Englewood Cliffs, N.J.

[52] SLUIS, W.M. (1993), A necessary condition for dynamic feedback linearization, *Systems & Control Letters*, **21**, 277–283.

[53] WONHAM, W.M. (1974), *Linear Multivariable Control: a Geometric Approach*, Springer-Verlag, New York.

[54] ZHARINOV, V.V. (1992), *Geometrical Aspects of Partial Differential Equations*, World Scientific, Singapore.

CHAPTER 5

On nonlinear digital control

S. MONACO AND D. NORMAND-CYROT

5.1 Introduction

The interest of digital control lies in the opportunity of using digital computers for controlling physical processes which usually admit continuous models.

Given a continuous-time plant, a digital controller can be set either by implementing an "equivalent" of a continuous controller by means of holding and sampling devices, or by designing a discrete-time controller based on the sampled model of the plant (Aström and Wittenmark, 1984; Hanselmann, 1987).

With respect to the first approach and in a linear context, several methods are available for discretizing a continuous filter. The most usual way in a nonlinear context is to use a holding system of zero-order on the sampling of the continuous-time controller. In this case, the performances of such a control scheme are often not satisfactory, since technological or practical constraints do not always allow very fast sampling.

In the second approach, the designer sets the controller starting from a sampled model of the plant. This is well-known and usually referred to as direct digital control in a linear context, where it is possible to satisfy more interesting control requirements, such as dead beat or minimum time. In addition, when dealing with systems including delays or non minimum-phase systems, one may benefit from a previous sampling procedure.

This brief recall of the known advantages of a purely discrete-time control design, when digital computers are used, especially concerns the linear context. In fact, when dealing with discrete-time or sampled linear systems, well-known design methods for controllers are based on linear algebra and matrix manipulations. The situation is rather different in the nonlinear context. Nonlinear discrete-time control theory is still more difficult to handle than continuous-time control and generally leads to solutions which are difficult to compute.

In the previous chapters, one may notice the extensive use of differential geometric tools for studying nonlinear continuous-time control systems. Such a mathematical framework is powerful for obtaining, at least locally, analysis and synthesis results as illustrated in several works (Byrnes and Isidori, 1985; 1988; Claude *et al.*, 1983; Isidori, 1985). In discrete-time, one may observe the lack of a rather universal mathematical domain. In particular, the difficulty of setting down local notions in a purely discrete-time case makes many control problems unsolvable.

To counterbalance this rather pessimistic analysis, let us note an increasing interest in discrete-time control theory, reinforced by the design of many recent promising concepts, results and techniques of proofs (Albertini and Sontag, 1993; Grizzle, 1993; Monaco and Normand-Cyrot, 1995). However, these studies suffer from giving theoretical results rather than constructive ones. This fact actually motivates the study of systems deduced from the sampling of continuous-time systems thus allowing the formulation of local concepts and easy iterative computations. More precisely, local concepts can be set down under sampling since integrating a differential equation during a small time interval generates a small state displacement. Moreover, the discrete-time dynamics thus obtained admits series expansions in powers of the sampling period and benefits from suitable geometric, algebraic or combinatoric properties which simplify their analysis. This explains the importance of research activity which studies structural and control systems' properties under sampling [2, 17, 20, 22, 24, 28, 30, 31]. The study of the control scheme by series expansion techniques renders control solutions iteratively computable in terms of polynomial approximations with respect to the sampling period.

In this book, some results in this area are presented. In particular, multirate sampling and control procedures are discussed hereafter in a nonlinear context. For a better homogeneity with the previous chapters the problem of the preservation of input–output linearization properties under digital control will especially be discussed. Input-state linearization under coordinates change and static state feedback will be deduced as a special case. The square multivariable case represented by a system with the same number of inputs and outputs and a non singular decoupling matrix will be easily deduced according to generalized notations.

Given a linear analytic continuous-time system satisfying the input–output or input-state feedback linearization properties, (cf. chapters 2 and 4), it has already been shown that the implementation, according to a zero-order holding system of the continuous-time control does not give satisfactory results. The control techniques here proposed here and introduced in Monaco and Normand-Cyrot (1988a), are based on the sampling of the input–output links that one intends to control. More precisely, this means that an overall sampled version of the full state dynamics is not always required.

Given a continuous-time control scheme, ensuring specific performances such as linearization, decoupling or tracking, a discrete-time control scheme which preserves these requirements at sampling instants is developed. These discrete-time techniques make use of a sampling at different rates on the control and measured variables. More precisely, the control variables are sampled faster than the state or output variables. The basic idea is thus very simple since controlling faster than measuring provides the possibility of obtaining more degrees of freedom on the control in order to fulfil additional specifications. In the present case, it will be shown that it is sufficient to control the output and a finite number of its time derivatives to preserve exactly at sampling instants the performance of an "ideal" continuous-time control scheme.

On these bases, it will be easier for the reader to understand why the same reasoning can be followed to design direct digital control schemes satisfying discrete-time requirements or to propose solutions to difficult continuous-time control problems. In fact, thanks to these extra degrees of freedom, one may control other state variables corresponding to physical process characteristics or may satisfy optimality or robustness criteria for the overall control system. These last aspects are not discussed in the present volume. The same benefits may hold when higher-order holding circuits are used for digital implementation (Barbot, 1989a, 1989b). However, even if the results are comparable, the technology required for their practical realization is still rather complicated.

To control an input–output link generally implies the cancellation of a part of the dynamics which may generate internal instability of the closed-loop control system. Such a problem occurs when dealing with input–output linearization [15, 17, 20, 22, 25, 27, 28, 30] and can be overcome, at least locally, if the system is minimum phase (Byrnes and Isidory, 1985). Under single rate sampling, this minimum phase property is lost for almost every sampling period. This fact is the nonlinear counterpart of the appearance of extra zeros (the sampling zeros in the transfer function) when usual sampling in terms of zero-order holding circuit is performed (Aström *et al.*, 1984). It will be shown how to deal with this problem in a nonlinear context and to maintain, at the sampling instants $t = k\delta$, the stability properties of the continuous-time process.

The present chapter is organized as follows. Section 5.2 introduces nonlinear sampling procedures for nonlinear forced dynamics. Two cases are studied, the usual single-rate case and the multirate case. Section 5.3 discusses the sampling of a continuous-time state feedback control providing a discrete-time solution, which preserves the requirements of the continuous-time control scheme at the sampling instants. How to substitute to an "ideal" continuous-time control scheme a discrete-time one in order to satisfy the same requirements at the sampling instants, is studied. In this approach, an essential role is played by the notion of relative degree which, as in the continuous-time case, generalizes to the nonlinear context the notion of excess of the poles over the zeros of a transfer function. In section 5.4 the problem is studied for systems with a relative degree which is equal to one. The result is generalized in section 5.5 to systems admitting with a relative degree higher than one and a solution is given in terms of multirate control.

5.2 Nonlinear sampling

5.2.1 The general case

Let us consider the following nonlinear continuous-time system denoted as Σ_c.

$$\Sigma_c \begin{cases} \dot{x}_c(t) = f(x_c(t); \underline{u}(t)) & (5.1) \\ y_c(t) = h(x_c(t)) & (5.2) \end{cases}$$

where the state x_c is defined on a open set $\mathbb{M}$ of $\mathbb{R}^n, \underline{u} \in \mathbb{R}^m, y_c \in \mathbb{R}$. The functions, $f(.,\underline{u}) : \mathbb{M} \times \mathbb{R}^m \longrightarrow \mathbb{R}^n$ and $h(x) : \mathbb{R}^n \longrightarrow \mathbb{R}$ are supposed to be analytic.

If one assumes the control vector $\underline{u}(t)$ constant over time intervals of amplitude δ, that is

$$\underline{u}(t+\tau) = \underline{u}(t) = \underline{u}(k) \quad \text{for } 0 \leq \tau < \delta, t = k\delta, k \geq 0,$$

the object of this paragraph is to build a discrete-time state space representation, denoted as Σ_d, which is equivalent to system Σ_c, from the point of view of their input-state behaviors at the sampling instants $t = k\delta$ for $k \geq 0$ (Monaco and Normand-Cyrot, 1985).

More precisely, if Σ_d denotes a discrete-time system defined by the difference equations

$$\Sigma_d \begin{cases} x_d(k+1) &= F^{\delta}(x_d(k), \underline{u}(k)) \quad (5.3) \\ y_d(k) &= h(x_d(k)) \quad (5.4) \end{cases}$$

where the state $x_d(k) \in \mathbb{M}$ and the function $F^{-}(.,.): \; [0, \delta_0[\times \mathbb{M} \times \mathbb{R}^m \longrightarrow \mathbb{R}^n$ is assumed analytic for δ_0 sufficiently small, the following definition is stated.

Definition 5.1

The system Σ_d is called the exact sampled equivalent of system Σ_c if, for a same initialization $x(0)$ and for constant controls over time intervals of amplitude δ, Σ_d and Σ_c have the same input-state behaviors, at the sampling instants $t = k\delta$; i.e. for $k \geq 1$ and $x_c(0) = x_d(0) = x(0)$ one has

$$x_c(k\delta) = x_d(k)$$

Remark 5.1

From the input-state coincidence, one deduces the input–output coincidence at the sampling instants $t = k\delta$; i.e. for $k \geq 0$ and $x_c(0) = x_d(0) = x(0)$ one has

$$y_c(k\delta) = h(x_c(k\delta)) = h(x_d(k)) = y_d(k)$$

Referring to the Lie derivatives or to the directional derivatives introduced in volume 2, let us introduce the Lie exponential series associated to any vector field $f(.)$ defined on M as

$$\mathrm{e}^{L_f} := \sum_{p \geq 0} \frac{1}{p!} L_f^p := 1 + L_f + \frac{1}{2!} L_f^2 + \cdots + \cdots + \frac{1}{p!} L_f^p + \cdots \quad (5.5)$$

where L_f^p represents the p times composition of the operator L_f. Id and 1 represent the identity function and the identity operator respectively.

Proposition 5.1

The sampled dynamics $F^{\delta} \; : \mathbb{M} \times \mathbb{R}^m \longrightarrow \mathbb{R}^n$ describing Σ_d is characterized for $\delta \in [0, \delta_0[$, by the exponential expansion

$$F^{\delta}(x_d(k), \underline{u}(k)) = \mathrm{e}^{\delta L_{f(\cdot;\underline{u}(k))}}(I_d)(x_d(k)) = \mathrm{e}^{\delta L_{f(\cdot;\underline{u}(k))}}(x_d(k)) \quad (5.6)$$

(When it is applied to the identity function one writes $\mathrm{e}^{L_f}(x)$ instead of e^{L_f} (Id) (x))

The proof is immediate. In fact, for $\delta \in [0, \delta_0[$, the Lie series

$$\mathrm{e}^{\delta L_{f(\cdot,\underline{u}(k))}}(x_d(k))$$

expresses the solution at time $t = (k+1)\delta$ of equation (5.1), initialized in $x_c(k\delta) = x_d(k)$, when the control is constant and equal to $\underline{u}(k)$. The exact solution (5.6) can be expanded either in powers of δ or $\underline{u}$.

The expansion with respect to δ, directly deduced from (5.5) allows the notion of approximated sampling to be stressed at order p which corresponds to truncating at order p in δ, the series (5.6). One obtains

$$e^{\delta L_{f(\cdot;\underline{u})}}(x_d(k)) := \left(1 + \delta L_{f(\cdot;\underline{u})} + \frac{\delta^2}{2!}L^2_{f(\cdot;\underline{u})} + \cdots + \frac{\delta^p}{p!}L^p_{f(\cdot;\underline{u})}\right)(x_d(k)) + \mathrm{O}(\delta^{p+1}) \quad (5.7)$$

($\mathrm{O}(\delta^{p+1})$ denotes the remaining terms of order δ^{p+1} at least).

An alternative method for computing the sampled dynamics (5.6) is to integrate over the sampling period δ, the Volterra series which characterizes the input-state behavior of (5.1) initialized in $x_d(k)$ and submitted to piecewise constant controls.

An exact calculus being generally difficult to perform, one considers in practice a truncated series of the type (5.7), its validity depending on the value of δ which is assumed small enough to ensure the convergence of (5.6).

Remark 5.2

Approximated sampling at order 1 *with respect to* δ *corresponds to the usual Euler sampling method, i.e. the time derivative is approximated as*

$$\dot{x}(t) = \lim_{\delta \to 0} \frac{x(t+\delta) - x(t)}{\delta}$$

For computational purposes, one notes that the generic term of order p appearing in the series (5.7) is nothing else but the time derivative at order p of the state variable $x_c(t)$ computed at time $t = k\delta$. The sampled dynamics (5.6) can thus be rewritten as

$$x_d(k+1) \quad := \quad F^\delta(x_d(k), \underline{u}(k)) \ := \sum_{p \geq 0} \frac{\delta^p}{p!} \overset{(p)}{x}_c(k\delta) \qquad (5.8)$$

with $$\overset{(p)}{x}_c(k\delta) \quad = \quad L^p_{f(\cdot;\underline{u}(k))}(x_d(k)).$$

The expansion of the sampled dynamics with respect to $\underline{u}$ can be iteratively deduced from the formula (5.6) after a preliminary Taylor type expansion of the continuous-time dynamics $f(x;\underline{u})$ with respect to $\underline{u}$.

As shown in Monaco and Normand-Cyrot (1985), the expansion with respect to $\underline{u}$ of $F^\delta(x;\underline{u})$ satisfies homogeneous recurrent expressions in terms of $f(x,\underline{u})$. In order to do so, combinatorial tools as well as functional expansions, closely related to the Baker-Campbell-Hausdorff formula, are extensively used. In the present case, we limit the study to linear analytic dynamics (5.1), (mainly described in these books) with a scalar control and we just recall the essential tools for computing F^δ. An effective computation makes use of formal calculus methods (Barbot, 1989a; 1989b).

Following the same procedure, it is possible to sample continuous-time dynamics driven by continuous-time control, thus obtaining sampled dynamics depending on the control and their time derivatives computed at the sampling instants (Monaco and Normand-Cyrot (1991b)).

Example 1

The following simple dynamics, defined on $\mathbb{R}^2$, illustrates the proposed method when the control acts in a nonlinear manner. Approximated sampling at order three is considered. Given the dynamics

$$\begin{cases} \dot{x}_1(t) &= x_1(t) + u_1^2(t)x_2(t) \\ \dot{x}_2(t) &= x_1(t)x_2(t) \end{cases}$$

Assuming the control $u_1(t)$ constant over time intervals of amplitude δ, the approximated dynamics at order three according to δ is given by the discrete-time difference equation

$$\begin{aligned}
x_1(k+1) &= x_1(k) + \delta[x_1(k) + u_1^2(k)x_2(k)] + \frac{\delta^2}{2}[x_1(k) + u_1^2(k)x_2(k) \\
&+ u_1^2(k)x_1(k)x_2(k)] + \frac{\delta^3}{3!}[x_1(k) + u_1^2(k)x_2(k) + 2u_1^2(k)x_1(k)x_2(k) \\
&+ u_1^4(k)x_2^2(k) + u_1^2(k)x_1^2(k)x_2(k)]
\end{aligned}$$

$$\begin{aligned}
x_2(k+1) &= x_2(k) + \delta x_1(k)x_2(k) + \frac{\delta^2}{2}[x_1(k)x_2(k) + u_1^2(k)x_2^2(k) \\
&+ x_1^2(k)x_2(k)] + \frac{\delta^3}{3!}[x_1(k)x_2(k) + u_1^2(k)x_2^2(k) + 4u_1^2(k) \\
&\qquad x_1(k)x_2^2(k)) + 3x_1^2(k)x_2(k) + x_1^3(k)x_2(k)]
\end{aligned}$$

5.2.2 *The linear analytic case*

Let us consider the SISO continuous-time system, Σ_c

$$\Sigma_c \begin{cases} \dot{x}_c(t) = f(x_c(t)) + g(x_c(t))u(t) & (5.9) \\ y_c(t) = h(x_c(t)) & (5.10) \end{cases}$$

where the state x_c is defined on an open set $\mathbb{M}$ of $\mathbb{R}^n, u \in \mathbb{R}$. The vector fields $f(x)$, $g(x) \to \mathbb{R}^n$ and the output function $h(x) : \mathbb{R}^n \to \mathbb{R}$ are supposed to be analytic.

When the controls are assumed constant over time intervals of amplitude δ, the exact sampled equivalent to (5.9) (5.10) is described by the nonlinear difference equations

$$\Sigma_d \begin{cases} x_d(k+1) = F^\delta(x_d(k), u(k)) & (5.11) \\ y_d(k) = h(x_d(k)) & (5.12) \end{cases}$$

The dynamics $F^\delta(x_d(k), u(k))$ admits, according to Proposition 5.1, the exponential representation

$$F^\delta(x_d(k), u(k)) = e^{\delta L_{f(.)} + \delta u(k) L_{g(.)}}(x_d(k)) \tag{5.13}$$

which can be expanded with respect to u as

$$F^\delta(x, u) = \sum_{i \geq 0} u^i F_i^\delta(x) \tag{5.14}$$

with

$$
\begin{aligned}
F_0^\delta(x) &= e^{\delta L_f}(x) = (1 + \delta L_f + \frac{\delta^2}{2!}L_f^2 + \cdots + \frac{\delta^p}{p!}L_f^p)(x) + O(\delta^{p+1}) \\
F_1^\delta(x) &= \left(\frac{e^{\delta L_f} - 1}{L_f}\right)\omega L_g(x) = \sum_{p\geq 1}\frac{\delta^p}{p!}L_f^{p-1}\omega L_g(x) \\
&= \delta g(x) + \frac{\delta^2}{2!}(L_f L_g(x) + L_g L_f(x)) + \frac{\delta^3}{3!}(L_f^2 L_g(x) + L_f L_g L_f(x) \\
&\qquad + L_g L_f^2(x)) + O(\delta^4) \\
F_2^\delta(x) &= \left(\frac{e^{\delta L_f} - 1 - \delta L_f}{L_f^2}\right)\omega L_g^2(x) = \sum_{p\geq 2}\frac{\delta^p}{p!}L_f^{p-2}\omega L_g^2(x) \\
&= \frac{\delta^2}{2!}L_g^2(x) + \frac{\delta^3}{3!}(L_f L_g^2(x) + L_g L_f L_g(x) + L_g^2 L_f(x)) + O(\delta^4) \\
F_3^\delta(x) &= \frac{\delta^3}{3!}L_g^3(x) + O(\delta^4)
\end{aligned}
$$

For $i > 1$, one has

$$
\begin{aligned}
F_i^\delta(x) &= \left(\frac{e^{\delta L_f} - 1 - \delta L_f - \cdots - \left(\frac{\delta^{i-1}L_f^{i-1}}{(i-1)!}\right)}{L_f^i}\right)\omega L_g^i(x) \qquad (5.15) \\
&= \sum_{p\geq i}\frac{\delta^p}{p!}L_f^{p-i}\omega L_g^i(x)
\end{aligned}
$$

The shuffle product (Ree, 1958), denoted as "ω", is defined in a recurrent way as follows

$$
\begin{aligned}
1\omega L_g &= L_g\omega 1 = L_g \\
L_f\omega L_g &= L_f L_g + L_g L_f \\
L_f^p\omega L_g^i &= L_f(L_f^{p-1}\omega L_g^i) + L_g(L_f^p\omega L_g^{i-1})
\end{aligned}
$$

Remark 5.3

To know the exact expression (5.15) of the F_i^δ, in terms of f and g, is not only suitable for explicitly computing the sampled dynamics. In fact, it enters into the study of the structural properties of the sampled model and in the understanding of whether or not one preserves the properties of a continuous-time control scheme under sampling (Monaco and Normand-Cyrot, 1995). In this sense, the proposed method is not strictly a numerical integration method, as the Runge-Kutta technique for example, even if it may satisfy this objective too.

Remark 5.4

Each function $F_p^\delta(x)$ is infinitely small of order δ^p. It results that the approximated sampled dynamics at order p with respect to δ is polynomial also with respect to u, of degree p at most.

Remark 5.5

Integrating successively over a period of amplitude δ, the Volterra kernels which specify the input-state behavior of (5.9), one recovers the exact sampled dynamics expanded according to the successive powers in u. The functions $F_p^\delta(x)$ satisfy in this case the integro-differential Poincaré formulae (Monaco and Normand-Cyrot, 1985).

Example 2

Let the linear analytic system be defined on $\mathbb{R}^2$ as

$$\begin{cases} \dot{x}_1(t) &= x_2(t) \\ \dot{x}_2(t) &= x_2(t) + u(t)x_1^2(t) \end{cases}$$

The approximated sampled dynamics at order 2 in δ is described by the equations

$$\begin{aligned} x_1(k+1) &= x_1(k) + \delta x_2(k) + \frac{\delta^2}{2}\left[x_2(k) + u(k)x_1^2(k)\right] + \mathrm{O}(\delta^3) \\ x_2(k+1) &= x_2(k) + \delta\left[x_2(k) + u(k)x_1^2(k)\right] \\ &\quad + \frac{\delta^2}{2}\left[x_2(k) + u(k)x_1^2(k) + 2u(k)x_1(k)x_2(k)\right] + \mathrm{O}(\delta^3) \end{aligned}$$

The link $\dot{x}_1(t) = x_2(t)$ is transformed into

$$\frac{\partial}{\partial \delta} x_1(k+1) = x_2(k+1)$$

5.2.2.1 Input–output sampling

It has previously been noted that the exact coincidence of the input-state behavior implies, at the sampling instants, the exact coincidence of the input–output behavior because of the functional link: $y = h(x)$. Conversely, input–output coincidence does not guarantee, *a priori*, an input-state coincidence.

Definition 5.2

Given system Σ_c, under the action of a constant control over time intervals of amplitude δ, the sampled system Σ_d is said to reproduce the input–output behavior of Σ_c, at sampling instants, if for any $x_c(0) = x_d(0)$, one verifies the equalities

$$y_c(k\delta) = h(x_c(k\delta)) = h(x_d(k)) = y_d(k) \text{ for } k \geq 0$$

5.2.3 *Multirate sampling*

5.2.3.1 The general case

Multirate sampling (of order m) is referred to as a sampling scheme in which the control variables are sampled faster (say m times) than the measured variables (the state).
This technique, which is usual in a linear context, is hereafter introduced in a nonlinear context and will be used to design nonlinear control schemes satisfying several objectives (precisely m).

Let us assume the single input $u(t)$, constant over subintervals of amplitude $\overline{\delta} = \dfrac{\delta}{m}$ and denote as $u_i(k)$ the value of $u(t)$ over $[k\delta + (i-1)\overline{\delta}, k\delta + i\overline{\delta}[$ for $i = 1, \ldots, m$

$$u_i(k) = u(k\delta + \tau) \text{ for } (i-1)\overline{\delta} \leq \tau < i\overline{\delta} \tag{5.16}$$

In this paragraph, a discrete-time state space representation, Σ_d^m, which is equivalent to system Σ_c, at the sampling instants $t = k\delta$, for $k \geq 0$, is proposed.

Let us denote Σ_d^m a discrete-time system with m controls of the form

$$\Sigma_d^m \left\{ \begin{array}{rcl} x_d(k+1) & = & (\mathcal{F}^\delta)^m(x_d(k), u_1(k), \ldots, u_m(k)) \quad (5.17) \\ y_d(k) & = & h(x_d(k)) \quad (5.18) \end{array} \right.$$

where $x_d(k) \in \mathbb{M}$ and where the function $(\mathcal{F}^-)^m : [0, \delta_0[\times\mathbb{M}\times\mathbb{R}^m \longrightarrow \mathbb{R}^n$ is supposed to be analytic. The following definition is stated.

Definition 5.3

The system Σ_d^m is referred to as the exact multirate sampled equivalent of order m to system Σ_c if, with the same initialization and constant controls over time intervals of amplitude $\overline{\delta} = \dfrac{\delta}{m}$, Σ_d^m and Σ_c exhibit the same input-state behaviors at the sampling instants $t = k\delta$, i.e., for $k \geq 1$ and $x_c(0) = x_d(0) = x(0)$ one verifies the equalities

$$x_c(k\delta) = x_d(k)$$

Lemma 5.1

The sampled dynamics $(\mathcal{F}^\delta)^m : \mathbb{M} \times \mathbb{R}^m \longrightarrow \mathbb{R}^n$ of system Σ_d^m admits the following exponential expansion

$$\begin{array}{rcl} \left(\mathcal{F}^\delta\right)^m (x_d(k), u_1(k), \ldots, u_m(k)) & = & F^{\overline{\delta}}(\ldots(F^{\overline{\delta}}(.), u_1(k)), \ldots), u_m(k))(x_d(k)) \\ & = & \mathrm{e}^{\overline{\delta} L_{f(.;u_1(k))}} \circ \cdots \circ \mathrm{e}^{\overline{\delta} L_{f(.;u_m(k))}}(x_d(k)) \end{array} \tag{5.19}$$

The expansion (5.19) is nothing else but the composition m times of exponential representations of the form (5.6), each one being parametrized by the successive controls $u_1(k), \ldots, u_m(k)$.

5.2.3.2 The linear analytic case

In this case the dynamics (5.19) takes the form

$$(\mathcal{F}^\delta)^m(x_d(k), u_1(k), \ldots, u_m(k)) = \mathrm{e}^{\overline{\delta} L_{f(.)} + \overline{\delta} u_1(k) L_{g(.)}} \circ \cdots \circ \mathrm{e}^{\overline{\delta} L_{f(.)} + \overline{\delta} u_m(k) L_{g(.)}}(x_d(k)) \tag{5.20}$$

and can be expanded according to the powers of $\overline{\delta}$ or according to the powers of the controls, $u_i(k)$, for $i = 1, \ldots, m$.

The expansion according to $\overline{\delta}$ and thus the approximated sampled dynamics of order p is deduced from (5.19) by substituting to each $F^{\overline{\delta}}(., u_i)$ its exponential expansion according to $\overline{\delta}$ and by adequately regrouping the terms of the same power.

With reference to the expansion according to the control variables, one substitutes to each $F^{\overline{\delta}}(., u_i)$ its expansion (5.14) according to $u_i^j, i = 1, \ldots, m; j \geq 0$ thus obtaining

$$\left(\mathcal{F}^{\delta}\right)^m (x, u_1, \ldots, u_m) = \sum_{i_1,\ldots,i_m \geq 0} u_1^{i_1} \ldots u_m^{i_m} F^{\overline{\delta}}_{i_1\ldots i_m}(x) \tag{5.21}$$

with $F^{\overline{\delta}}_{i_1\ldots i_m}(x) = F^{\overline{\delta}}_{i_1}\circ \ldots \circ F^{\overline{\delta}}_{i_m}(x)$ and where any $F^{\overline{\delta}}_i$ admits an expansion of the form (5.15) according to $\overline{\delta}$.

More precisely, for the first terms one obtains

$$\begin{aligned}
\left(\mathcal{F}^{\delta}\right)^m (x, u_1, \ldots, u_m) = \; & \left(F^{\overline{\delta}}_0\right)^m (x) + \sum_{i=1}^{m} u_i \left(F^{\overline{\delta}}_0\right)^{i-1} \circ F^{\overline{\delta}}_1 \circ \left(F^{\overline{\delta}}_0\right)^{m-i} (x) \\
& + \sum_{i=1}^{m} u_i^2 \left(F^{\overline{\delta}}_0\right)^{i-1} \circ F^{\overline{\delta}}_2 \circ \left(F^{\overline{\delta}}_0\right)^{m-i} (x) + \\
& + \sum_{\substack{i,j=1 \\ i<j}}^{m} u_i u_j \left(F^{\overline{\delta}}_0\right)^{i-1} \circ F^{\overline{\delta}}_1 \circ \left(F^{\overline{\delta}}_0\right)^{j-i-1} \circ F^{\overline{\delta}}_1 \circ \left(F^{\overline{\delta}}_0\right)^{m-j} (x) \\
& + \mathrm{O}(u_i u_j u_k)
\end{aligned}$$

with $(F^{\overline{\delta}}_0)^m = F^{\overline{\delta}}_0 \circ \ldots \circ F^{\overline{\delta}}_0 = F^{\delta}_0 (m \text{ times})$.

5.3 Nonlinear sampling of a continuous-time control scheme

The notion of relative degree associated to the input–output behavior of a nonlinear discrete-time system, as introduced in Monaco and Normand-Cyrot (1986), characterizes the delay occurring between the influence of the control on the output. Hereafter, the definition is recalled on the basis of sampled systems of the form (5.3, 5.4).

Definition 5.4

Σ_d has a relative degree r if

(i) For any $i = \{1, \ldots, m\}$ $\quad \dfrac{\partial}{\partial u_i} h\circ (F^{\delta}_0)^k \circ F^{\delta}(.; \underline{u})(x) = 0$

$0 \leq k \leq r-2 \quad \forall (x, \underline{u}) \in \mathbb{M} \times \mathbb{R}^m$

(ii) There exists $i \in \{1, \ldots, m\}$ *such that* $\dfrac{\partial}{\partial u_i} h\circ (F^{\delta}_0)^{r-1} \circ F^{\delta}(.; \underline{u})(x) \neq 0$

$\forall (x, \underline{u}) \in D(\mathbb{M} \times \mathbb{R}^m)$, *an open dense subset of* $\mathbb{M} \times \mathbb{R}^m$.

Remark 5.6

$y_d(k+r)$ is the first output of the system Σ_d, initialized in $x(k)$, under the influence of one of the controls $u_i(k)$, i.e. the control u_i influences the output y with a delay of r steps.

The following lemma is an immediate consequence of the definition of Σ_d.

Lemma 5.2

Given a SISO linear analytic continuous-time system, Σ_c, with relative degree r, the relative degree associated to the exact sampled equivalent system Σ_d is equal to 1 almost everywhere.

The proof of this lemma is directly deduced from the expansion according to $\delta(\delta \in [0, \delta_0[)$ of the output of the system Σ_d as

$$y_d(k+1) = h(x_d(k+1)) = h(x(k)) + \sum_{j \geq 1} \frac{\delta^j}{j!} \overset{(j)}{y_c}(k\delta) \tag{5.22}$$

where $\overset{(j)}{y_c}(k\delta)$ represents the jth output derivative of $y_c(t)$, computed at time $t = k\delta$, for constant controls $u(t)$ over sampling intervals.

As the rth output derivative y_c is the first output derivative depending on the control, this means that in (5.22), the first term depending on u is of order r according to δ. One obtains

$$\frac{\partial}{\partial u} \overset{(r)}{y_c}(k\delta) = \frac{\delta^r}{r!} L_g L_f^{r-1}(h)(x(k))$$

Among the consequences of lemma (5.2), one notes hereafter the impossibility to reproduce, by means of a discrete-time feedback designed on the single rate sampled equivalent system, the input–output behavior of a continuous-time feedback system with a relative degree greater than one. Another difficulty is the appearance of unstable zeros which generate the closed-loop system instability, *i.e.* the zero dynamics is not preserved under sampling. These difficulties justify the introduction of two digital control strategies depending on the relative degree of the initial continuous-time system.

Given a continuous-time linear analytic system Σ_c, and a linear analytic feedback γ_c, defined as

$$\gamma_c(x, v) = \alpha_c(x) + v\beta_c(x) \tag{5.23}$$

where $\alpha_c(x) : \mathbb{M} \longrightarrow \mathbb{R}$ and $\beta_c(x) : \mathbb{M} \longrightarrow \mathbb{R}$ are analytic functions, β_c non singular and $v \in \mathbb{R}$ an external input, one notes as $\tilde{\Sigma}_c$ the system modified by γ_c. It satisfies the dynamics

$$\tilde{\Sigma}_c \begin{cases} \dot{x}_c(t) &= \tilde{f}(x_c(t)) + \tilde{g}(x_c(t))v(t) \\ &= f(x_c(t)) + g(x_c(t))\alpha_c(x_c(t)) + g(x_c(t))\beta_c(x_c(t))v(t) \quad (5.24) \\ y_c(t) &= h(x_c(t)) \quad (5.25) \end{cases}$$

The stated problem is how to implement the continuous-time control scheme (5.2) without modifying its performances, *i.e.* how to maintain, at least at the sampling instants, the performances of the continuous-time control scheme $\tilde{\Sigma}_c$, making use of a discrete-time control scheme?

The most usual technique would be to consider a zero-order holding system which corresponds to maintaining the continuous-time control scheme γ_c constant, over time intervals of the amplitude of the sampling period. However, such an implementation can only give satisfactory results when the sampling period is sufficiently small, otherwise it may generate internal unstability.

In a linear context, several methods are available, like higher-order holding circuits or the design of a discrete-time control scheme on the basis of the sampled system (Hanselmann, 1987). This last approach requires a good validity for the sampled model.

In a nonlinear context, it may be difficult to compute with accuracy the sampled equivalent Σ_d of a continuous-time dynamics Σ_c, as well as the technological realization of higher-order holding circuits (Barbot, 1989b) may be unrealistic. Hereafter, we focus on the design of piecewise constant control schemes satisfying, at the sampling instants, the requirements fixed on the input–output continuous-time behavior while preserving its stability.

More precisely, the problem discussed in this paragraph is how to find a discrete-time feedback γ_d, which, when applied to the continuous-time system Σ_c, enables the reproduction, at the sampling instants, of the input–output behavior of $\tilde{\Sigma}_c$ (*i.e.* the reference with internal stability)?

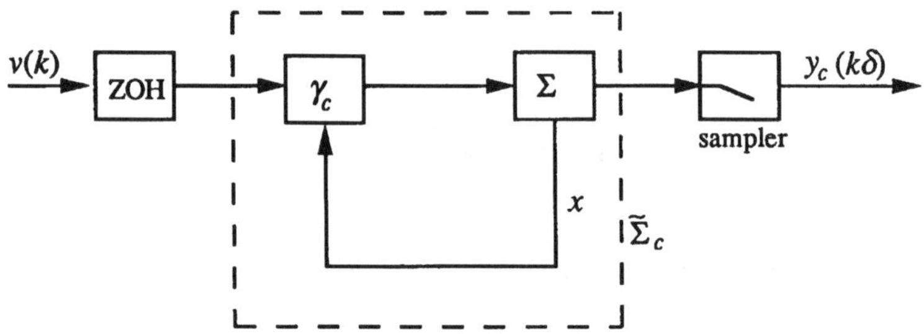

Figure 5.1 : Continuous-time feedback system under sampling (reference).

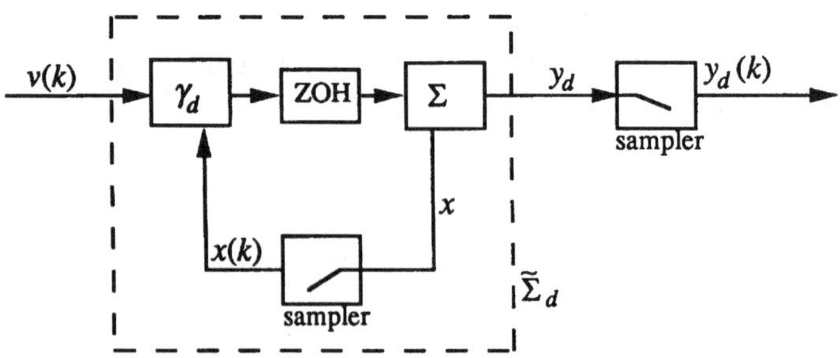

Figure 5.2 : Sampled system under the feedback action γ_d.

Denoting as $\gamma_d(x, v)$ any analytic discrete-time feedback law $\gamma_d \ : \mathbb{M} \times \mathbb{R} \longrightarrow \mathbb{R}$, such that $\dfrac{\partial \gamma_d}{\partial v} \neq 0$ and $\tilde{\Sigma}_d$ the feedback sampled system associated to Σ_d, one obtains the closed-loop dynamics

$$\tilde{\Sigma}_d \begin{cases} x_d(k) = \tilde{F}^\delta(x_d(k), v(k)) = e^{\delta(L_f + \gamma_d(x_d(k), v(k))L_g)}(x_d(k)) & (5.26) \\ y_d(k+1) = h(x_d(k)) & (5.27) \end{cases}$$

Definition 5.5

The system $\tilde{\Sigma}_d$ is said to reproduce, at each step, the input–output behavior of $\tilde{\Sigma}_c$, if for $x_c(0) = x_d(0)$ and for any $k \geq 0$ one verifies the equalities $y_c(k\delta) = y_d(k)$.

The following definition corresponds to a weaker concept of input–output reproduction. In fact, the two control systems $\tilde{\Sigma}_c$ and $\tilde{\Sigma}_d$ being, at the same state at time $t = k\delta$, one requires the output reproduction, at time $t = (k+1)\delta$ only.

Definition 5.6

The feedback system $\tilde{\Sigma}_d$ is said to reproduce, step by step, the input–output behavior of $\tilde{\Sigma}_c$, if for any $k \geq 0$ such that $x_d(k) = x_c(k\delta) = x(k)$ one verifies the equality $y_c((k+1)\delta) = y_d(k+1)$.

In the sequel, the external control $v(t)$, appearing in (5.23), is generally assumed to be constant over time intervals of amplitude δ

$$v(t) = v(k) \text{ for } t \in [k\delta, k\delta + \delta[\text{ for } k \geq 0. \tag{5.28}$$

The following lemma is an immediate consequence of (5.5) and (5.6).

Lemma 5.3

The system $\tilde{\Sigma}_d$ reproduces, at each step, the input–output behavior of $\tilde{\Sigma}_c$, submitted to external controls $v(t)$ constant over the time intervals if and only if, for $x_c(0) = x_d(0)$, the following equality is satisfied for any $k \geq 0$

$$h_\circ e^{\delta(L_f + \gamma_d(x_d(k), v(k))L_g)}(x_d(k)) = e^{\delta(L_{\tilde{f}} + v(k)L_{\tilde{g}})}\, h(x_c(k\delta)) \tag{5.29}$$

The next proposition is deduced from the implicit function theorem.

Proposition 5.2

Given a SISO linear analytic continuous-time system Σ_c, with relative degree r, and a non singular feedback law γ_c with piecewise constant external control v, there exists a non singular discrete-time feedback law γ_d such that the feedback system $\tilde{\Sigma}_d$ reproduces, step by step, the input–output behavior $\tilde{\Sigma}_c$.

Proof

When $x_c(k\delta) = x_d(k) = x(k)$, then the right-hand side of equality (5.29), which corresponds to the reference to be reproduced, admits an expansion according to δ of the form

$$e^{\delta(L_{\tilde{f}} + v(k)L_{\tilde{g}})}\, h(x(k)) := \sum_{i=0}^{r-1} \frac{\delta^i}{i!} L^i_{\tilde{f}} h(x(k)) + \frac{\delta^r}{r!} \overline{S}_{x(k)}(\delta, v(k)) \tag{5.30}$$

where $\overline{S}_x(\delta, v)$ is the infinite series defined by

$$\overline{S}_x(\delta, v) := \sum_{i\geq 0} \frac{\delta^i r!}{(r+i)!}(L_{\tilde{f}} + vL_{\tilde{g}})^{i+1} L_f^{r-1} h(x) = \sum_{i\geq 0} \frac{\delta^i r!}{(r+i)!} \overset{(i+r)}{y_c} \tag{5.31}$$

Since the discrete-time feedback law is, by definition, constant over the sampling intervals, the left-hand side of (5.29) admits, according to δ the expansion

$$\begin{aligned} h_\circ \tilde{F}^\delta(x(k), v(k)) &\coloneqq e^{\delta(L_f + \gamma_d(x(k),v(k))L_g)} h(x(k)) \\ &\coloneqq \sum_{i=0}^{r-1} \frac{\delta^i}{i!} L_f^i h(x(k)) + \frac{\delta^r}{r!} S_x(k)(\delta; \gamma_d(x(k), v(k))) \end{aligned} \tag{5.32}$$

where $S_x(\delta, \gamma_d)$ is the infinite series defined by

$$S_x(\delta, \gamma_d) := \sum_{i\geq 0} \frac{\delta^i r!}{(r+i)!}(L_f + \gamma_d L_g)^{i+1} L_f^{r-1} h(x) \tag{5.33}$$

From these notations, when $x_d(k) = x_c(k\delta) = x(k)$, it results that the equality (5.29) can be rewritten as the equality of series

$$S_x(k)(\delta, \gamma_d(x(k), v(k))) = \overline{S}_{x(k)}(\delta, v(k)) \tag{5.34}$$

or equivalently as

$$(L_f + \gamma_d L_g) L_f^{r-1} h(x) + \sum_{i\geq 1} \frac{\delta^i r!}{(r+i)!}(L_f + \gamma_d L_g)^{i+1} L_f^{r-1} h(x) = \overset{(r)}{y_c} + \sum_{i\geq 1} \frac{\delta^i r!}{(r+i)!} \overset{(i+r)}{y_c} \tag{5.35}$$

which admits for $\delta \in [0, \delta_0[$, a solution of the form

$$\gamma_d(x, v) = \gamma_{d0}(x, v) + \sum_{i\geq 1} \delta^i \gamma_{di}(x, v) \tag{5.36}$$

if and only if the condition

$$L_g L_f^{r-1}(h)(x) \neq 0 \tag{5.37}$$

is satisfied in $x(k)$.

In fact, setting $\delta = 0$ in (5.35), one verifies that the discrete-time feedback, γ_{d0} (zero-order holding of γ_c, over the interval $[k\delta, (k+1)\delta[$) is a solution, *i.e.*, one verifies the equality

$$(L_f + \gamma_{d0} L_g) L_f^{r-1} h(x(k)) = \overset{(r)}{y_c}(k) = (L_{\tilde{f}} + vL_{\tilde{g}}) L_f^{r-1} h(x(k))$$

This implies

$$\gamma_{d0}(x(k), v(k)) = \gamma_c(x(k), v(k)) \tag{5.38}$$

For $\delta \in [0, \delta_0[$, then $\overline{S}_{x(k)}(\delta, v(k))$ belongs to a neighborhood of $\overline{S}_{x(k)}(0, v(k))$ and the equality (5.34) admits a solution of the form (5.36) if and only if

$$\left.\frac{\partial S_x}{\partial u}\right|_{\substack{\delta = 0 \\ u = \gamma_c(x, v(k))}} = L_g L_f^{r-1} h(x) \neq 0 \tag{5.39}$$

which is satisfied because of the relative degree definition.

5.3.1 *Computation of the digital feedback law.*

The explicit computation of γ_d can be performed according to two procedures. The first, a direct one, makes reference to formal series reversion as proposed in Barbot (1989b), Gröbner (1973), and Monaco and Normand-Cyrot (1987), in the context of discrete-time systems. More precisely, one computes

$$\gamma_d(x, v) = S_x^{-1}\left(\delta, \overline{S}_x(\delta, v)\right) \tag{5.40}$$

where $S_x^{-1}(\delta, X)$ represents the reverse series associated to $S_x(\delta, X)$, *i.e.*

$$S_x\left(\delta, S_x^{-1}(\delta, X)\right) = X$$

In the second procedure, one takes into account the δ-dependency of the series $S_x(\delta, v)$ and $\overline{S}_x(\delta, v)$ and performs an iterated computation of the expansion according to δ of the exact solution γ_d.

Substituting the expansion (5.36) to γ_d into (5.35) and comparing homogeneous terms in the power of δ, one can show that any $\gamma_{di}(x, v)$ is a solution of an algebraic equation. Such a solution exists because of condition (5.39).

Arguing thus, the first terms $\gamma_{d0}, \gamma_{d1}, \gamma_{d2}$, are solutions of the following equations

$$(L_f + \gamma_{d0} L_g) L_f^{r-1} h(x) = (L_f + \gamma_c L_g) L_f^{r-1} h(x) \tag{5.41}$$

$$\begin{aligned} &\gamma_{d1} L_g L_f^{r-1} h(x) + \tfrac{1}{r+1}(L_f + \gamma_{d0} L_g)^2 L_f^{r-1} h(x) \\ &\qquad = \tfrac{1}{r+1}(L_f + \gamma_c L_g)^2 L_f^{r-1} h(x) \end{aligned} \tag{5.42}$$

$$\begin{aligned} &\gamma_{d2} L_g L_f^{r-1} h(x) + \tfrac{1}{r+1}\gamma_{d1}(L_f + \gamma_{d0} L_g) L_g L_f^{r-1} h(x) \\ &\qquad + \tfrac{1}{r+1}\gamma_{d1} L_g (L_f + \gamma_{d0} L_g) L_f^{r-1} h(x) \\ &\qquad + \tfrac{1}{(r+1)(r+2)}(L_f + \gamma_{d0} L_g)^3 L_f^{r-1} h(x) \\ &\qquad = \tfrac{1}{(r+1)(r+2)}(L_f + \gamma_c L_g)^3 L_f^{r-1} h(x) \end{aligned} \tag{5.43}$$

From this, one deduces

$$\begin{aligned} \gamma_{d0}(x(k), v(k)) &= \gamma_c(x(k), v(k)) \\ \gamma_{d1}(x(k), v(k)) &= \tfrac{1}{r+1}(L_f + \gamma_c L_g)(\gamma_c)(x(k), v(k)) \\ &= \tfrac{1}{r+1}\dot{\gamma}_c(x(k), v(k)) \end{aligned} \tag{5.44}$$

$$\begin{aligned} &\gamma_{d2}(x(k), v(k)) L_g L_f^{r-1} h(x(k)) = \\ &\tfrac{1}{(r+1)(r+2)}\ddot{\gamma}_c(x(k), v(k)) L_g L_f^{r-1} h(x(k)) \\ &-\tfrac{1}{(r+1)^2(r+2)}\dot{\gamma}_c(x(k), v(k))(L_f + \gamma_{d0} L_g) L_g L_f^{r-1}(h)(x(k), v(k)) \\ &-\tfrac{1}{(r+1)^2(r+2)}\dot{\gamma}_c(x(k), v(k)) L_g (L_f + \gamma_{d0} L_g) L_f^{r-1}(h)(x(k), v(k)) \end{aligned} \tag{5.45}$$

Remark 5.7

Except when the relative degree is equal to 1*, Proposition (5.2) does not ensure the input–output reproduction at each step. This will be discussed hereafter by distinguishing the cases of* $r = 1$ *and* $r > 1$.

A preliminary result concerning the application of such a discrete-time control procedure is proposed in chapter 6.

5.4 Preservation of the input–output linearization under digital feedback – the case of $r = 1$

In this section, one considers a SISO linear analytic continuous-time system of the form (5.9, 5.10), with a relative degree equal to 1 $(L_g h(x) \neq 0)$.
It can easily be verified that the linearizing continuous-time feedback law, given by the expression

$$\gamma_c(x(t), v(t)) = [L_g h(x(t))]^{-1}(-L_f h(x(t)) + v(t)), \tag{5.46}$$

ensures to the closed-loop, $\tilde{\Sigma}_c$, the linear input–output link

$$\dot{y}_c(t) = v(t) \tag{5.47}$$

Remark 5.8

Various choices for the external control $v(t)$ *enable us to satisfy various input–output objectives such as regulation, stabilization or reference trajectory tracking.*

In the following the external control $v(t)$ is assumed to be constant over time intervals of amplitude δ

$$v(t) = v(k) \text{ for } t \in [k\delta, k\delta + \delta[\text{ with } k \geq 0$$

Theorem 5.1

Given a SISO linear analytic continuous-time system Σ_c*, with relative degree* $r = 1$*, modified by the linearizing feedback law* γ_c *(5.46), then there exists a discrete-time feedback law* γ_d *such that the closed-loop system* $\tilde{\Sigma}_d$ *preserves, at each step, the input–output behavior of* $\tilde{\Sigma}_c$.

Proof

It directly follows from Proposition 5.2 since the objective to be reproduced (*i.e.* the right-hand side of (5.29)) depends on the ouput and on constant external controls only.
More precisely, in this case the series $\overline{S}_x(\delta, v)$ defined in (5.31) reduces to

$$\overline{S}_x(\delta, v(k)) = v(k) \tag{5.48}$$

because $v(k)$ is assumed to be constant.
The series $S_x(\delta, \gamma_d)$, defined in (5.33), can be rewritten as

$$S_x(\delta, \gamma_d) = (L_f + \gamma_d L_g)h(x) + \sum_{i \geq 1} \frac{\delta^i}{(i+1)!}(L_f + \gamma_d L_g)^{i+1} h(x) \tag{5.49}$$

and the equality (5.29) can be rewritten as

$$h(x)+\delta(L_f+\gamma_d L_g)h(x)+\sum_{i\geq 1}\frac{\delta^{i+1}}{(i+1)!}(L_f+\gamma_d L_g)^{i+1}h(x)=h(x)+\delta v(k) \quad (5.50)$$

According to Proposition (5.2) the feedback γ_d is computed to satisfy (5.50). Let us show by recursion that the input–output reproduction at each step is satisfied too. Starting from the same initial condition, *i.e.* $x_d(0)=x_c(0)=x(0)$, one immediately verifies, at time $t=0$, the equality

$$y_d(0)=y_c(0)=h(x(0)).$$

The discrete-time feedback $\gamma_d(x(0),v(0))$, being designed to satisfy (5.50) (step by step reproduction), *i.e.*, at time $t=\delta$, the equality

$$y_d(1)=h(x_d(1))=y_c(\delta)=h(x(0))+\delta v(0)$$

one obtains also the equality

$$y_d(1)=y_d(0)+\delta v(0)$$

with $x_d(1)\neq x_c(\delta)$. The discrete-time feedback $\gamma_d(x_d(1),v(1))$ being designed to satisfy (5.50) at time $t=2\delta$, *i.e.* is the equality

$$y_d(2)=h(x_d(1))+\delta v(1)$$

one deduces from $y_d(1)=y_c(\delta)$ the equality

$$y_d(2)=y_c(2\delta)=y_d(0)+\delta v(0)+\delta v(1)$$

Iterating the reasoning with $x_d(k)\neq x_c(k\delta)$, and assuming that the input–output reproduction holds at time $t=k\delta$, *i.e.*

$$y_d(k)=y_c(k\delta)=h(x_d(k))=y_d(0)+\sum_{i=0}^{k-1}\delta v(i)$$

the discrete-time feedback $\gamma_d(x_d(k),v(k))$ being designed to satisfy (5.50), *i.e.*

$$y_d(k+1)=h(x_c(k\delta))+\delta v(k)$$

one deduces the equality

$$y_d(k+1)=y_c((k+1)\delta)=y_d(0)+\sum_{i=0}^{k}\delta v(i) \quad (5.51)$$

This concludes the proof of Theorem 5.1.

Corollary

(i) Writing $v(k)=\dfrac{-y(k)}{\delta}$, the discrete-time feedback law γ_d brings to zero the output in one step and keeps it to zero at each sampling instant. This corresponds to minimum time regulation as discussed in Glad (1987) and Monaco and Normand-Cyrot (1987).

(ii) Writing

$$v(k) = \frac{(y_M(k+1) - y_M(k)) + (c_0 + 1)(y_M(k) - y(k))}{\delta}$$

one satisfies after one step only, the output error trajectory $(y_M(k+1) - y(k+1)) + c_0(y_M(k) - y(k)) = 0$, *which is then maintained at the sampling instants.*

(iii) Writing $v(k) = \dfrac{-y(k) - c_0 y(k)}{\delta}$, *one obtains in one step only, the output dynamics* $y(k+1) + c_0 y(k)$, *which is then maintained at the sampling instants.*

(iv) The invariance of the Ker dh *distribution is ensured after one step i.e. given two states* $x_1(0)$ *and* $x_2(0)$ *such that* $h(x_1(0)) = h(x_2(0))$ *then, because of (5.51),*

$$h(x_1(k)) = h(x_2(k)) \text{ with } v(k) = 0 \text{ for any } k \geq 0$$

Remark 5.9

From part (iv) of the corollary, one deduces that $\gamma_d(x, 0)$ *is the discrete-time analogue of the continuous-time feedback defining the zero dynamics of* Σ_c. *Indeed, for* $v(k) = 0$ *and* $x(0)$ *such that* $h(x(0)) = 0$, *the feedback* $\gamma_d(x, 0)$ *maintains, at each step, the ouput at zero:* $h(x(k)) = h(x(0)) = 0$ *for any* $k \geq 1$. *It results that, at least locally, the minimum phase property of the continuous-time dynamics is preserved and internal stability is maintained.*

Remark 5.10

When the external control $v(t)$ *is not constant but depends on* $y(t)$ *and possibly on other external variables, which are constant over the sampling periods, the same control technique can be applied in order to satisfy the input–output reproduction at each step. In this case, one has to replace the right-hand side of (5.50) by the expansion*

$$h(x) + \delta \overline{S}_x(\delta, v(k)) = h(x) + \delta v(k) + \sum_{i \geq 1} \frac{\delta^{i+1}}{(1+i)!} \overset{(i)}{v}(k)$$

where $\overset{(i)}{v}(k)$ *represents the* i*th time derivative of the control* $v(t)$, *computed at time* $t = k\delta$. *For example, for* $v(t) = -\lambda h(x(t))$, *one obtains*

$$h(x) + \delta \overline{S}_x(k)(\delta, v(k)) = \mathrm{e}^{-\lambda\delta}\, h(x(k))$$

5.4.1 *Computation of the digital feedback law*

It suffices to reconsider the formulae (5.41), (5.42) and (5.43) when $r = 1$. For the first terms one computes

$$\begin{aligned}
\gamma_{d0}(x, v) &= \gamma_c(x, v) = (L_g h(x))^{-1} (-L_f h(x) + v) \\
\gamma_{d1}(x, v) &= -\tfrac{1}{2} (L_g h(x))^{-1} (L_f + \gamma_{d0} L_g)^2 h(x) \\
\gamma_{d2}(x, v) &= -\tfrac{1}{2} (L_g h(x))^{-1} \gamma_{d1}((L_f + \gamma_{d0} L_g)\omega L_g) h(x) \\
&\quad - \tfrac{1}{3!} (L_g h(x))^{-1} (L_f + \gamma_{d0} L_g)^3\, h(x).
\end{aligned}$$

5.5 Preservation of the input–output linearization under discrete-time feedback – the case of $r > 1$

Given a SISO linear analytic continuous-time system Σ_c, with relative degree r, it is easy to verify that the linearizing feedback γ_c defined by

$$\gamma_c(x, v) = (L_g L_f^{r-1} h(x))^{-1}(-L_f^r h(x) + v) \tag{5.52}$$

ensures for the closed-loop system Σ_c the linear input–output link

$$\overset{(r)}{y} = v \tag{5.53}$$

The possibility to reproduce, step by step, under discrete-time control design, the input–output behavior of Σ_c, has been shown in Proposition 5.2.
In the present case, the continuous-time linearizing feedback being given by (5.52), the series $\overline{S}_x(\delta, v)$, defined in (5.31) reduces to

$$\overline{S}_x(\delta, v(k)) = \sum_{i\geq 0} \frac{\delta^i}{(i+1)!} \overset{(i+1)}{y_c}(k) = v(k) \tag{5.54}$$

because $v(k)$ is supposed to be constant.
The series $S_x(\delta, \gamma_d)$, defined in (5.33), is equal to

$$S_x(\delta, \gamma_d) = (L_f + \gamma_d L_g) L_f^{r-1} h(x) + \sum_{i\geq 1} \frac{\delta^i r!}{(r+i)!} (L_f + \gamma_d L_g)^{i+1} L_f^{r-1} h(x) \tag{5.55}$$

The step by step input–output reproduction is satisfied when γ_d is a solution of equality (5.29) or equivalently of

$$\begin{aligned}\sum_{i=0}^{r-1} \frac{\delta^i}{i!} L_f^i h(x) + \frac{\delta^r}{r!}(L_f + \gamma_d L_g) L_f^{r-1} h(x) + \sum_{i\geq 1} \frac{\delta^{i+r}}{(r+i)!}(L_f + \gamma_d L_g)^{i+1} L_f^{r-1} h(x) \\ = \sum_{i=0}^{r-1} \frac{\delta^i}{i!} L_f^i h(x) + \frac{\delta^r}{r!} v(k)\end{aligned} \tag{5.56}$$

The purpose of this paragraph is to propose a generalization of Theorem 5.1 to the case of $r \geq 1$.

Let us first note that if the equality (5.56) is satisfied at time $t = (k+1)\delta$ when $x_d(k) = x_c(k\delta) = x(k)$, *i.e.* if

$$h(x_d(k+1)) = h(x_c((k+1)\delta)),$$

it will generally not be satisfied at time $t = (k+2)\delta$, because nothing ensures the equality

$$L_f^i h(x_d(k+1)) = L_f^i h(x_c((k+1)\delta)) \quad \text{for } i = 1, \ldots, r-1.$$

It results that the discrete-time feedback designed to satisfy (5.56) does not enable the reproduction, at each step, of the continuous-time control scheme. The solution proposed hereafter is performed to ensure the reproduction, at the sampling instants, of the input–output behaviors associated to the r functions $L_f^{i-1} h(x)$ for $i = 1, \ldots, r$.

To do so, one considers the multirate sampled system of order r denoted as Σ_d^r and defined by system (5.17, 5.18), completed by the $r-1$ output derivatives $L_f^i h(x)$ for $i = 1, \ldots, r-1$. Arguing thus, one obtains a MIMO square system (r inputs and r outputs).

Given r discrete-time feedback laws

$$\gamma_d^i(x, v) \ : \mathbb{M} \times \mathbb{R} \longrightarrow \mathbb{R} \text{ for } i = 1, \ldots, r \tag{5.57}$$

let us denote $\tilde{\Sigma}_d^r$ as the multirate sampled system of order r submitted to the feedback action γ_d^i.

Assuming in (5.52) the external control $v(t)$, constant over time intervals of amplitude δ

$$v(t) = v(k), k\delta \leq t < (k+1)\delta \qquad \text{for } k \geq 0 \tag{5.58}$$

One deduces from the implicit function theorem the next proposition.

Proposition 5.3

Given a SISO linear analytic continuous-time system Σ_c, with relative degree r, and the linearizing non singular feedback γ_c defined in (5.52), there exist r non singular discrete-time feedback laws γ_d^i, $(i = 1, \ldots, r)$, such that the closed-loop system $\tilde{\Sigma}_d^r$ reproduces, step by step, the r input–output behaviors of $\tilde{\Sigma}_c$.

Proof

When $x_d(k) = x_c(k\delta) = x(k)$, the same reasoning as previously (for $r = 1$) allows us to assert that the step by step reproduction of the r outputs is ensured if there exist r feedback laws satisfying the following r equalities

$$\mathrm{e}^{\bar{\delta} L_{f(.,\gamma_d^1(k))}} \circ \ldots \circ \mathrm{e}^{\bar{\delta} L_{f(.,\gamma_d^r(k))}} (L_f^i h)(x(k)) =$$

$$\sum_{j=0}^{r-1-i} \frac{\delta^j}{j!} L_f^{(i+j)} h(x(k)) + \frac{\delta^{r-i}}{(r-i)!} \overline{S}^i_{x(k)}(\delta, v(k)) \tag{5.59}$$

where $\overline{S}^i_{x(k)}(\delta, v(k))$ is reduced to

$$\overline{S}^i_{x(k)}(\delta, v(k)) = v(k) \qquad \text{for } i = 0, \ldots, r-1 \tag{5.60}$$

The discrete-time feedback γ_d^i being, by definition, constant over the sampling intervals of amplitude $\bar{\delta}$, the left-hand side of (5.59) can be expanded according to $\bar{\delta}$ for $i = 0, \ldots, r-1$, as

$$\mathrm{e}^{\bar{\delta} L_{f(.,\gamma_d^1(k))}} \circ \ldots \circ \mathrm{e}^{\bar{\delta} L_{f(.,\gamma_d^r(k))}} (L_f^i h)(x(k)) =$$

$$\sum_{j=0}^{r-1-i} \frac{\delta^j}{j!} L_f^{i+j} h(x(k)) + \frac{\delta^{r-i}}{(r-i)!} S^i_{x(k)}(\bar{\delta}, \gamma_d^1, \ldots, \gamma_d^r) \tag{5.61}$$

where each $S^i_{x(k)}(\bar{\delta}, \gamma_d^1, \ldots, \gamma_d^r)$ represents an infinite series organized according to the powers of $\bar{\delta}$.

With these notations, the equality (5.59) can be rewritten as the equality of the r series

$$S^i_{x(k)}(\bar{\delta}, \gamma^1_d, \ldots, \gamma^r_d) = v(k) \qquad \text{for } i = 1, \ldots, r. \tag{5.62}$$

which admits r solutions $\gamma^i_d(x, v) : \mathbb{M} \times \mathbb{R} \longrightarrow \mathbb{R}$ analytic for $i = 1, \ldots, r$, of the form

$$\gamma^i_d(x, v) = \gamma^i_{d,0}(x, v) + \sum_{j \geq 1} \delta^j \gamma^i_{d,j}(x, v) \tag{5.63}$$

if and only if

$$\det \left[\frac{\partial}{\partial u_j} \left[S^i_{x(k)}(\bar{\delta}, u_1 \ldots, u_r) \right] \begin{array}{rcl} i & = & 1, \ldots, r \\ j & = & 1, \ldots, r \\ \bar{\delta} & = & 0 \\ \underline{u} & = & \gamma_{d0} \end{array} \right] \neq 0 \tag{5.64}$$

A simple computation shows that condition (5.64) is reduced to $L_g L_f^{r-1}(h)(x) \neq 0$ which is satisfied because of the relative degree definition.

Theorem 5.2

Given a SISO linear analytic continuous-time system Σ_c, with relative degree r, and the non singular linearizing feedback γ_c defined in (5.52) with piecewise constant external controls v, there exist r discrete-time non singular feedback laws γ^i_d such that the closed-loop system $\tilde{\Sigma}^r_d$ reproduces, at each step, the r input–output behaviors of $\tilde{\Sigma}_c$ when $x_c(0) = x_d(0)$.

Proof

It is directly deduced from Proposition 5.3 because the right-hand side of equality (5.59), *i.e.* the references to reproduce, depends on the output function, on its first $r-1$ derivatives and on the external piecewise constant control $v(k)$ only.

Input–output reproduction at each step will be shown arguing in a recurrent manner. For, one first assumes the initial condition: $x_d(0) = x_c(0) = x(0)$.

At time $t = 0$, one immediately verifies

$$L^i_f h(x_d(0)) = L^i_f h(x_c(0)) = L^i_f h(x(0)) \qquad \text{for } i = 0, \ldots, r-1.$$

The r discrete-time feedback $\gamma^i_d(x(0), v(0))$ being designed in order to satisfy (5.59), that is at time $t = \delta$, the r equalities

$$\begin{aligned} L^i_f h(x_d(1)) &= \sum_{j=0}^{r-1-i} \frac{\delta^j}{j!} L^{i+j}_f h(x_d(0)) + \frac{\delta^{r-i}}{(r-i)!} v(0) \\ &= \sum_{j=0}^{r-1-i} \frac{\delta^j}{j!} L^{i+j}_f h(x(0)) + \frac{\delta^{r-i}}{(r-i)!} v(0) \end{aligned}$$

one immediately obtains

$$L_f^i h(x_d(1)) = L_f^i h(x_c(1)) \qquad \text{for } i = 0, \ldots, r-1.$$

Iterating the reasoning with $x_d(k) \neq x_c(k)$, but assuming that the input–output reproduction holds at time $t = k\delta$, *i.e.*, for $i = 0, \ldots, r-1$

$$L_f^i h(x_d(k)) = L_f^i h(x_c(k\delta)) = \sum_{j=0}^{r-1-i} \frac{\delta^j}{j!} L_f^{i+j} h(x(0)) + \sum_{p=0}^{k-1} \frac{\delta^{r-i}}{(r-i)!} v(p)$$

The discrete-time feedback $\gamma_d^i(x_d(k), v(k))$ being designed to satisfy (5.59), that is

$$L_f^i h(x_d(k+1)) = \sum_{j=0}^{r-1-i} \frac{\delta^j}{j!} L_f^{i+j} h(x_d(k)) + \frac{\delta^{r-i}}{(r-i)!} v(k)$$

one obtains the equality

$$L_f^i h(x_d(k+1)) = L_f^i h(x_c(k+1)) = \sum_{j=0}^{r-1-i} \frac{\delta^j}{j!} L_f^{i+j} h(x(0)) + \sum_{p=0}^{k} \frac{\delta^{r-i}}{(r-i)!} v(p)$$

and the proof of Theorem 5.2 is complete.

Corollary

(i) *The discrete-time feedback laws $\gamma_d^i(x_d(k), v(k))$ preserve, at the sampling instants, the invariance of the distribution* $\mathrm{Ker}\left\{\mathrm{d}h, \ldots, \mathrm{d}L_f^{r-1}h\right\}$ *associated to $\tilde{\Sigma}_c$. (i.e. given two states $x_1(0)$ and $x_2(0)$ such that $L_f^i h(x_1(0)) = L_f^i h(x_2(0))$ for $i = 0, \ldots, r-1$, then $L_f^i h(x_1(k)) = L_f^i h(x_2(k))$ for $i = 0, \ldots, r-1$ and for any $k \geq 1$).*

(ii) *Admitting now different constant values for $v(k)$ over the r small time intervals of amplitude $\bar{\delta} = \dfrac{\delta}{r}$, (i.e. $v(t) = v_i(k)$ for $t \in [k\delta + (i-1)\bar{\delta}, k\delta + i\bar{\delta}[, i = 1, \ldots, r)$, one can constraint the output and its first $r-1$ derivatives to zero in one step of amplitude δ, and then maintain them at zero at each sampling instant.*

Remark 5.11

From part (ii) of the corollary one deduces that the feedback laws $\gamma_d^i(x, 0)$ are the discrete-time analogues of the continuous-time feedback law which enables to define the zero dynamics. More precisely, for $v(k) = 0$ for any $k \geq 0$ and for $x(0)$ such that $L_f^i h(x(0)) = 0$ for $i = 0, \ldots, r-1$, then the feedback laws $\gamma_d^i(x, 0)$ maintain at zero, at each sampling instants, the output and its first $r-1$ derivatives: $L_f^i h(x(k)) = 0$, for $k \geq 0$ and $i = 0, \ldots, r-1$.

5.5.1 Computation of the discrete-time feedback laws

It is carried out as in the previous case ($r = 1$) according to simple expansions with respect to δ.

The computation is here proposed when $r = 2$, allowing different constant values for the external control over the small time intervals $\overline{\delta} = \delta/2$. This gives more degrees of freedom to the digital scheme which can fulfil extra requirements with respect to the continuous-time scheme. The case of $v(k + \frac{1}{2}) = v(k)$ is then easily deduced.

Starting again from the SISO continuous-time system Σ_c, modified by the linearizing feedback law (5.52)

$$\gamma_c(x, v) = (L_g L_f h(x))^{-1}(-L_f^2 h(x) + v) = \alpha(x) + \beta(x)v \tag{5.65}$$

which leads to the linear input–output link

$$\ddot{y} = v \tag{5.66}$$

One designs, as previously explained, the square multirate sampled system of order 2

$$\left\{\begin{array}{rcll} x(k+1) & = & \mathcal{F}_2^\delta(x(k); u_2(k); u_1(k)) & (5.67) \\ y_1(k) & = & h(x(k)) & (5.68) \\ y_2(k) & = & L_f h(x(k)) & (5.69) \end{array}\right.$$

with

$$\mathcal{F}_2^\delta(x(k), u_2, u_1) = \mathrm{e}^{\overline{\delta}(L_f + u_1 L_g)}\,\mathrm{e}^{\overline{\delta}(L_f + u_2 L_g)}(x(k)) \quad \text{where} \quad \overline{\delta} = \delta/2$$

From (5.59), the reproduction at each step will be ensured by the two discrete-time feedback laws $\gamma_d^1(x, v), \gamma_d^2(x, v)$ of the form

$$\gamma_d^i\left(x, v(k), v\left(k + \tfrac{1}{2}\right)\right) = \gamma_{d0}^i\left(x, v(k), v\left(k + \tfrac{1}{2}\right)\right) + \sum_{j \geq 1} \delta^j \gamma_{dj}^i\left(x, v(k), v\left(k + \tfrac{1}{2}\right)\right) \tag{5.70}$$

which, substituted to $u_1(k)$ and $u_2(k)$, in (5.67), are solutions of the following series equalities

$$\begin{aligned} & h \circ \tilde{\mathcal{F}}_2^\delta\left(x(k), v(k), v\left(k + \tfrac{1}{2}\right)\right) \\ & = \mathrm{e}^{\overline{\delta}(L_f + (\alpha + \beta v(k)) L_g)} \circ \mathrm{e}^{\overline{\delta}(L_f + (\alpha + \beta v(k + \frac{1}{2})) L_g)}\, h(x(k)) \\ & = h(x(k)) + \delta L_f h(x(k)) + \tfrac{1}{2}\overline{\delta}^2\left(3v(k) + v\left(k + \tfrac{1}{2}\right)\right) \end{aligned}$$

with

$$\overline{S}_x^1\left(\overline{\delta}, v(k), v\left(k + \tfrac{1}{2}\right)\right) = \tfrac{1}{4}\left(3v(k) + v\left(k + \tfrac{1}{2}\right)\right)$$

and

$$\begin{aligned} & L_f h \circ \tilde{\mathcal{F}}_2^\delta\left(x(k), v(k), v\left(k + \tfrac{1}{2}\right)\right) \\ & = \mathrm{e}^{\overline{\delta}(L_f + (\alpha + \beta v(k)) L_g)} \circ \mathrm{e}^{\overline{\delta}(L_f + (\alpha + \beta v(k + \frac{1}{2})) L_g)}\, L_f h(x(k)) \\ & = L_f h(x(k)) + \overline{\delta}\left(v(k) + v\left(k + \tfrac{1}{2}\right)\right) \end{aligned}$$

with

$$\overline{S}_x^2\left(\overline{\delta}, v(k), v\left(k+\tfrac{1}{2}\right)\right) = \tfrac{1}{2}\left(v(k) + v\left(k+\tfrac{1}{2}\right)\right)$$

According to the previous notations, one can rewrite the two series equalities as

$$h\circ\tilde{\mathcal{F}}_2^\delta\left(x(k), v(k), v\left(k+\tfrac{1}{2}\right)\right) = h(x(k)) + \delta L_f h(x(k)) + \tfrac{1}{2}\overline{\delta}^2 S_x^1\left(\overline{\delta}, \gamma_d^1, \gamma_d^2\right) \quad (5.71)$$

with

$$\begin{aligned} S_x^1\left(\overline{\delta}, \gamma_d^2, \gamma_d^1\right) &= \tfrac{3}{4}\left(L_f + \gamma_d^1 L_g\right) L_f h + \tfrac{1}{4}\left(L_f + \gamma_d^2 L_g\right) L_f h \\ &\quad + \tfrac{1}{6}\delta\left(L_f + \gamma_d^1 Lg\right)^2 L_f h + \tfrac{\delta}{8}\left(L_f + \gamma_d^1 L_g\right)\left(L_f + \gamma_d^2 L_g\right) L_f h \\ &\quad + \tfrac{1}{24}\delta\left(L_f + \gamma_d^2 L_g\right)^2 L_f h + \mathrm{O}\left(\delta^2\right) \end{aligned}$$

and

$$L_f h\circ\tilde{\mathcal{F}}_2^\delta\left(x(k), v\left(k+\tfrac{1}{2}\right), v(k)\right) = L_f h(x(k)) + \overline{\delta} S_x^2\left(\delta, \gamma_d^2, \gamma_d^1\right) \quad (5.72)$$

with

$$\begin{aligned} S_x^2\left(\overline{\delta}, \gamma_d^1, \gamma_d^2\right) &= \tfrac{1}{2}\left(L_f + \gamma_d^1 L_g\right) L_f h + \tfrac{1}{2}\left(L_f + \gamma_d^2 L_g\right) L_f h \\ &\quad + \tfrac{1}{8}\delta\left(L_f + \gamma_d^1 L_g\right)^2 L_f h + \tfrac{1}{8}\delta\left(L_f + \gamma_d^2 L_g\right)^2 L_f h \\ &\quad + \tfrac{1}{4}\delta\left(L_f + \gamma_d^1 L_g\right)\left(L_f + \gamma_d^2 L_g\right) L_f h + \mathrm{O}(\delta^2) \end{aligned}$$

Substituting the expansions (5.70) to the feedback laws γ_d^j for $j = 1, 2$ into the equalities (5.71-72), and then regrouping and equating the terms of the same power in δ, one can compute the solutions γ_{di}^j for $i = 0, 1$ and $j = 1, 2$.

To compute γ_{d0}^j for $j = 1, 2$, one obtains the two equations

$$\begin{aligned} \tfrac{3}{4}\left(L_f + \gamma_{d0}^1 L_g\right) L_f h + \tfrac{1}{4}\left(L_f + \gamma_{d0}^2 L_g\right) L_f h &= \tfrac{3}{4}v(k) + \tfrac{1}{4}v\left(k+\tfrac{1}{2}\right) \\ \tfrac{1}{2}\left(L_f + \gamma_{d0}^1 L_g\right) L_f h + \tfrac{1}{2}\left(L_f + \gamma_{d0}^2 L_g\right) L_f h &= \tfrac{1}{2}v(k) + \tfrac{1}{2}v\left(k+\tfrac{1}{2}\right) \end{aligned}$$

which can be satisfied by considering

$$\gamma_{d0}^1 = \gamma_c(., v(k)) \quad \text{and} \quad \gamma_{d0}^2 = \gamma_c\left(., v\left(k+\tfrac{1}{2}\right)\right)$$

To compute γ_{d1}^j for $j = 1, 2$, one considers the two equations

$$\begin{aligned} &\left(\tfrac{3}{4}\gamma_{d1}^1 + \tfrac{1}{4}\gamma_{d1}^2\right) L_g L_f h + \tfrac{1}{6}\left(L_f + \gamma_{d0}^1 L_g\right)^2 L_f h \\ &\quad + \tfrac{1}{8}\left(L_f + \gamma_{d0}^1 L_g\right)\left(L_f + \gamma_{d0}^2 L_g\right) L_f h + \tfrac{1}{24}\left(L_f + \gamma_{d0}^2 L_g\right)^2 L_f h = 0 \end{aligned}$$

and

$$\begin{aligned} &\tfrac{1}{2}\left(\gamma_{d1}^1 + \gamma_{d1}^2\right) L_g L_f h + \tfrac{1}{4}\left(L_f + \gamma_{d0}^1 L_g\right)\left(L_f + \gamma_{d0}^2 L_g\right) L_f h \\ &\quad + \tfrac{1}{8}\left(L_f + \gamma_{d0}^1 L_g\right)^2 L_f h + \tfrac{1}{8}\left(L_f + \gamma_{d0}^2 L_g\right)^2 L_f h = 0 \end{aligned}$$

which can be satisfied by setting

$$\begin{aligned}
\gamma_{d1}^{1} &= -(L_gL_fh)^{-1}\left(\tfrac{5}{24}\left(L_f+\gamma_{d0}^{1}L_g\right)^2 L_fh - \tfrac{1}{24}\left(L_f+\gamma_{d0}^{2}L_g\right)^2 L_fh\right)\\
\gamma_{d1}^{2} &= -(L_gL_fh)^{-1}\left(\tfrac{1}{24}\left(L_f+\gamma_{d0}^{1}L_g\right)^2 L_fh + \tfrac{7}{24}\left(L_f+\gamma_{d0}^{2}L_g\right)^2 L_fh\right) + \\
&\quad -\tfrac{1}{2}(L_gL_fh)^{-1}\left(L_f+\gamma_{d0}^{1}L_g\right)\left(L_f+\gamma_{d0}^{2}L_g\right)L_fh
\end{aligned}$$

Remark 5.12

For $v(k) = v\left(k+\frac{1}{2}\right)$, one finds $\overline{S}^{1}_{x(k)}(\delta, v(k)) = \overline{S}^{2}_{x(k)}(\delta, v(k)) = v(k)$ and thus

$$\begin{aligned}
\gamma_{d0}^{1}(x(k), v(k)) &= \gamma_{d0}^{2}(x(k), v(k)) = \gamma_c(x(k), v(k))\\
\gamma_{d0}^{1}(x(k), v(k)) &= \tfrac{1}{6}\dot{\gamma}_c(x(k), v(k))\\
\gamma_{d0}^{2}(x(k), v(k)) &= \tfrac{5}{6}\dot{\gamma}_c(x(k), v(k))
\end{aligned}$$

that is

$$\begin{aligned}
\gamma_{d}^{1}(x(k), v(k)) &= \gamma_c(x(k), v(k)) + \tfrac{1}{6}\delta\dot{\gamma}_c(x(k), v(k)) + O(\delta^2)\\
\gamma_{d}^{2}(x(k), v(k)) &= \gamma_c(x(k), v(k)) + \tfrac{5}{6}\delta\dot{\gamma}_c(x(k), v(k)) + O(\delta^2)
\end{aligned} \tag{5.73}$$

The sampled system under the feedback action (5.73), exhibits the following two input–output behaviors

$$h(x_d(k+1)) = h(x_d(k)) + \delta L_fh(x_d(k)) + \frac{\delta^2}{2}v(k) + O(\delta^4)$$

and
$$L_fh(x_d(k+1)) = L_fh(x_d(k)) + \delta v(k) + O(\delta^3)$$

Thus, when $x_d(k) = x_c(t = k\delta) = x(k)$, the input–output reproduction at time $t = (k+1)\delta$ is verified up to a pre-fixed order of approximation with respect to δ, *i.e.*,

$$\begin{aligned}
h(x_c((k+1)\delta)) &= h(x_d(k+1)) + O(\delta^4)\\
L_fh(x_c(k+1)\delta)) &= L_fh(x_d(k+1)) + O(\delta^3)
\end{aligned}$$

The input–output reproduction at each step is then easily deduced, *i.e.* if $x_c(0) = x_d(0) = x(0)$, one obtains for $k \geq 0$ the equalities

$$h(x_c(k\delta)) = h(x_d(k)) + O(\delta^4)$$

and

$$L_fh(x_c(k\delta)) = L_fh(x_d(k)) + O(\delta^3)$$

It can also easily be verified that the transfer function associated to the input–output link $h \longleftrightarrow v$ of the discrete-time system (5.67-68-69), under the feedback action (5.73), is

$$H(z) = \frac{y_1}{v} = \frac{\delta^2(z+1)}{2(z-1)^2} + \mathrm{O}(\delta^4)$$

which is the sampled version of $H(s) = \dfrac{1}{s^2}$ up to an error in $\mathrm{O}(\delta^4)$.

Remark 5.13

Choosing

$$v(k) = \frac{-4h - 3\delta L_f h}{\delta^2} \quad \text{and} \quad v\left(k + \tfrac{1}{2}\right) = \frac{4h + \delta L_f h}{\delta^2}$$

one brings to zero the output h and its first time derivative $L_f h$ in one step of amplitude δ. Then, they are maintained at zero by setting $v = 0$.

Remark 5.14

It can be shown (Monaco and Normand-Cyrot, 1987) that the remaining dynamics, i.e. the zero dynamics associated both with the closed-loop continuous-time system under sampling and with the r-multirate closed-loop system when $v(k) = 0$, coincide up to an approximation of order r in δ, i.e. the minimum phase property and the internal stability are preserved at least up to an error of order $r + 1$ according to δ under such digital control schemes.

Remark 5.15

When $r = n$, one achieves the preservation of the full linearization property, at the sampling instants, under n-multirate feedback control. It results that when dealing with input-state linearization, one can proceed in the same way with respect to some dummy output function having a continuous-time relative degree equal to n.

5.6 Conclusion

The aim of this chapter was to set the bases enabling the digital implementation of feedback control schemes designed in the continuous-time domain without deteriorating the perfomances, at least at the sampling instants. SISO linear analytic continuous-time systems under linearizing feedback laws have been widely investigated. Depending on the relative degree associated with the continuous-time system ($r = 1$ or $r > 1$), single rate or multirate feedback control schemes have been proposed. These methods can easily be generalized to MIMO systems thus enabling the design of discrete-time control schemes preserving input–output decoupling properties, for example. In this multivariable case, the main difficulty being the choice of the multirate orders on each control action, the minimization of optimality criteria may be performed. Let us also note that the possibility of extra degrees of freedom on the control variables enables to be fulfilled, several requirements such as energy minimization, or technological constraints. With respect to the most usual digital implementation of continous-time control schemes by means of zero-order holding circuits, the present approach maintains the control objectives but also the minimum phase property and thus the local stability of the closed-loop system. With respect to higher-order holding circuits, the present approach is still based on the design of piecewise constant controls which are simple from a technological point of view. An important advantage of this digital feedback conception is also to increase notably the possible sampling period, so that the frequency for variable measures is much lower with respect to a continuous-time conception. This can be of practical interest when fixed sampling period or control frequency are imposed by the process or technological constraints. It is also important to stress that, in practice, approximated solutions are available in the sense that the solutions which are expressed as δ-dependent series can be truncated at a certain order. This considerably enlarges the applicability of the proposed method which is comparable to a continuous-time conception regarding the control computing time. In addition, because of specific structural properties of the continuous-time dynamics, it may occur that the series manipulated are finite and reduce to polynomials. This is very attractive because it allows the design of an exact digital control scheme on the basis of exact sampled models.

This chapter is introductory in the sense that several extensions of these methods are possible such as digital stabilization of controllable continuous-time systems, adaptive multirate control with parameter uncertainties and digital control of perturbed dynamics. The interested reader may consult the authors' references both for a theoretical presentation and the discussion of case studies. More precisely, concerning applications, benefits have been obtained for the stabilization of oscillatory responses due to flexibles modes in space or robotics domains (Georgiou *et al.*, 1992). This is mainly related to the fact that multirate techniques take into account the inter sampling behaviors. In relation with the possibility to synchronize sampling and measurement or control frequencies, improvements have been noted for mimicking gas jets control modes in aeronautics (Monaco and Normand-Cyrot, 1989) or pulse modulations in electrotechnology (Djemaï and Barbot, 1993; Georgiou *et al.*, 1992). Another interesting and simple application refers to the steering of mobile robots between two configurations (Chelouah *et al.*, 1993; Monaco and Normand-Cyrot, 1992). In this case, because exact sampling is possible, an exact multirate steering is achieved. This also applies to more general nonholonomic dynamics present in mechanics, space and robotics.

5.7 Bibliography

[1] ALBERTINI, F. and SONTAG, E.D. (1993), Discrete time transitivity and accessibility: analytic systems, *SIAM J. Cont. and Opt.*, **31**, 6, 1599–1627.

[2] ARAPOSTATHIS, A., JAKUBCZYK, B., LEE, H.G., MARCUS, S.I. and SONTAG, E.D. (1989), The effect of sampling on linear equivalence and feedback linearization, *Syst. and Contr. Letters*, **13**, 373–381.

[3] ASTRÖM, K.T. and WITTENMARK, B. (1984), *Computer Controlled Systems: Theory and Design*, Prentice-Hall.

[4] ASTRÖM, K.T., HAGANDER, P. and STERNBY, J. (1984), Zeros of sampled systems, *Automatica*, **20**, 21–30.

[5] BARBOT, J.P. (1989), A computer aided design for sampling a nonlinear system, *Applied Algebra, Algebraic Algorithms and Error Correcting Codes*, Lect. Notes in Comp. Sc., **357**, T. Mora (ed), 74–88.

[6] BARBOT, J.P. (1989), Méthodes de calcul appliquées aux systèmes non linéaires sous échantillonnage, Doctorate Thesis, Université de Paris Sud, Laboratoire des Signaux et Systèmes CNRS-ESE.

[7] BARBOT, J.-P., PANTALOS, N., MONACO, S. and NORMAND-CYROT, D. (1991), *Discretization schemes for nonlinear singularly perturbed systems*, Proc. 30th IEEE-CDC, Brighton, UK, 443–448.

[8] BYRNES, C. and ISIDORI, A. (1985), *A survey on recent developments in nonlinear control theory*, Proc. IFAC. Symp., SYROCO, Barcelone, 287–291.

[9] BYRNES, C. and ISIDORI, A. (1988), Local stabilization of critically minimum phase nonlinear systems, *Syst. and Cont. Letters*, **11**, 9–17.

[10] CHELOUAH, K., DI GIAMBERARDINO, P., MONACO, S. and NORMAND-CYROT, D. (1993), *Digital control of nonholonomic systems, two case studies*, Proc. 32nd IEEE-CDC, San Antonio, USA, 2664–2669.

[11] CLAUDE, D., FLIESS, M. and ISIDORI, A. (1983), Immersion directe et par bouclage d'un système non-linéaire dans un linéaire, *C.R. Acad. Sc., Paris*, **1**, 237–240.

[12] DJEMAÏ and M., BARBOT, J.-P. (1993), *Singularly perturbed methods for digital control of a synchronous motor*, Proc. IEEE Conf. on Syst. Man. and Cyb., Touquet, France.

[13] GEORGIOU, G., DI GENNARO, S., MONACO, S. and NORMAND-CYROT, D. (1991), *On the nonlinear adaptive control of a flexible spacecraft*, First ESA Int. Conf. on Spacecraft, Guidance, Navigation and Control Systems, ESTEC, Noordwijk.

[14] GEORGIOU, G., K. CHELOUAH, K., MONACO, S. and NORMAND-CYROT, D. (1992), *Nonlinear multirate adaptive control of a synchronous motor*, Proc. 31st IEEE-CDC, Tucson, USA, 3523–3528.

[15] GLAD, S.T. (1987), Output dead-beat control for nonlinear systems with one zero at infinity, *Syst. and Cont. Letters*, **9**, 249–255.

[16] GRIZZLE, J.W. (1993), A linear algebraic framework for the analysis of discrete-time nonlinear systems, *SIAM J. Cont. and Opt.*, **31**, 4, 1026–1044.

[17] GRIZZLE, J.W. and KOKOTOVIC, P.V. (1988), Feedback linearization of sampled-data systems, *IEEE Trans. on Automatic Control*, **33**, 857–859.

[18] GRÖBNER, W. (1973), *Serie di Lie e Loro Applicazioni*, Cremonese Ed., Rome.

[19] HANSELMANN, A. (1987), Implementation of digital controllers: a survey, *Automatica*, **23**, 7–32.

[20] HSU, P. and SASTRY, S.S. (1987), *The effect of discretized feedback in a closed-loop system*, Proc. 26th IEEE-CDC, Los Angeles, 1518–1523.

[21] ISIDORI, A. (1985), Nonlinear control systems, an introduction, *Lect. Notes on Cont. and Info. Sci.*, **72**, Springer-Verlag.

[22] LEE, H.G., ARASPOTATHIS, A. and MARCUS, S.I. (1987), *Remarks on discretization and linear equivalence of continuous-time nonlinear systems*, Proc. 26th IEEE-CDC, Los Angeles, 1783–1785.

[23] MONACO, S. and NORMAND-CYROT, D. (1985), *On the sampling of a linear analytic control system*, Proc. 24th IEEE-CDC, Fort Lauderdale, 1457–1462.

[24] MONACO, S. and NORMAND-CYROT, D. (1986), *On the linearizing feedback in nonlinear sampled data control scheme*, Proc. 25th IEEE-CDC, Athens, 2056–2060.

[25] MONACO, S. and NORMAND-CYROT, D., (1987), *Minimum phase nonlinear discrete-time systems and feedback stabilization*, Proc. 26th IEEE-CDC, Los Angeles.

[26] MONACO, S. and NORMAND-CYROT, D. (1988), *Multirate digital control*, Rep. D.I.S. University of Rome “La Sapienza”.

[27] MONACO, S. and NORMAND-CYROT, D. (1988), *Sur la commande digitale d’un système non linéaire à déphasage minimal*, Proc. 8th Int. Conf. Analysis and Optimization of Systems, Lect. Notes in Cont. and Info Sci., A. Bensoussan and J.L. Lions (eds), III, Springer-Verlag, Berlin, 193–205.

[28] MONACO, S. and NORMAND-CYROT, D. (1988), Zero dynamics of sampled nonlinear systems, *Syst. and Cont. Letters*, **11**, 229–234.

[29] MONACO, S. and NORMAND-CYROT, D. (1989), *Multirate three axis attitude stabilization of spacecrafts*, Proc. 28th IEEE-CDC, Tampa, USA.

[30] MONACO, S. and NORMAND-CYROT, D. (1991), Multirate sampling and zero dynamics, from linear to nonlinear, in *Nonlinear Synthesis*, C.I. Byrnes and A. Kurzhansky, (eds), Progress in Systems and Control Theory, Birkhäuser, Boston, 200–214.

[31] MONACO, S. and NORMAND-CYROT, D. (1991), *On the sampling of time delayed nonlinear systems*, Proc. ECC -91, Grenoble, Hermès, Paris, 556–560.

[32] MONACO, S. and NORMAND-CYROT, D. (1992), *An introduction to motion planning under multirate digital control*, Proc. 31st IEEE-CDC, Tucson, USA, 1780–1785.

[33] MONACO, S. and NORMAND-CYROT, D. (1995), A unifying representation for nonlinear discrete-time and sampled dynamics, *J. of Mathematics, Systems, Estimation and Control*, summary, **5**, 1, 103–106.

[34] REE, R. (1958), Lie elements and an algebra of shuffles, *Ann. of Maths*, **68**, 210–220.

CHAPTER 6

Applications of input–ouput linearization and decoupling under static state feedback

D. THOMASSET, E. RICHARD, B. CARON, S. SCAVARDA, A. CHARARA

Three physical processes – a chemical reactor, an electropneumatic servodrive and an active magnetic bearing – are considered. For each process, this chapter describes the design of linearizing control schematized by the scheme shown in figure 6-1

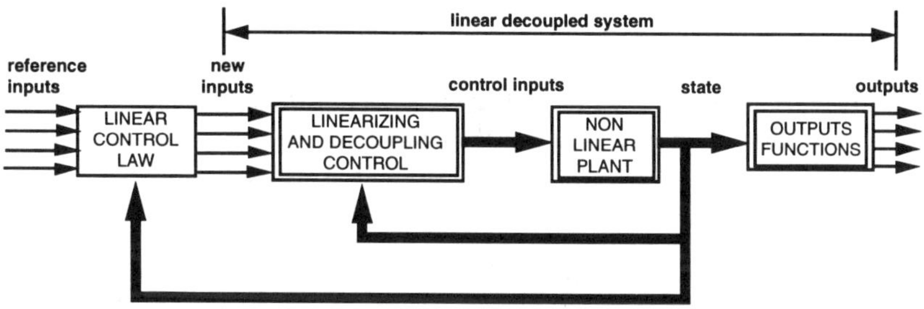

Figure 6.1: Control structure with two loops.

Only the continuous-time synthesis of each internal loop is presented. Then, a discrete-time form of this control law is obtained. The design of the external loops is achieved, directly in discrete-time, using classical linear methods (pole placement, optimal control, robust control, *etc.*)

6.1 Chemical reactor

6.1.1 Presentation of the studied pilot plant

The process investigated was a neutralization reaction of a strong acid (HCl, concentration C_A) by addition of a strong base (NaOH, concentration, C_B) in an aqueous solution.

The acid flow-rate Q_A and the base flow-rate Q_B can be modulated by means of two gates. These gates are controlled by two local PID in order to eliminate the effect of the hysteresis. A pH meter gives the H^+ concentration and the volume V in the reactor is measured by means of a pressure sensor. A microcomputer controls the plant and gives the reference inputs of acid and base flow-rate local PID controllers.

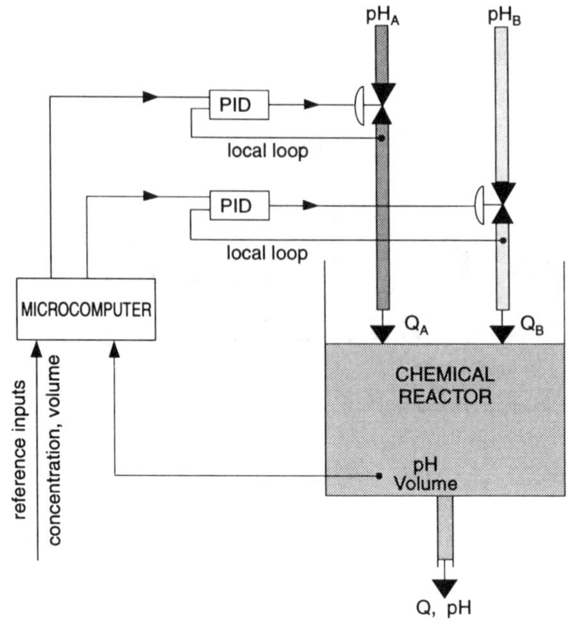

Figure 6.2: Neutralization chemical reactor.

The neutralization is assumed to be instantaneous, with sufficient agitation in the reactor. The electroneutrality of the solution gives the equation

$$\frac{\mathrm{d}\,[C(t)V(t)]}{\mathrm{d}t} = C_A Q_A(t) - C_B Q_B(t) - C(t)Q(t) \tag{6.1}$$

where $Q(t)$ and $C(t)$ represent, respectively, the reactor output flow-rate and the concentration.

A second equation is obtained from a volume balance

$$\frac{\mathrm{d}V(t)}{\mathrm{d}t} = Q_A(t) + Q_B(t) - Q(t) \tag{6.2}$$

If we consider a constant opening of the output gate, the reactor output flow rate may be expressed by

$$Q(t) = \hat{\alpha}\sqrt{V(t)} \quad \text{with } \hat{\alpha} \text{ being the gate opening coefficient} \tag{6.3}$$

Equation (6.1) may be rewritten as

$$C(t)\frac{\mathrm{d}V(t)}{\mathrm{d}t} + V(t)\frac{\mathrm{d}C(t)}{\mathrm{d}t} = C_A Q_A(t) - C_B Q_B(t) - C(t)Q(t)$$

then, using (6.2) and (6.3), the nonlinear state model of the reactor may be obtained:

$$\begin{cases} \dfrac{\mathrm{d}C(t)}{\mathrm{d}t} = \dfrac{1}{V(t)}\left[C_A Q_A(t) - C_B Q_B(t) - C(t)\left(Q_A(t) + Q_B(t)\right)\right] \\ \dfrac{\mathrm{d}V(t)}{\mathrm{d}t} = Q_A(t) + Q_B(t) - \hat{\alpha}\sqrt{V(t)} \end{cases} \tag{6.4}$$

6.1.1.1 Validity of this model

In fact, there is no concentration sensor and so its control uses a pH meter. Therefore, the pH measure depends on the temperature, the weak acid concentrations, the dielectric constant of the solution, the solvent density and the salt effect (Jutila, 1983). Differences between the process and the proposed model occur for concentrations greater than 0,001 mol/l (pH 3) (Paskiewicz, 1987). To take these phenomena into account, heuristic laws could be used (Gustafson and Waller, 1983) with the measurement of the CO_2 concentration. However, this measurement is never done on the industrial process. Therefore, the dynamical model (6.4) of the neutralization reactor may be used for the acid concentrations of the aqueous solution (pH 0-3). The effects of these previous problems will be discussed in section 6.1.5.

6.1.2 Input-output linearization and decoupling by static state feedback

The nonlinear model of the chemical reactor may be expressed in the form

$$\begin{cases} \dot{x} &= f(x) + \sum_{i \in m} g_i(x) u_i \\ y_1 &= C(t) \\ y_2 &= V(t) \end{cases} \tag{6.5}$$

Q_A and Q_B are the two controls and C and V are the two outputs. Thus, it is a square model ($m = 2$). In the basis $\frac{\partial}{\partial C}, \frac{\partial}{\partial V}$ the vector fields of this system are:

$$\begin{aligned} f &= \begin{bmatrix} 0 \\ -\hat{\alpha}\sqrt{V(t)} \end{bmatrix} \\ g_1(x) &= \begin{bmatrix} \frac{1}{V(t)}(C_A - C(t)) \\ 1 \end{bmatrix} \\ g_2(x) &= \begin{bmatrix} -\frac{1}{V(t)}(C_B + C(t)) \\ 1 \end{bmatrix} \end{aligned} \tag{6.6}$$

The state

$$x = \begin{bmatrix} C(t) \\ V(t) \end{bmatrix}$$

is in a two-dimensional manifold locally equivalent to $\mathbb{R}^+ \times \mathbb{R}^{*+}$. The equilibrium point $f(x) = 0$ such that $V(t) = 0$ (reactor empty without controls) is also a singular point for the vector fields g_1 and g_2. Therefore, this point is not a practical working point.

To compute the characteristic indexes ρ_1, ρ_2 and the linearizing control, we use the Lie derivatives:

$$\begin{aligned} L_f &= \sum_{i=1}^{n} f_i(x)\frac{\partial}{\partial x_i} = -\hat{\alpha}\sqrt{V(t)}\frac{\partial}{\partial V} \\ L_{g_1} &= \sum_{i=1}^{n} g_{1i}(x)\frac{\partial}{\partial x_i} = \frac{C_A - C(t)}{V(t)}\frac{\partial}{\partial C} + \frac{\partial}{\partial V} \\ L_{g_2} &= \sum_{i=1}^{n} g_{2i}(x)\frac{\partial}{\partial x_i} = -\frac{C_B + C(t)}{V(t)}\frac{\partial}{\partial C} + \frac{\partial}{\partial V} \end{aligned} \tag{6.7}$$

Now, we compute the Lie derivatives of the output function.

$$\begin{aligned} L_{g_1}h_1(x) &= \frac{C_A - C(t)}{V(t)} \quad ; \quad & L_{g_2}h_1(x) &= -\frac{C_B + C(t)}{V(t)} \\ L_{g_1}h_2(x) &= 1 \quad ; \quad & L_{g_2}h_2(x) &= 1 \end{aligned}$$

The characteristic indexes ρ_1 and ρ_2 are equal to 1 and the decoupling matrix $\Delta(x)$ may be expressed as

$$\Delta(x) = \begin{bmatrix} L_{g_1}L_f^{\rho_1-1}h_1(x) & L_{g_2}L_f^{\rho_1-1}h_1(x) \\ L_{g_1}L_f^{\rho_2-1}h_2(x) & L_{g_2}L_f^{\rho_2-1}h_2(x) \end{bmatrix} = \begin{bmatrix} \dfrac{C_A - C(t)}{V(t)} & -\dfrac{C_B + C(t)}{V(t)} \\ 1 & 1 \end{bmatrix} \tag{6.8}$$

The singular point $V(t) = 0$ being outside the physical domain, the rank of the $\Delta(x)$ matrix is equal to two in the whole physical domain (because $C_A + C_B > 0$). There is no singularity.

The inverse of $\Delta(x)$ is

$$\Delta^{-1}(x) = \begin{bmatrix} \dfrac{V(t)}{C_A + C_B} & \dfrac{C_B + C(t)}{C_A + C_B} \\ \dfrac{-V(t)}{C_A + C_B} & \dfrac{C_A - C(t)}{C_A + C_B} \end{bmatrix} \tag{6.9}$$

Then, we obtain the linearizing control: $u = \alpha(x) + \beta(x)w$ with

$$\alpha(x) = -\Delta^{-1}(x)\begin{bmatrix} L_f^{\rho_1}h_1(x) \\ L_f^{\rho_2}h_2(x) \end{bmatrix} = -\Delta^{-1}(x)\begin{bmatrix} 0 \\ -\hat{\alpha}\sqrt{V(t)} \end{bmatrix}$$

which is equal to

$$\alpha(x) = \begin{bmatrix} \hat{\alpha}\sqrt{V(t)}\left(\dfrac{C_B + C(t)}{C_A + C_B}\right) \\ \hat{\alpha}\sqrt{V(t)}\left(\dfrac{C_A - C(t)}{C_A + C_B}\right) \end{bmatrix} \tag{6.10}$$

and

$$\beta(x) = \Delta^{-1}(x) \tag{6.11}$$

The two control laws versus the new inputs $w_1(t)$ and $w_2(t)$ may be obtained in the form

$$\begin{aligned} Q_A(t) &= \frac{\hat{\alpha}\sqrt{V}(C_B + C(t))}{C_A + C_B} + \frac{V(t)}{C_A + C_B} w_1(t) + \frac{C_B + C(t)}{C_A + C_B} w_2(t) \\ Q_B(t) &= \frac{\hat{\alpha}\sqrt{V}(C_A - C(t))}{C_A + C_B} - \frac{V(t)}{C_A + C_B} w_1(t) + \frac{C_A - C(t)}{C_A + C_B} w_2(t) \end{aligned} \tag{6.12}$$

Then, the closed-loop system is equal to two SISO integrators

$$\begin{cases} \dfrac{dC(t)}{dt} = w_1(t) \\ \dfrac{dV(t)}{dt} = w_2(t) \end{cases} \tag{6.13}$$

Notice that, in our case the dimension of the unobservable submanifold obtained by the application of this control law is equal to zero ($n - \Sigma\rho_i = 0$).

6.1.3 *Experimental results*

6.1.3.1 Nonlinear control law

Using the results obtained by Monaco and Normand-Cyrot (chapter 5), a discrete-time control law can be computed. Therefore, in our case the sampling period chosen is very small, then the control designed in continuous-time is implemented in discrete-time on the microcomputer without any correcting terms. Figure 6.3 represents the control structure of the chemical reactor.

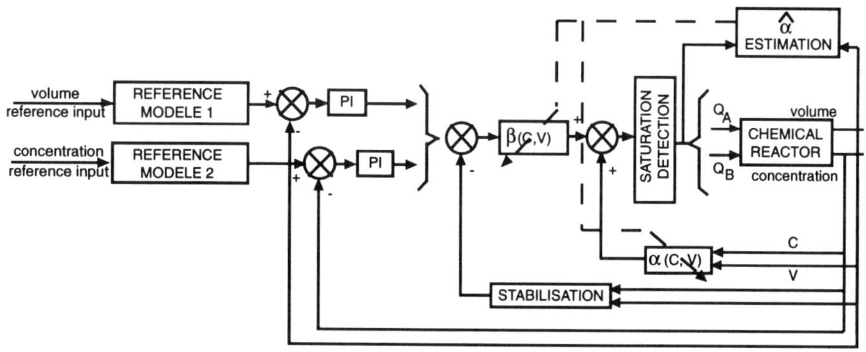

Figure 6.3: Chemical reactor control scheme.

The inner loop performs the linearization and the decoupling. The asymptotic stability of each SISO subsystem is obtained by means of a second loop (pole placement, optimal quadratic control, *etc.*). The effect of control inputs saturation is included in the control law and the on-line estimation of the $\hat{\alpha}$ coefficient is also used.

6.1.3.2 Practical results

Figure 6.4 shows the performances obtained for a starting phase of the plant corresponding to the reference inputs $V = 22$ l and pH $= 2.7$ with this control scheme. These results can be qualified as rather satisfactory because they are obtained when the plant is started. These figures also show the decoupling effect of the control: a changing value of the concentration set point weakly affects the reactor volume. The results shown in figure 6.5 are obtained for smaller variations of reference inputs and show the decoupling effect on the concentration when the volume input reference is modified. In this case, a pH variation can be noticed, then the plant outputs are not totally decoupled, which is due to the modeling approximations previously mentioned.

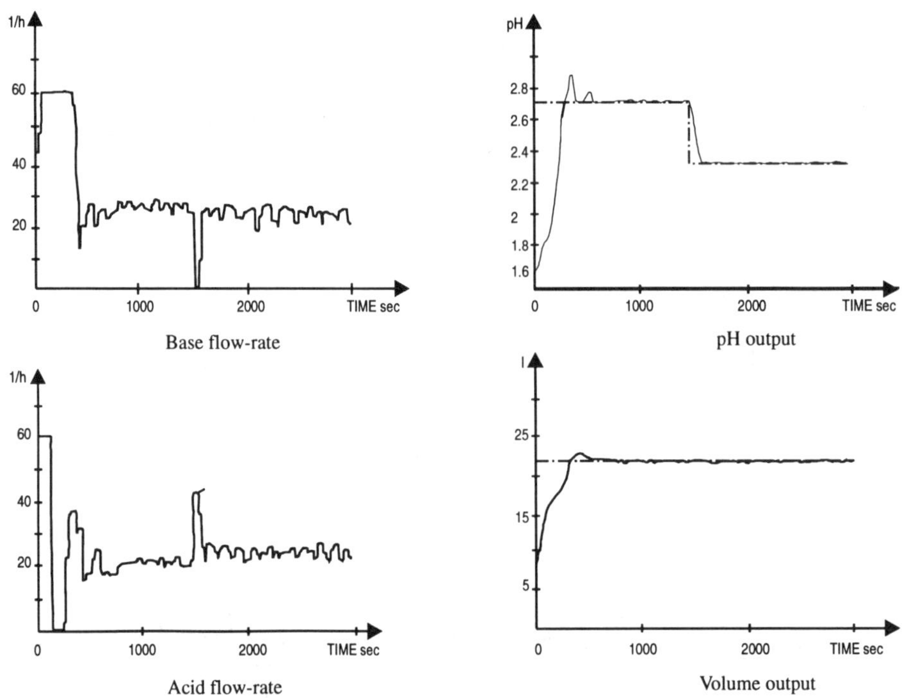

Figure 6.4: pH and volume variations during the starting phase.

6.1.4 Discussion

The nonlinear control law operates satisfactorily over a wide range of working conditions and this is the main result of this example. Nevertheless, the validity of the proposed model is reduced (*cf.* section 6.1.2). For long-period experiments (longer than 1-2 hours), the unmodeled phenomena lead to the modification of the system closed-loop behavior as shown in figure 6.6.

To solve this problem, fuzzy PI controllers are implemented on each SISO subsystem obtained by means of the inner loop. The presented results point out the improvements obtained with this method (Paszkiewicz, 1987).

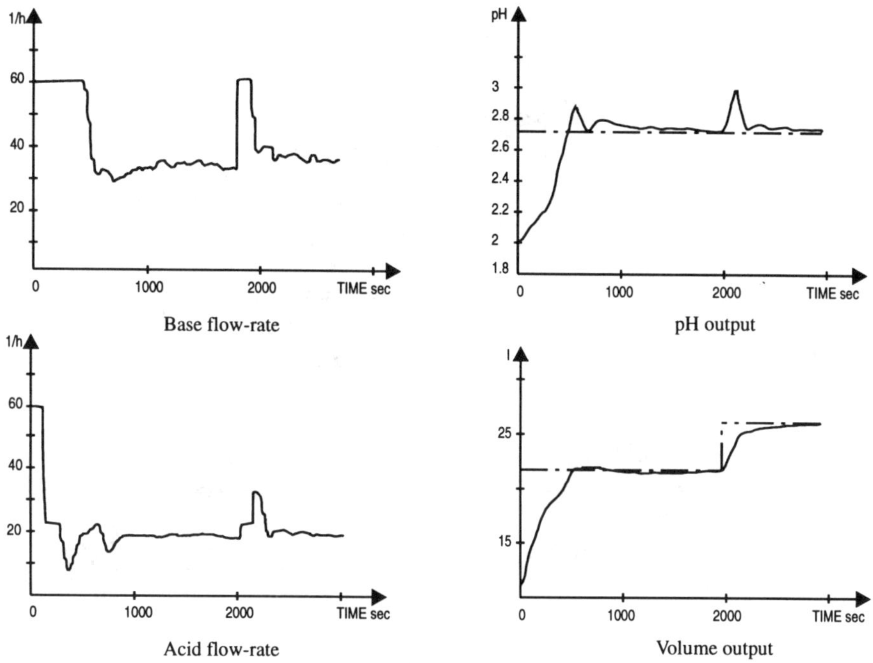

Figure 6.5: pH and volume responses.

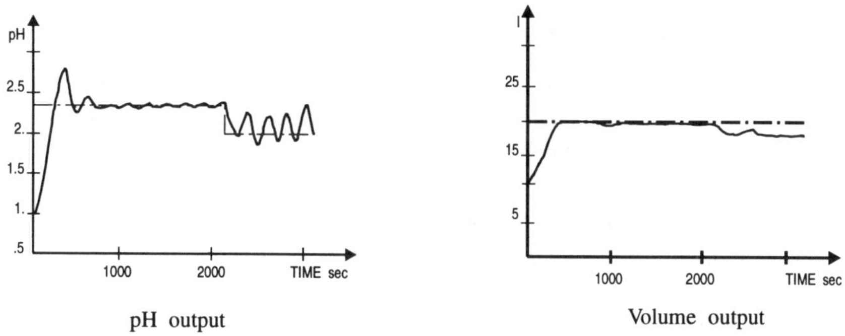

Figure 6.6: pH and volume responses.

6.2 Position control of electropneumatic actuators

6.2.1 The experimental plant

The system under consideration (fig. 6.7) is composed of a rotary vane actuator with an inertial load (the resistive torque depends on the position), an interface, p controlled by the current i_p, an interface, n controlled by the current i_n, the required sensors and a microcomputer. The two control variables i_p and i_n are assumed to be equal when one five-way interface is used.

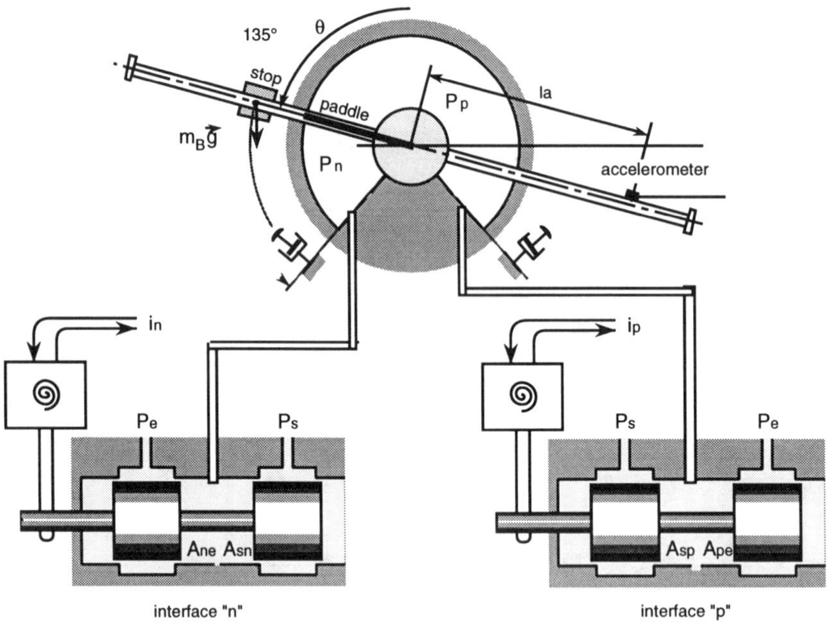

Figure 6.7: Technological scheme of the studied system.

6.2.2 *Nonlinear model*

6.2.2.1 Hypothesis and state model

The electropneumatic system model can be obtained by using the three laws: the mass flow rate through a restriction; the pressure behavior in an open chamber with variable volume; and the mechanics of the system.

(*i*) the expression of the mass flow rate through a restriction can be written in the form (6.14): (Comolet, 1979; Mc Cloy and Martin, 1980). If the flow is turbulent, the mass flow rate is positive when P_1 is greater than P_2. A is the orifice area. For the interface restrictions, this area is a function of the control currents. $D(P_1, P_2)$ depends on the chamber pressures. This function can be deduced from thermodynamic laws (Comolet, 1979) including a correcting coefficient to take into account the orifice shape or experimental data. In this last case, the function is tabulated.

$$\dot{m} = A \cdot D(P_1, P_2) \tag{6.14}$$

(*ii*) To establish the corresponding expression, for the pressure behavior in an open chamber with variable volume, the following assumptions are usually made: the pressure and the temperature are homogeneous in the chamber, and the air is a perfect gas. Two kinds of models can be obtained: the first one uses a mass balance and an internal energy balance (Shearer, 1956; Kain andWartelle, 1973; Jebar, 1977) the second one uses a mass balance and an evolution law (Andersen, 1976; Burrows, 1972; Zalmanzon, 1965). In the first approach, the temperature is a state variable and a thermal transfer law is necessary. Therefore, we prefer the second kind of model with a polytropic process (k coefficient). (Andersen, 1976). We also assume that the temperature changes are neglected with respect to the mean temperature Ts and these assumptions lead to

$$\frac{\mathrm{d}P}{\mathrm{d}t} = \frac{krT_s}{V} \cdot \left(\sum_i \dot{m}_i - \frac{P}{rT_s} \cdot \frac{\mathrm{d}V}{\mathrm{d}t} \right) \tag{6.15}$$

with r the perfect gas constant, V the chamber volume, P the chamber pressure and $\sum_i \dot{m}_i$: total flows entering and leaving the chamber (positive if entering).

(*iii*) The mechanical equation is

$$K_c. (P_p - P_n) - C_r(\theta) = J.\frac{\mathrm{d}^2\theta}{\mathrm{d}t^2} \tag{6.16}$$

with K_c the actuator constant torque, $C_r(\theta)$ the resistive torque function of the position and J the inertia.

(*iv*) With the mass flow rate distribution (fig. 6.8), the flow rate law, the pressure law (6.15) for each chamber and the mechanical equation (6.16), we obtain the state space equations

$$\begin{aligned} \frac{\mathrm{d}P_p}{\mathrm{d}t} &= \frac{krT_s}{(V_0 + K_c.\theta)} \left[\dot{m}_p(i_p, P_p) - \dot{m}_v(P_p, P_n) - \frac{P_p}{rT_s} K_c \Omega \right] \\ \frac{\mathrm{d}P_n}{\mathrm{d}t} &= \frac{krT_s}{(V_0 - K_c.\theta)} \left[\dot{m}_n(i_n, P_n) + \dot{m}_v(P_p, P_n) + \frac{P_n}{rT_s} K_c \Omega \right] \\ \frac{\mathrm{d}\Omega}{\mathrm{d}t} &= \frac{1}{J} \left[K_c(P_p - P_n) - C_r(\theta) \right] \\ \frac{\mathrm{d}\theta}{\mathrm{d}t} &= \Omega \end{aligned} \tag{6.17}$$

with the following notations:

$$\begin{aligned} V_p(\theta) &= V_0 + K_c\theta \qquad V_n(\theta) = V_0 - K_c\theta \\ \dot{m}_p(i_p, P_p) &= A_{sp}(i_p) D(P_s, P_p) - A_{pe}(i_p) D(P_p, P_e) \\ \dot{m}_n(i_n, P_n) &= A_{sn}(i_n) D(P_s, P_n) - A_{ne}(i_n) D(P_n, P_e) \end{aligned}$$

where V_0 is the sum of the piping volume and each actuator chamber in the central position ($\theta = 0$).

The two interface currents i_p and i_n are the control variables.

6.2.2.2 Adaptation of the nonlinear state model to control design inputs

To obtain an explicit control law, a linear model with respect to the control inputs is necessary. In our case, a justified assumption (flow rate leakages independent of the interface currents i_p and i_n) leads to two new control variables A_p^* and A_n^* which appear linearly in the state equations. To introduce these new control variables a decomposition of each orifice area is made. An orifice area is assumed to be the sum of a constant leakage area (denoted A_{fp} or A_{fn}) and a control area (denoted A_p^* or A_n^*) modulated by the corresponding current (i_p or i_n). Then, the input actuator mass flow rates can be written

$$\begin{aligned} \dot{m}_p(i_p, P_p) &= -\dot{m}_{Fp}(P_p) + \varphi_p\left(P_p, \mathrm{sgn}\left(A_p^*\right)\right).A_p^* \\ \dot{m}_n(i_n, P_n) &= -\dot{m}_{Fn}(P_n) - \varphi_n\left(P_n, \mathrm{sgn}\left(A_n^*\right)\right) \cdot A_n^* \end{aligned} \tag{6.18}$$

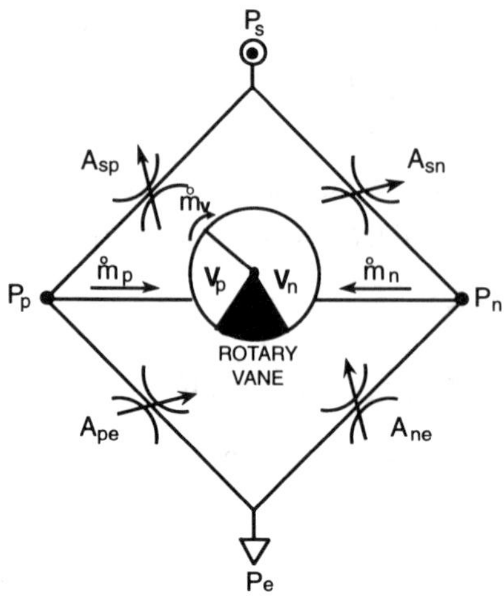

Figure 6.8: Bridge lay-out of a pneumatic stage.

with the notations

$$A_p^* = \psi_p(i_p) = \begin{cases} A_{sp}(i_p) - A_{fp} & \text{if} \quad i \geq i_{p0} \\ -(A_{pe}(i_p) - A_{fp}) & \text{if} \quad i < i_{p0} \end{cases}$$

$$\varphi_p\left(P_p, \text{sgn}\left(A_p^*\right)\right) = \begin{cases} D(P_s, P_p) & \text{if} \quad A_p^* \geq 0 \\ D(P_p, P_e) & \text{if} \quad A_p^* < 0 \end{cases}$$

where the leakage area A_{fp} and the current i_{p0} are obtained by solving the equation :

$$A_{sp}(i_{p0}) = A_{pe}(i_{p0}) = A_{fp}.$$

$$A_n^* = \psi_n(i_n) = \begin{cases} A_{ne}(i_n) - A_{fn} & \text{if} \quad i_n \geq i_{n0} \\ -A_{sn}(i_n) - A_{fn} & \text{if} \quad i_n < i_{n0} \end{cases}$$

$$\varphi_n\left(P_n, \text{sgn}\, A_n^*\right) = \begin{cases} D(P_n, P_e) & \text{if} \quad A_n^* \geq 0 \\ D(P_s, P_n) & \text{if} \quad A_n^* < 0 \end{cases}$$

Similarly the leakage area A_{fn} and the current i_{n0} of the other interface are given by the solution of

$$A_{sn}(i_{n0}) = A_{ne}(i_{n0}) = A_{fn}$$

The servovalve leakage mass flow rate can be expressed as

$$\dot{m}_{Fx} = A_{Fx}.\left(D\left(P_x, P_e\right) - D\left(P_s, P_x\right)\right)$$

$$x = p, n$$

The previous model (6.17) simplification introduces the functions φ_p and φ_n depending on the the sign of A_p^* and A_n^*. However, this non derivability of φ_p and φ_n is on the control variables and then introduces no restrictions. Using (6.18) in the state equation (6.17), gives the model

$$\dot{x} = f(x) + g_1(x)A_p^* + g_2(x)A_n^*.$$

With a state vector $x = [P_p P_n \Omega\, \theta]$ and in the basis $\left(\frac{\partial}{\partial P_p}\frac{\partial}{\partial P_n}\frac{\partial}{\partial \Omega}\frac{\partial}{\partial \theta}\right)$, the coordinate of the vector fields f, g_1 and g_2 are

$$f = \begin{bmatrix} \frac{krT_s}{V_p(\theta)}\left[-\dot{m}_v(P_p, P_n) - \dot{m}_{Fp}(P_p) - \frac{P_p}{rT_s}K_c\Omega\right] \\ \frac{krT_s}{V_n(\theta)}\left[\dot{m}_v(P_p, P_n) - \dot{m}_{Fn}(P_n) + \frac{P_n}{rT_s}K_c\Omega\right] \\ \frac{1}{J}\left[K_c(P_p - P_n) - C_r(\theta)\right] \\ \Omega \end{bmatrix}$$

$$g1 = \begin{bmatrix} \frac{krT_s}{V_p(\theta)}\varphi_p\left(P_p, \operatorname{sgn}\left(A_p^*\right)\right) \\ 0 \\ 0 \\ 0 \end{bmatrix} \qquad g2 = \begin{bmatrix} 0 \\ -\frac{krT_s}{V_n(\theta)}\varphi_n\left(P_n, \operatorname{sgn}\left(A_n^*\right)\right) \\ 0 \\ 0 \end{bmatrix} \tag{6.19}$$

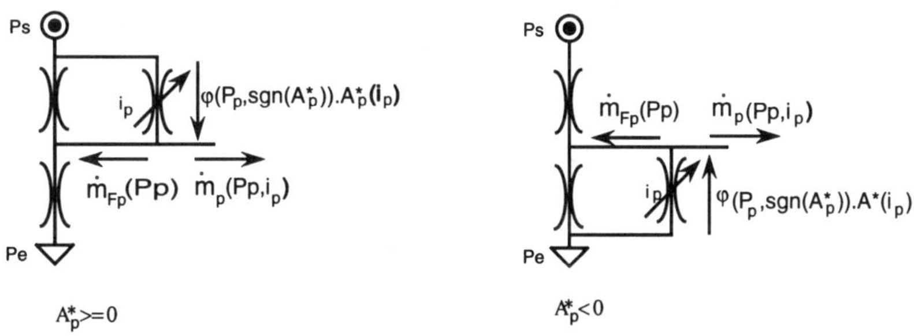

Figure 6.9: Decomposition of the interface "p" flow rates.

6.2.3 *Control law synthesis*

This section now considers an input–output linearizing control for a plant with only one five-way servovalve and a linearizing and decoupling control for the same plant with two three-way servovalves interfaces.

6.2.3.1 SISO case

6.2.3.1.1 Control design

In this case, symmetrical assumptions lead to $A^* = A_p^* = A_n^* = \psi(i)$. Then there is only one control variable and consequently one control vector field: $g = g1 + g2$. The theoretical result (*cf.* input–output decoupling, Chapter 2) gives the existence conditions and the synthesis control method. The Lie differential operators L_f and L_g are

$$\begin{aligned} L_f &= \frac{krT_s}{V_p}\left[-\dot{m}_v - \dot{m}_{Fp} - \frac{P_p}{rT_s}K_c\Omega\right]\frac{\partial}{\partial P_p} + \frac{krT_s}{V_n}\left[\dot{m}_v - \dot{m}_{Fn} + \frac{P_n}{rT_s}K_c\Omega\right]\frac{\partial}{\partial P_n} \\ &\quad + \frac{1}{J}\left[K_c(P_p - P_n) - C_r(\theta)\right]\frac{\partial}{\partial \Omega} + \Omega\frac{\partial}{\partial \theta} \qquad (6.20) \\ L_g &= \frac{krT_s}{V_p}\varphi_p\left(P_p, \mathrm{sgn}\left(A^*\right)\right)\frac{\partial}{\partial P_p} - \frac{krT_s}{V_n}\varphi_n\left(P_n, \mathrm{sgn}\left(A^*\right)\right)\frac{\partial}{\partial P_n} \end{aligned}$$

The Lie derivatives of the output function $y = h(x) = \theta$ along the vector field f are given by:

$$\begin{aligned} L_f h &= \Omega & L_g h &= 0 \\ L_f^2 h &= \frac{1}{J}\left[K_c\left(P_p - P_n\right) - C_r\left(\theta\right)\right] & L_g L_f h &= 0 \\ L_g L_f^2 h &= \Delta\left(P_p, P_n, \theta, \mathrm{sgn}(A^*)\right) \neq 0 \end{aligned}$$

A characteristic index ρ equal to 3 is obtained, then the A^* variable directly influences the third position time derivative (the jerk). The decoupling matrix $\Delta(x)$ (scalar in this case) becomes:

$$\Delta(P_p, P_n, \theta, \mathrm{sgn}A^*) = \frac{K_c.krT_s}{J}\left[\frac{\varphi_p(P_p, \mathrm{sgn}A^*)}{V_p(\theta)} + \frac{\varphi_n(P_n, \mathrm{sgn}A^*)}{V_n(\theta)}\right] \qquad (6.21)$$

The vector of the natural feedbacks is calculated by:

$$\Delta_0(x) = L_f^3 h = -a.\left[d_f\left(P_p, P_n, \theta\right) + d_u\left(P_p, P_n, \theta, \Omega\right) + d_r\left(\theta, \Omega\right)\right] \qquad (6.22)$$

with the notations

$$\begin{aligned} \mathrm{d}_f\left(P_p, P_n, \theta\right) &= \dot{m}_v\left(P_p, P_n\right)\left[\frac{V_o}{V_p(\theta)} + \frac{V_o}{V_n(\theta)}\right] \\ &\quad + \left[\dot{m}_{F_p}\left(P_p\right)\frac{V_o}{V_p(\theta)} - \dot{m}_{F_n}\left(P_n\right)\frac{V_o}{V_n(\theta)}\right] \\ \mathrm{d}_u\left(P_p, P_n, \theta, \Omega\right) &= \frac{K_c}{rT_s}\left[P_p\frac{V_o}{V_p(\theta)} + P_n\frac{V_o}{V_n(\theta)}\right]\Omega \\ \mathrm{d}_r(\theta, \Omega) &= \frac{V_o}{krT_s}\cdot\frac{1}{K_c}\cdot\frac{\mathrm{d}C_r}{\mathrm{d}\theta}\cdot\Omega \\ a &= \frac{K_c}{J}\cdot\frac{krT_s}{V_o} \end{aligned}$$

where V_0 is an arbitrary reference volume. Then the control law has the form

$$\begin{aligned} A^* &= \frac{1}{L_g L_f^2 h}\left[-L_f^3 h + v\right] = \alpha\left(x\right) + \beta\left(x\right)v = -\Delta^{-1}(x)\Delta_0(x) + \Delta^{-1}(x)v \\ A^* &= \frac{a}{\Delta\left(P_p, P_n, \theta, \mathrm{sgn}\,A^*\right)}\left[\mathrm{d}_f\left(P_p, P_n, \theta\right) + \mathrm{d}_u\left(P_p, P_n, \theta, \Omega\right) + \mathrm{d}_r\left(\theta, \Omega\right) + \frac{v}{a}\right] \qquad (6.23) \end{aligned}$$

By using the reciprocal function $i = \psi^{-1}\left(A^*\right)$, the input servovalve current can be calculated.

6.2.3.1.2 *Physical interpretation*

In the expression of A^* (6.23), all the terms inside the right-hand side bracket have the dimension of a mass flow rate. The triple integrator input v is the jerk, the unit of the parameter "a" is a jerk under mass flow rate. Each term of the right-hand side bracket can be interpreted as follows:

- ***The matrix decoupling*** $\Delta(P_p, P_n, \theta, \text{sgn}(A^*))$
  This expression depends on the sign of the input A^*. Then, there exist two different operating modes for $A^* > 0$ and $A^* < 0$. Notice that a continuous transition between the two operating modes will appear if the change of flow type is taken into account in the vicinity of the middle position of the interface spool. In this case, an algebraic equation can be solved in order to obtain the control.
- ***Mass flow rate due to the rotary vane movement*** $\mathrm{d}u(P_p, P_n, \theta, \Omega)$
  This term, proportional to the velocity, introduces a mass flow rate which compensates, on the effort, the effect of the actuator chamber volume changes.
- ***Mass flow rate due to the variable perturbation torque*** $\mathrm{d}_r(\theta, \Omega)$
  This term takes into account the influence, on the jerk, of the perturbation torque changes.
- ***Compressibility mass flow rate*** $\mathrm{d}_c = v/a$
  This part of the mass flow rate is used for the acceleration variations.
- ***Leakage mass flow rate*** $\mathrm{d}f$
  This term takes into account the internal actuator leakage mass flow rate and the leakage mass flow rate of the servovalve flow stage. During the piston movement, this term is negligible. At the equilibrium, this leakage mass flow rate compensation introduces an input current i_0 in order to obtain the differential pressure which is necessary to balance the perturbation torque. Without this compensation, this current i_0 is obtained with a static position error.

6.2.3.1.3 *The validity domain*

Singular points

The decoupling matrix (scalar in this case) $\Delta(x)$ is generally positive. However, for some particular systems and under exceptionnal conditions, this coefficient can be canceled or have a sign change, due to the inversion of the mass flow rate in one or two of the variable orifices of the Wheatstone bridge. This problem occurs during the decelerating phase with very big inertial loads. In this case, the interface is closed when $\Delta(x)$ is negative in order to get through this singularity.

Unobservable

If the characteristic index is less than the space manifold dimension, then, a one-dimensional submanifold is obtained in closed-loop. Pressure P_p can be chosen to span this submanifold.

A diffeomorphism is used in order to obtain normal form of the closed-loop. Taking as new state variables, the output and its two first derivatives along the vector field f completed by an independent function, leads to a local diffeomorphism. Then the chosen diffeomorphism is

$$\theta = h \qquad \Omega = L_{\mathrm{f}}h \qquad \gamma = L_{\mathrm{f}}^2 h \qquad P_p = P_p \tag{6.24}$$

It is easy to verify that this previous transformation is a local diffeomorphism. Therefore, these new coordinates do not lead to the classical normal form because the control input A^* appears in the pressure state equation $P_p(L_g P_p \neq 0)$. In these new coordinates, the closed-loop state equations become

$$\Sigma_L \begin{cases} \dfrac{d\theta}{dt} &= \Omega \\ \dfrac{d\Omega}{dt} &= \gamma \\ \dfrac{d\gamma}{dt} &= v \end{cases} \tag{6.25}$$

$$\Sigma_{NL} \; \frac{dP_p}{dt} = \frac{krT_s}{V_n V_p} \cdot \frac{1}{\frac{\varphi_p}{V_p} + \frac{\varphi_n}{V_n}} \cdot \Big\{ \dot{m}_v . (\varphi_p - \varphi_n) - \dot{m}_{F_n}\varphi_p - \dot{m}_{F_p}\varphi_n$$
$$+ (\varphi_p P_n - \varphi_n P_p) \frac{K_c}{rT_s} \cdot \Omega + \varphi_p \frac{V_n}{V_o} \left[\frac{V_o}{krT_s} \cdot \frac{1}{K_c} \frac{dC_r}{d\theta} \Omega + \frac{v}{a} \right] \Big\}$$

with $\quad P_p - P_n = [J\gamma + C_r(\theta)/K_c]$

Figure 6.10 points out the linearizing part and the unobservable part due to the control law. The asymptotic stability at the first-order of the zero dynamic (pressure P_p) can be proved with simplifying assumptions (Richard, 1990) .

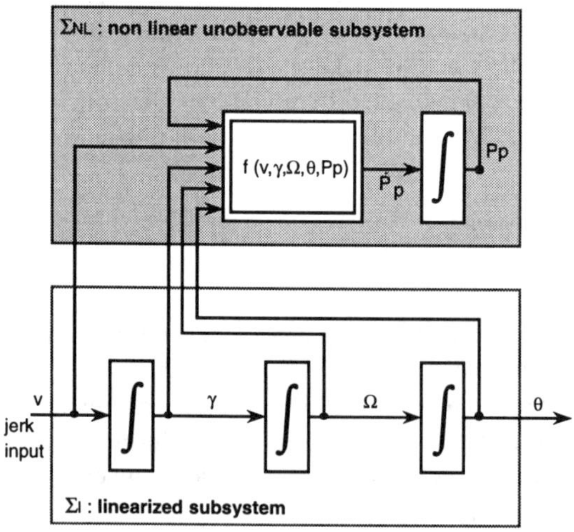

Figure 6.10: Schematic representation of the closed-loop system.

6.2.3.2 Multivariable case

6.2.3.2.1 The control design

Now a pressure/ position decoupling is investigated: the first output function $h_1(x)$ is P_p and the second output function $h_2(x)$ is the position θ. For the model (6.19) described in section 6.2.2.2, theoretical results on linearization and decoupling are used (*cf.* Chapter 2) in order to design a control law. The Lie differential operators are

$$\begin{aligned} L_f &= \frac{krT_s}{V_p}\left[-\dot{m}_v - \dot{m}_{Fp} - \frac{P_p}{rT_s}K_c\Omega\right]\frac{\partial}{\partial P_p} + \frac{krT_s}{V_n}\left[\dot{m}_v - \dot{m}_{Fn} + \frac{P_n}{rT_s}K_c\Omega\right]\frac{\partial}{\partial P_n} \\ &\quad + \frac{1}{J}\left[K_c\left(P_p - P_n\right) - C_r\left(\theta\right)\right]\frac{\partial}{\partial \Omega} + \Omega\frac{\partial}{\partial \theta} \\ L_{g1} &= \frac{krT_s}{V_p}\varphi_p\left(P_p, \operatorname{sgn}\left(A^*\right)\right)\frac{\partial}{\partial P_p} \quad L_{g2} = -\frac{krT_s}{V_n}\varphi_n\left(P_n, \operatorname{sgn}\left(A^*\right)\right)\frac{\partial}{\partial P_n} \end{aligned} \tag{6.26}$$

For the output h_1, the characteristic index ρ_1 is equal to one and for the second output h_2, the characteristic index ρ_2 is equal to three. In a closed-loop, a triple integrator for the position and an integrator for the pressure are obtained. Therefore, there is no unobservable submanifold . The decoupling matrix $\Delta(x)$ is given by

$$\begin{aligned} \Delta(x) &= \begin{bmatrix} L_{g_1}L_f^0 h_1 & L_{g_2}L_f^0 h_1 \\ L_{g_1}L_f^2 h_2 & L_{g_2}L_f^2 h_2 \end{bmatrix} \\ &= \begin{bmatrix} \dfrac{krT_s}{V_p(\theta)}\varphi_p(P_p, \operatorname{sgn}(A_p^*)) & 0 \\ \dfrac{K_c}{J}\dfrac{krT_s}{V_p(\theta)}\varphi_p(P_p, \operatorname{sgn}(A_p^*)) & \dfrac{K_c}{J}\dfrac{krT_s}{V_n}\varphi_n(P_n, \operatorname{sgn}(A_n^*)) \end{bmatrix} \end{aligned} \tag{6.27}$$

and the vector of the natural feedbacks $\Delta_0(x)$ is given by

$$\begin{aligned} \Delta_0(x) &= \begin{bmatrix} L_f^1 h_1 \\ L_f^3 h_2 \end{bmatrix} \\ &= \begin{bmatrix} \dfrac{-krT_s}{V_p}\cdot\left(\dot{m}_{Fp} + \dot{m}_v + \dfrac{P_p}{rT_s}\cdot K_c\Omega\right) \\ -k\dfrac{K_c^2}{J}\left(\dfrac{P_p}{V_p} + \dfrac{P_n}{V_n}\right)\Omega - krT_s\dfrac{K_c}{J}\left(\dfrac{\dot{m}_v + \dot{m}_{Fp}}{V_p} + \dfrac{\dot{m}_v - \dot{m}_{Fn}}{V_n}\right) - \dfrac{\Omega}{J}\dfrac{\mathrm{d}C_r}{\mathrm{d}\theta} \end{bmatrix} \end{aligned} \tag{6.28}$$

The control law is then obtained:

$$\begin{bmatrix} A_p^* \\ A_n^* \end{bmatrix} = \Delta^{-1}(x).\left[-\Delta_0(x) + \begin{bmatrix} v_1 \\ v_2 \end{bmatrix}\right] \tag{6.29}$$

$$\begin{bmatrix} i_p \\ i_n \end{bmatrix} = \begin{bmatrix} \psi_p^{-1}\left(A_p^*\right) \\ \psi_n^{-1}\left(A_n^*\right) \end{bmatrix} \tag{6.30}$$

where v represents the input vector of the linearized and decoupled system, the components v_1 and v_2 are, respectively, the first derivative reference pressure P_p and the jerk reference.

6.2.3.2.2 *Physical interpretation*

The relation (6.29) may be expressed in the form

$$A^*_p\varphi_p = \underbrace{\dot{m}_{F_p} + \dot{m}_v}_{\substack{\text{leakage mass}\\ \text{flow rates}}} + \underbrace{\frac{P_p}{rT_s}K_c\Omega}_{\substack{\text{mass flow rate}\\ \text{due to}\\ \text{the vane movement}}} + \underbrace{\frac{V_p}{krT_s}v_1}_{\substack{\text{compressibility mass}\\ \text{flow rate}}} \tag{6.31}$$

$$A^*_n\varphi_n = \underbrace{\dot{m}_{F_n} + \dot{m}_v}_{\substack{\text{leakage mass}\\ \text{flow rates}}} + \underbrace{\frac{P_n}{rT_s}K_c\Omega}_{\substack{\text{mass flow rate}\\ \text{due to}\\ \text{the vane movement}}} + \underbrace{\frac{V_n}{krT_s}\cdot\frac{1}{K_c}\frac{\mathrm{d}C_r}{\mathrm{d}\theta}\Omega}_{\substack{\text{compressibility mass}\\ \text{flow rate}\\ \text{due to the variable}\\ \text{perturbation torque}}} + \underbrace{\frac{V_n}{krT_s}\frac{1}{K_c}\left[Ju_2 - K_c u_1\right]}_{\substack{\text{compressibility mass flow rate}\\ \text{due to the jerk}}} \tag{6.32}$$

For the p chamber, the control input v_1 consists of a compensation of the leakage mass flow rate and the mass flow rate due to the piston movement. The third term corresponds to the compressibility mass flow rate $\frac{V_p}{krT_s}v_1$.

For the n chamber the servovalve opening leads to a compensation of the leakage mass flow rates and the mass flow rate due to the piston movement. The last two terms consist of a compressibility mass flow rate associated with the dependence on the resistive torque with respect to the position and a compressiblity mass flow rate for the required jerk. The singularities of the decoupling matrix are due to the inversion of the flow in one of the chambers $(\varphi_p = 0\ \varphi_n = 0)$.

6.2.4 *Experimental results in the SISO case*

The linearizing control law leads to a triple integrator input–output behaviour, then it is necessary to stabilize this triple integrator. This is obtained with a discrete-time external loop which is achieved for positioning and for tracking.

6.2.4.1 Step responses

The control of the linearized system is achieved with a classical discrete-time pole placement method. Figure 6.11 shows the control scheme with the linearizing inner loop and the outer loop which imposes the closed-loop dynamical behavior. For the computer implementation the compensating terms d_r and d_f have been neglected.

Results obtained both with linear and nonlinear control are presented in order to point out the nonlinear control improvements. The linear control is designed in order to correspond

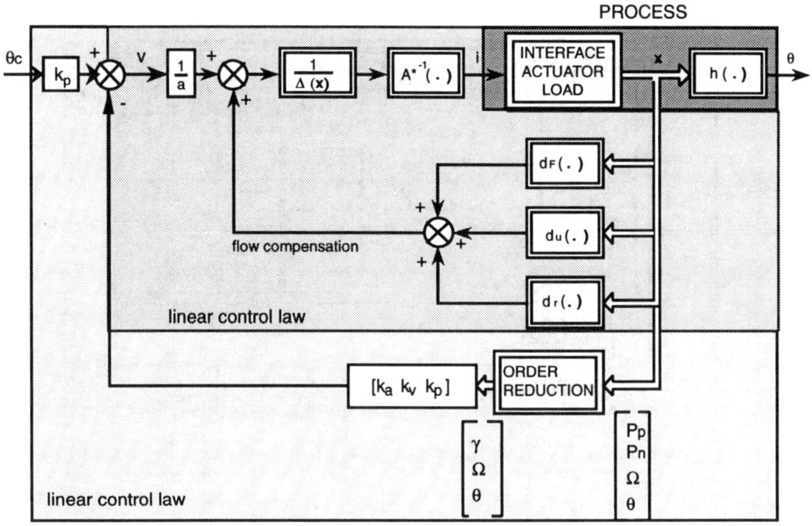

Figure 6.11: Control law scheme.

to the linearization of the nonlinear control law in equilibrium conditions at the middle stroke. Figure 6.12 shows a step response from $-10°$ to $0°$ obtained with the two control laws and the figure 6.13 corresponds to a step response from $-115°$ to $-105°$. The two control laws lead to the same results in the vicinity of the middle stroke. However at the end stroke oscillations appear with the linear control while nonlinear control gives a good behavior which is approximatively the same as the behavior obtained at the middle stroke.

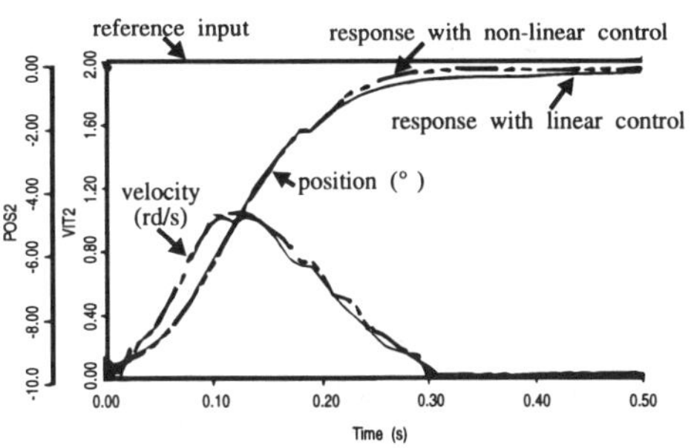

Figure 6.12:
Step response of the position
$-10° \longrightarrow 0°$.

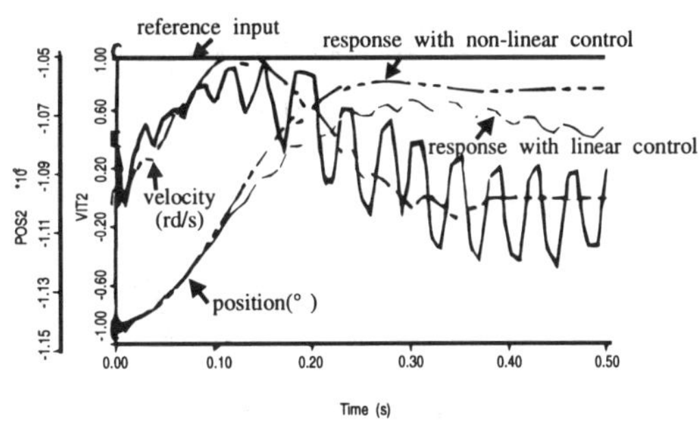

Figure 6.13:
Step response of the position
$-115° \longrightarrow -105°$.

6.2.4.2 Tracking

The tracking control scheme is shown in figure 6.14. Two different behaviors in positioning and in regulation can be obtained with this kind of control. The reference trajectory imposes the positioning behavior and the state feedback gains are fixed by the trajectory error dynamic.

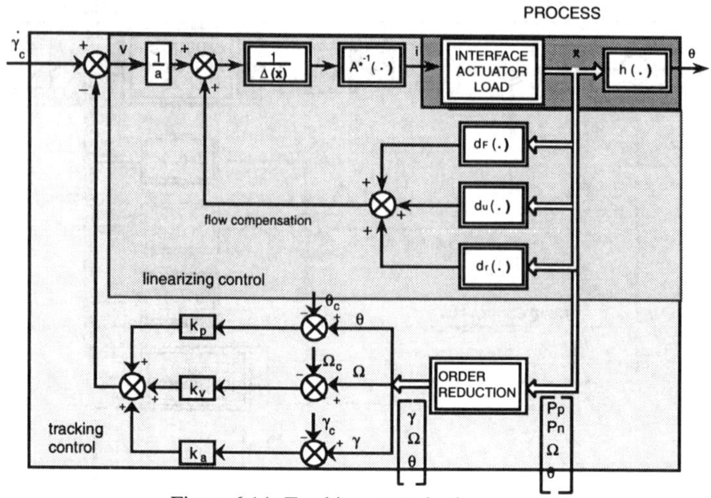

Figure 6.14: Tracking control scheme.

Figure 6.15 shows the tracking results obtained from $-100°$ to $0°$. A good harmony is obtained between reference and real trajectories and almost the same result is given by the linear control. However for angles of almost $-110°$, the linear control presents oscillations as in the case of the step response.

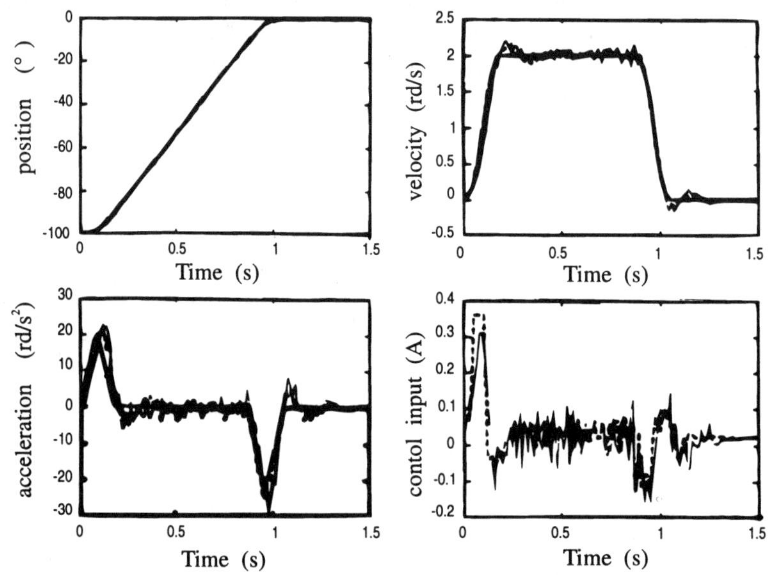

Figure 6.15: Tracking.

6.2.5 Simulated results in the MIMO case

For each decoupled subsystem, a pole placement state feedback is used for the outer loop. Figure 6.16 shows a step response of 20° for two different pressure references: 4.10^5 Pa and 2.10^5 Pa. For the pressure reference $P_p = 4.10^5$ Pa, an important overshoot comes from the fact that only a small differential pressure $DP = 10^5$ Pa is available during the slow down phase. For the second reference pressure value, a small differential pressure is available $DP = 10^5$ Pa during the speedup phase which leads to a long setting time. For these two simulations, the pressure and the position are decoupled.

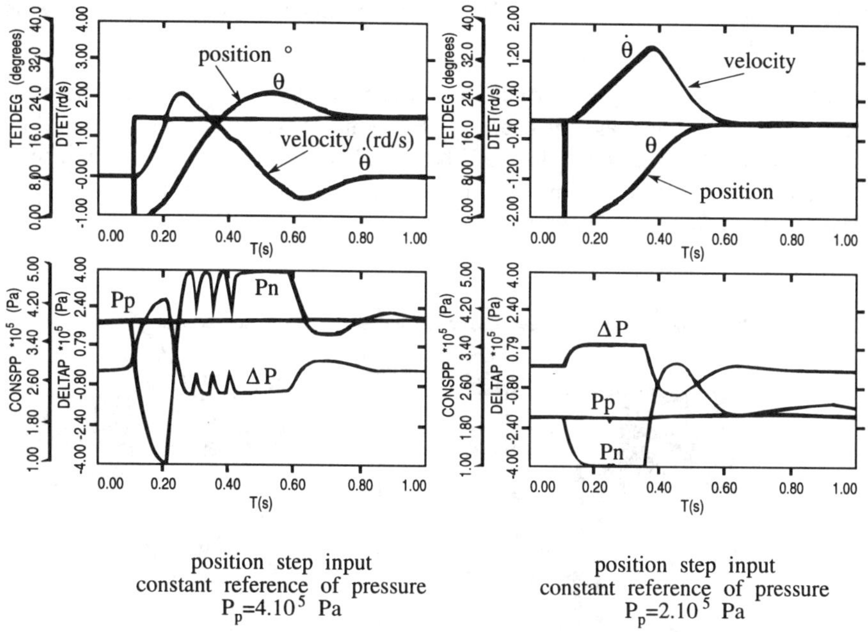

Figure 6.16: Pressure/position decoupling.

6.2.6 *Conclusion*

As shown by the experimental results, nonlinear control is active in larger range of operating conditions than the linear one. Nonlinear control needs two pressure sensors (one speed sensor and one position sensor) while linear control obtained with a reduced tangent linearized model uses only three sensors. Moreover, physical interpretation of the nonlinear control law increases the knowledge of the system.

6.3 Position control of an active magnetic bearing

6.3.1 *Description of the process*

A magnetic bearing is a device where guidance forces between the fixed part (stator) and the moving part (rotor) are of a magnetic nature. The two parts are held in position without contact. Therefore, it is possible to reach a high rotation speed in hard conditions (vacuum, high temperature), with a theoretical non limited lifetime. Magnetic bearings are of two kinds: passive or active.

– With *passive magnetic bearings* positioning is performed by permanent magnets or by soft magnetic materials. The centering forces are due to the material shape of the bearing (Yonnet, 1980).

– With *active magnetic bearings* magnetic fields are created by electromagnets supplied by external power, so the centering forces depend only on a control law (Habermann and Liard, 1979).

The prototype (fig. 6.17) is an inertial wheel suspended by a passive magnetic bearing and centred by two active magnetic bearings. The centering (radial displacements along x and y) is performed by the active magnetic bearing. The passive active magnetic bearing controls three degrees of freedom: axial displacement along z and rotations ϕ_x and ϕ_y. The last degree of freedom is the rotation ϕ_z.

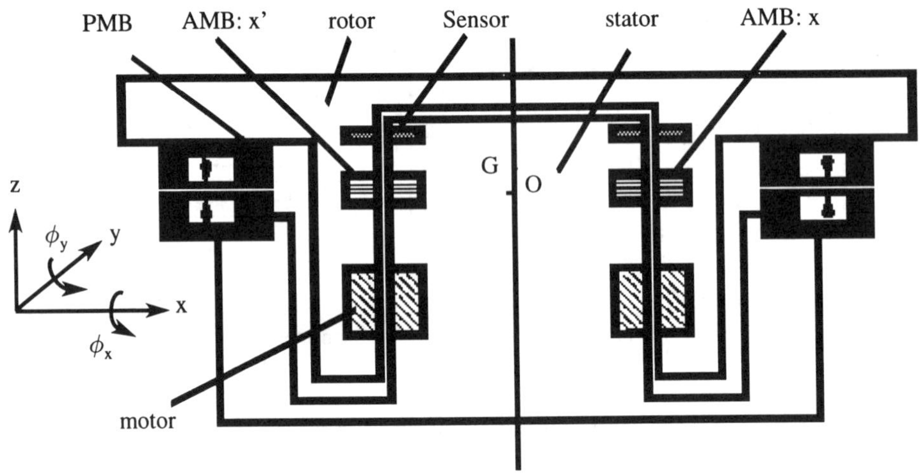

Figure 6.17: Prototype.

6.3.2 The magnetic bearing model

The process is made up from three parts (fig. 6.18):

1. The rotor.
2. The sensors.
3. The actuators formed by electromagnets that allow the creation of magnetic fields.

We make the following assumptions:

– The rotor is rigid.

– The axial displacement (along z) is only due to the passive magnetic bearing.

– Momentums due to rotations ϕ_x and ϕ_y are neglected.

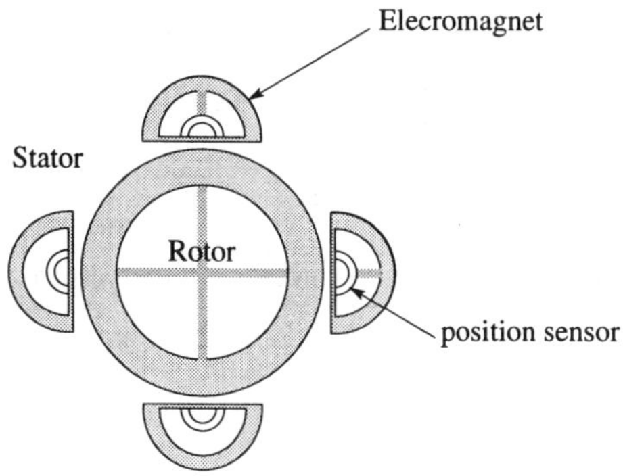

Figure 6.18: Active magnetic bearing.

– Rotations ϕ_x and ϕ_y are small, so they can be approximated by their first-order Taylor expansion, *i.e.* $\alpha = \sin(\phi_x)$ and $\beta = \sin(\phi_y)$.

The model is then given by Charara (1992):

$$
\begin{aligned}
m\ddot{x} &= \frac{\text{sign}(i_x)F_1 e_0^2 i_x^2}{(e_0 - \text{sign}(i_x)\Delta e_x)^2} + F_{px} = F_x + F_{px} \\
m\ddot{y} &= \frac{\text{sign}(i_y)F_1 e_0^2 i_y^2}{(e_0 - \text{sign}(i_y)\Delta e_y)^2} + F_{py} = F_y + F_{py} \\
I\ddot{\alpha} + J\omega_0\dot{\beta} &= l_1(F_y + F_{py}) \\
I\ddot{\beta} - J\omega_0\dot{\alpha} &= -l_1(F_x + F_{px}) \\
\frac{\mathrm{d}i_x}{\mathrm{d}t} &= -\frac{\text{sign}(i_x)\Delta\dot{e}_x i_x}{(e_0 - \text{sign}(i_x)\Delta e_x)} - \frac{R}{\lambda}(e_0 - \text{sign}(i_x)\Delta e_x)i_x \\
&\quad + \frac{1}{\lambda}(e_0 - \text{sign}(i_x)\Delta e_x)E_x \\
\frac{\mathrm{d}i_y}{\mathrm{d}t} &= -\frac{\text{sign}(i_y)\Delta\dot{e}_y i_y}{(e_0 - \text{sign}(i_y)\Delta\dot{e}_y)} - \frac{R}{\lambda}(e_0 - \text{sign}(i_y)\Delta e_y)i_y \\
&\quad + \frac{1}{\lambda}(e_0 - \text{sign}(i_y)\Delta e_y)E_y
\end{aligned}
$$

with (index j representing axis x or axis y)

Δe_j: variation of the air-gap and e_0 the nominal air-gap: $e_j = e_0 + \Delta e_j$.

x and y: coordinates of the gravity center G of the rotor.

l: distance between G and the sensor ($\Delta e_x = x - \beta l$ and $\Delta e_y = y + \alpha l$).

α and β: rotation angles around axes x and y.

i_j: current in the electromagnet j.

E_j: applied voltage on each coil j.

F_j: electromagnetic force produced by the electromagnet j.

ω_0: rotation speed of the rotor.

m: mass of the rotor.

F_{px} and F_{py}, perturbation forces (due to unbalance) given by: $ma_0\omega_0^2 \sin(\omega_0 t)$ and $ma_0\omega_0^2 \cos(\omega_0 t)$.

a_0: eccentricity.

l_1: length OG.

O: center of the active bearing.

μ_0: permeability of air.

S: surface of a pole of the actuator coil.

n: number of turns of the coil.

R: resistor of the coil.

$c = \cos \pi/8$: correction factor due to the shape of the coil.

I and J: radial and axial inertia momentums.

$F_1 = (\mu_0 n^2 Sc)/(4e_0^2)$

The first two equations are obtained by calculating the acceleration of the rotor along x and y. The next two equations are obtained by calculating the acceleration equations of the rotor around x and y. The two last equations are due to the electrical behavior of the coil. The sign function appears because we use only one coil at the same time on one axis. For example, in figure 6.17, if coil x' along axis x is supplied, then coil x is not. When a change of the direction of the force is desired, coil x is supplied instead of coil x', then the air-gap changes immediately if the rotor is not centred. The sign function describes this behavior. The dynamic of the process is then nonlinear and non analytical. To test the reliability of this model, a linear compensator has been calculated on each axis using the describing function approach of the $(\mathrm{sign(i)i^2})$ nonlinearity. The other nonlinearities and the coupling effects are neglected. This allows the stabilization of the process in order to compare the closed-loop response of the model to the closed-loop response of the process (Charara, 1992). We have obtained a similar behavior, so the proposed model will be used to calculate a noninteracting control.

6.3.3 *Control of the process*

Due to the behavior of the force which is proportional to sign(i)i^2 (if other nonlinearities are neglected), this model is not controllable if it is linearized around the equilibrium point $i = 0$ (fig. 6.19). To control the process, one usually supplies all the coils with a constant current i_0 (Maslen *et al.*, 1989; Matsumara and Nakagawa, 1986). This current introduces an equilibrium point in order to obtain a linear system around this equilibrium. This entails that the magnetic field generated by the coils cancel each other. If the control law needs a force along axis x, for example, a current $i = i_0 + \Delta i$ is applied to coil x and a current $i = i_0 - \Delta i$ is applied to coil x', so the mean consumption on each axis is equal to the maximum current.

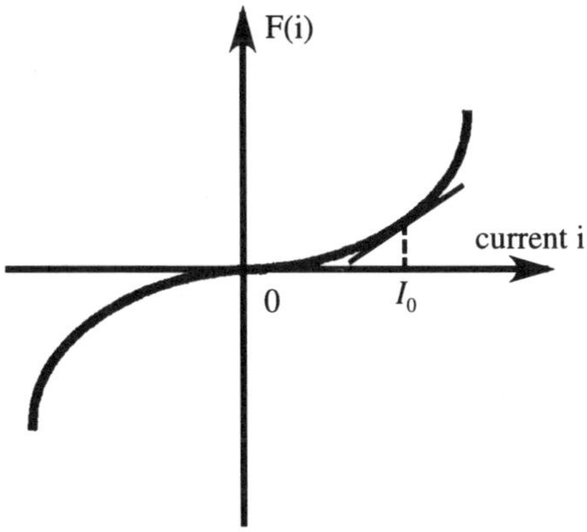

Figure 6.19: Static characteristic of the force in function of the current in the coil.

We now propose a control law without constant current in the coils. In that case, only one coil is fed at the same time depending on the desired direction of the force. This induces the sign function in the model. Without constant current in the coils, it is not possible to obtain a controllable linear model, so we propose to use an exact input–output linearization to control the process.

We can also remark that it is possible to use the currents in the coil as the control input instead of the voltages. However, in that case, the control does not appear linearly in the model, and this reduces the possibilities of a choice for the control algorithm.

6.3.3.1 Modification of the model

The model cannot be directly used to design a linearizing control because the sign function is non analytical. To implement the control, the sign function is then approximated by an analytical one

$$\mathrm{sign(i)} \simeq \frac{2}{\pi}\mathrm{atan}(\gamma \mathrm{i}) = \mathrm{q(i)}$$

where γ is chosen such as to have a good approximation of the sign function. It is important to see that $q(i)$ is always multiplied by i, so it is possible to differentiate the function $q(i)i$ around $i = 0$.

6.3.3.2 Expression of the control law.

The proposed model is transformed into an analytical one with

$$\begin{aligned}
\xi^t &= [x, y, \alpha, \beta, \dot{x}, \dot{y}, \dot{\alpha}, \dot{\beta}, i_x, i_y] && \text{state space vector.}\\
u^t &= [E_x, E_y] = [u_1(t), u_2(t)] && \text{control vector.}\\
h^t &= [x, y] = [h_1(t), h_2(t)] && \text{output vector.}
\end{aligned}$$

Let

$$e_x = e_0 - q(i_x)\Delta e_x; \quad e_y = e_0 - q(i_y)\Delta e_y$$

and

$$f(\xi) = \begin{pmatrix} \dot{x} \\ \dot{y} \\ \dot{\alpha} \\ \dot{\beta} \\ \dfrac{q(i_x)F_1 e_0^2 i_x^2}{m e_x^2} + \dfrac{F_{px}}{m} \\ \dfrac{q(i_y)F_1 e_0^2 i_y^2}{m e_y^2} + \dfrac{F_{py}}{m} \\ \dfrac{l_1 q(i_y)F_1 e_0^2 i_y^2}{I e_y^2} + l_1\dfrac{F_{py}}{I} - \dfrac{J\omega_0\dot{\beta}}{I} \\ -\dfrac{l_1 q(i_x)F_1 e_0^2 i_x^2}{I e_x^2} - l_1\dfrac{F_{px}}{I} + \dfrac{J\omega_0\dot{\alpha}}{I} \\ -\dfrac{q(i_x)\Delta\dot{e}_x i_x}{e_x} - \dfrac{R}{\lambda}e_x i_x \\ -\dfrac{q(i_y)\Delta\dot{e}_y i_y}{e_y} - \dfrac{R}{\lambda}e_y i_y \end{pmatrix} \qquad g(\xi) = \begin{pmatrix} 0 & 0 \\ 0 & 0 \\ 0 & 0 \\ 0 & 0 \\ 0 & 0 \\ 0 & 0 \\ 0 & 0 \\ 0 & 0 \\ \dfrac{1}{\lambda}e_x & 0 \\ 0 & \dfrac{1}{\lambda}e_y \end{pmatrix}$$

The model is then given by

$$\left|\begin{aligned}
\dot{\xi} &= f(\xi) + g(\xi)u \\
y &= h(\xi) = \begin{pmatrix} 1 & 0 & 0 & 0 & 0 & 0 & 0 & 0 & 0 & 0 \\ 0 & 1 & 0 & 0 & 0 & 0 & 0 & 0 & 0 & 0 \end{pmatrix}\xi
\end{aligned}\right.$$

In order to obtain the characteristic numbers $\rho_i (i = 1, 2)$, let us take the Lie derivatives

$$L_f h_1 = \xi_5 = \dot{x} \quad L_f h_2 = \xi_6 = \dot{y} \quad L_{g_k} L_f h_i = 0 \quad (i = 1, 2; k = 1; 2)$$

$$\begin{aligned}
L_{g_1} L_f^2 h_1 &= \frac{F_1 e_0^2}{m\lambda}\frac{\dot{q}(i_x)i_x^2 + 2q(i_x)i_x}{(e_0 - q(i_x)\Delta e_x)} + \frac{2F_1 e_0^2}{m\lambda}\frac{\dot{q}(i_x)q(i_x)i_x^2}{(e_0 - q(i_x)\Delta e_x)^2} \\
&\simeq \frac{F_1 e_0^2}{m\lambda}\frac{q(i_x)i_x}{(e_0 - q(i_x)\Delta e_x)} = \Delta_{11}(\xi) \neq 0 \implies \rho_1 = 3 \quad (\text{because } \dot{q}(i_x)i_x \simeq 0)
\end{aligned}$$

Using the same approximation it follows that

$$L_{g_2}L_f^2h_2 = 2\frac{F_1e_0^2}{m\lambda}\frac{q(i_y)i_y}{(e_0 - q(i_y)\Delta e_y)} = \Delta_{22}(\xi) \neq 0 \Longrightarrow \rho_2 = 3$$

Using this control law, each output is controlled through a triple integrator.

The decoupling matrix is given by

$$\Delta(x) = \begin{pmatrix} \Delta_{11} & 0 \\ 0 & \Delta_{22} \end{pmatrix}$$

This matrix is not invertible when currents i_x and i_y are equal to zero. In that case, the rotor is at equilibrium position and the currents are equal to zero. This problem is solved by introducing a small current (1mA) in the coils. This current induces a force equal to $10^{-7}N$, that could be neglected (Charara, 1992). With this current, the denominator of the expression of the control law cannot be equal to zero. Moreover, we use 12 bits ± 10 A analog to digital converters (for the current measurement), so i_x and i_y have a discretizing step of 5mA (compare with 1mA).

The control law is then given by

$$u_i(t) = \frac{-L_f^3h_i + \tilde{v}_i}{\Delta_{ii}} \qquad \text{for } i = (1,2)$$

where $\tilde{v}_i(i = 1,2)$ are the new inputs. Then we have

$$\begin{aligned} u_1(t) &= Ri_x + \left(\frac{m\lambda}{2F_1e_0^2}\right)\frac{(e_0 - q(i_x)\Delta e_x)}{q(i_x)i_x + 0.001}\tilde{v}_1 \\ u_2(t) &= Ri_y + \left(\frac{m\lambda}{2F_1e_0^2}\right)\frac{(e_0 - q(i_y)\Delta e_y)}{q(i_y)i_y + 0.001}\tilde{v}_2 \end{aligned}$$

In order to stabilize the two triple integrators, new inputs $\tilde{v}_i(i = 1,2)$ are taken as

$$\begin{aligned} \tilde{v}_1(t) &= \sigma_{12}\ddot{x}_g + \sigma_{11}\dot{x}_g + \sigma_{10}x_g + k_{r1}v_1 \\ \tilde{v}_2(t) &= \sigma_{22}\ddot{y}_g + \sigma_{21}\dot{y}_g + \sigma_{20}y_g + k_{r2}v_2 \end{aligned}$$

The new inputs, after linearization and stabilization, are v_1 and v_2. In our case, these inputs are equal to zero because the aim of the control is to center the rotor perfectly. After nonlinear feedback, the new state vector dimension is given by

$$\sum_{i=1}^{2}\rho_i = 6$$

The new state space vector is given by

$$\eta^t = [x, \dot{x}, \ddot{x}, y, \dot{y}, \ddot{y}] = [\eta_1, \eta_2, \eta_3, \eta_4, \eta_5, \eta_6]$$

The dimension of the state vector before linearization was equal to 10, so we have obtained an unobservable subspace of dimension 4.

6.3.3.3 Stability of the unobservable subspace

To apply the control, we have to prove the stability of the unobservable part of the dynamic. If α and β and their derivatives are neglected, the unobservable subspace has a zero dimension, so we can use these 4 state components to describe the unobservable subspace.

The observable part with the stabilizing feedback is given by

$$\begin{cases} \dot{\eta}_1 &= \eta_2 \\ \dot{\eta}_2 &= \eta_3 \\ \dot{\eta}_3 &= -\sigma_{12}\eta_3 - \sigma_{11}\eta_2 - \sigma_{10}\eta_1 + k_{r1}v_1 + \dfrac{\dot{F}_{px}}{m} \\ \dot{\eta}_4 &= \eta_5 \\ \dot{\eta}_5 &= \eta_6 \\ \dot{\eta}_6 &= -\sigma_{22}\eta_6 - \sigma_{21}\eta_5 - \sigma_{20}\eta_4 + k_{r2}v_2 + \dfrac{\dot{F}_{px}}{m} \end{cases}$$

The unobservable part is given by

$$\begin{cases} \dot{z}_1 &= z_2 \\ \dot{z}_2 &= \dfrac{ml_1}{I}\eta_6 - \dfrac{J\omega}{I}z_4 \\ \dot{z}_3 &= z_4 \\ \dot{z}_4 &= -\dfrac{ml_1}{I}\eta_3 + \dfrac{J\omega}{I}z_2 \end{cases}$$

When the linear part reaches the equilibrium point, all $\eta_i(i = 1, ..., 6)$ are equal to zero. The asymptotic behavior of the unobservable part is then given by

$$\begin{cases} \dot{z}_1 &= z_2 \\ \dot{z}_2 &= -\dfrac{J\omega}{I}z_4 \\ \dot{z}_3 &= z_4 \\ \dot{z}_4 &= \dfrac{J\omega}{I}z_2 \end{cases}$$

The solution of these differential equations is:

$$\begin{cases} z_1(t) &= \dfrac{c_1}{A}\sin(At) - \dfrac{c_2}{A}\cos(At) + c3 \\ z_2(t) &= c_1\cos(At) + c_2\sin(At) \\ z_3(t) &= -\dfrac{c_1}{A}\cos(At) - \dfrac{c_2}{A}\sin(At) + c_4 \\ z_4(t) &= c_1\sin(At) - c_2\cos(At) \end{cases}$$

where: $A = \dfrac{J\omega}{I}$ and $c_i(i = 1, \ldots, 4)$ are constants. To prove the Lyapunov's stability of the unobservable subspace, let us give the following proposition:

Proposition 6.1

If for $t = t_0, z_i(t) = b_i, \forall(i = 1, \ldots, 4)$, *then*

$$\forall\varepsilon > 0, \exists\theta/b_i \text{ belongs to the ball } B(0,\theta)$$
$$\Longrightarrow z_i \text{ belongs to the ball } B(0,\varepsilon), \forall t > t_0, \forall i = (1, \ldots, 4).$$

It is sufficient to take $\theta = \dfrac{A}{A+1}\varepsilon$ *in order to assume Lyapunov's stability of the unobservable part (Charara, 1992).*

We can remark that the preceding equations are only available at the equilibrium point. Then, if the model is perfect and without disturbances, the unobservable subspace is an oscillator. All simulations have shown (with and without a nominal plant) that outside the equilibrium point, the unobservable subspace is always stable.

6.3.4 Digital implementation

The preceding continuous control law has to be implemented digitally, so it is necessary to discretize it. We use the approach proposed by Monaco and Normand-Cyrot (Chapter 5). Let the nonlinear model of be as

$$\frac{\mathrm{d}\xi}{\mathrm{d}t} = L_{f+u(t)g}\xi(t)$$

We do a p-order Taylor expansion of the model

$$\xi[k+1] = \xi[k] + \delta\left(\frac{\mathrm{d}\xi}{\mathrm{d}t}\right)_{\xi[k]} + \cdots + \left(\frac{\delta^p}{p!}\right)\left(\frac{\mathrm{d}^p\xi}{\mathrm{d}t^p}\right)_{\xi[k]} \quad \text{with } \delta \text{ the sampling period.}$$

In the case where we stop at order zero, we obtain

$$u_1[k] = Ri_x[k] + \left(\frac{m\lambda}{2F_1e_0^2}\right)\frac{(e_0 - q(i_x)\Delta e_x[k])}{q(i_x)i_x[k] + 0.001}\tilde{v}_1[k]$$

and

$$u_2[k] = Ri_y[k] + \left(\frac{m\lambda}{2F_1e_0^2}\right)\frac{(e_0 - q(i_y)\Delta e_y[k])}{q(i_y)i_y[k] + 0.001}\tilde{v}_2[k]$$

In the case where we stop at order one, we obtain

$$\hat{u}_1[k] = u_1[k] + \delta\left(\frac{3mR}{2F_1e_0^2}\frac{[e_0 - q(i_x)\Delta e_x]^2}{q(i_x)i_x + 0.001}\right) + \frac{R^2}{\lambda}[e_0 - q(i_x)\Delta e_x]i_x - R\frac{\Delta\dot{e}_x q(i_x)i_x}{[e_0 - q(i_x)\Delta e_x]}$$

We obtain similar results for $\hat{u}_2[k]$.

6.3.5 Experimental results

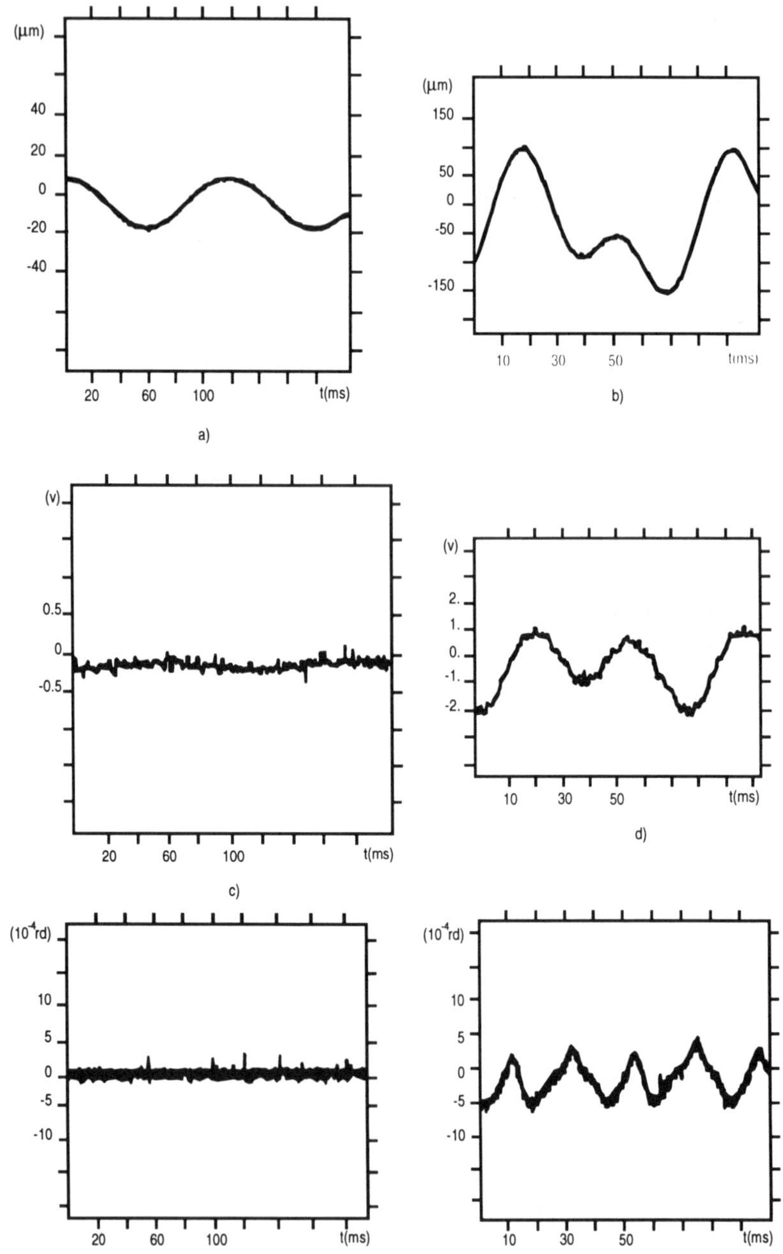

Figure 6.20: Linear control a) variation of position x around a nominal air-gap, f=0 Hz. b) variation of x, f=25 Hz. c) variation of the control, f=0 Hz. d) variation of the control, f=25 Hz. e) variation of β, f=0 Hz. f) variation of β, f=25 Hz.

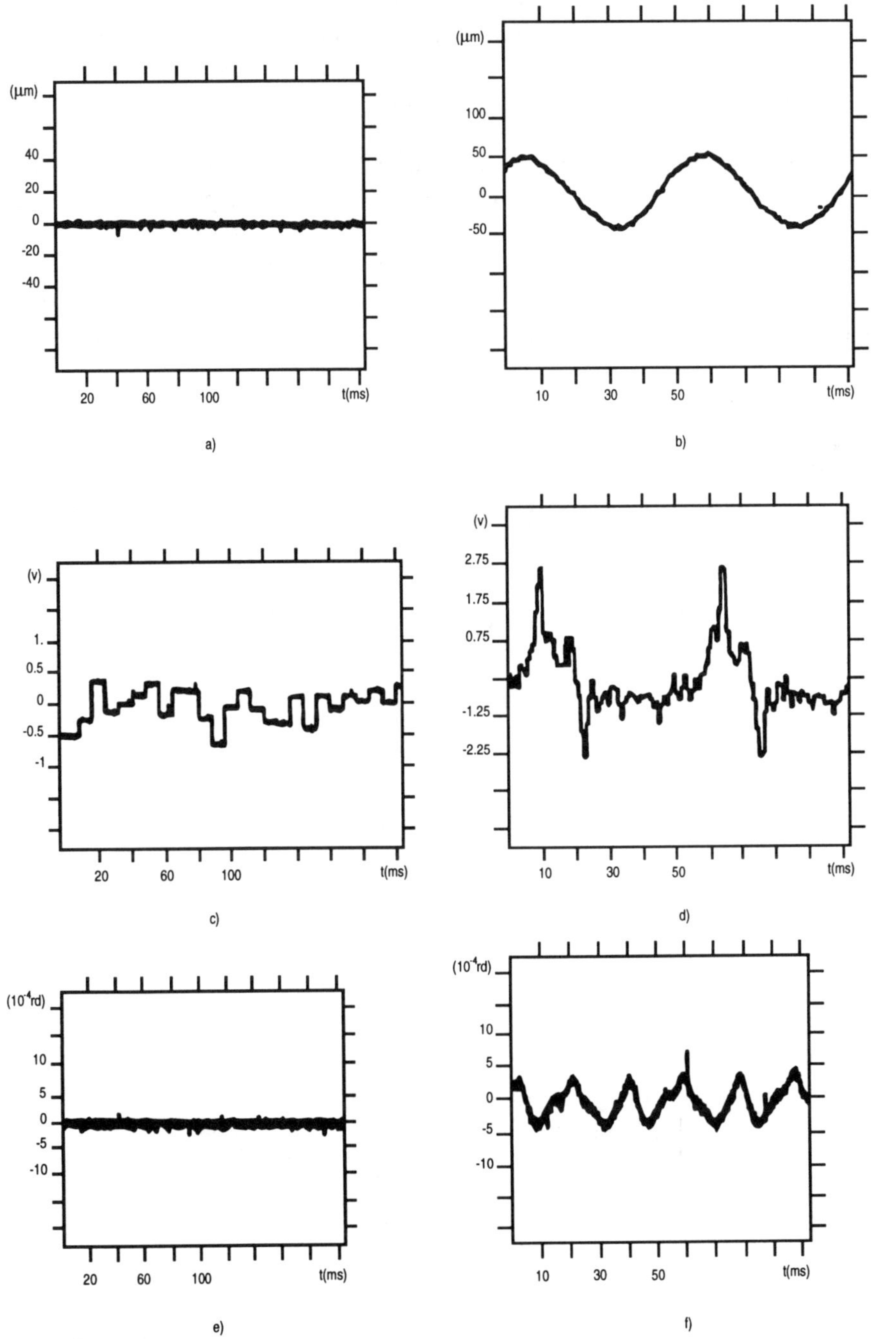

Figure 6.21: Nonlinear control a) variation of position x around a nominal air-gap, f=0 Hz. b) variation of x, f=25 Hz. c) variation of the control, f=0 Hz. d) variation of the control, f=25 Hz. e) variation of β, f=0 H. f) variation of β, f=25 Hz.

Figures 6.20 and 6.21 show trials for two frequencies (f) of a rotor rotation. The sampling period is taken equal to 800 microseconds. Figure 6.20 shows the performance in the case of a linear control using the describing function approach taking into account only the static characteristic of the force as a function of the current ($F = \mathrm{sign(i)i}^2$). Figure 6.21 shows the performance of the nonlinear control under the same conditions. The stabilizing feedback of the triple integrator is here done with a double lead controller. The stabilizing state space feedback has not been tested due to the lack of computation time. We can remark the good performance of the nonlinear control.

Figure 6.22 shows, in simulation, the comparison between the linear control (simple lead compensator using the describing function approach) and the whole nonlinear control using the stabilizing state space feedback. We can remark that the performance is even better.

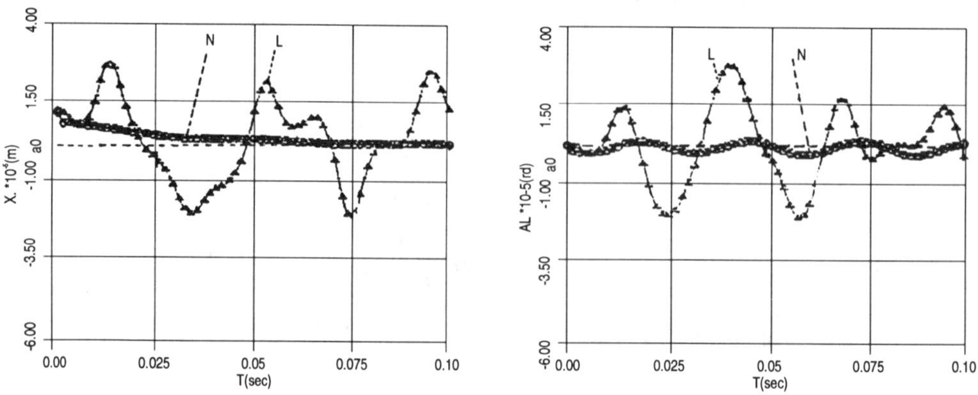

Figure 6.22: Comparison between linear (L) and nonlinear (N) control with state space feedback a) displacement along x b) coupling variable α.

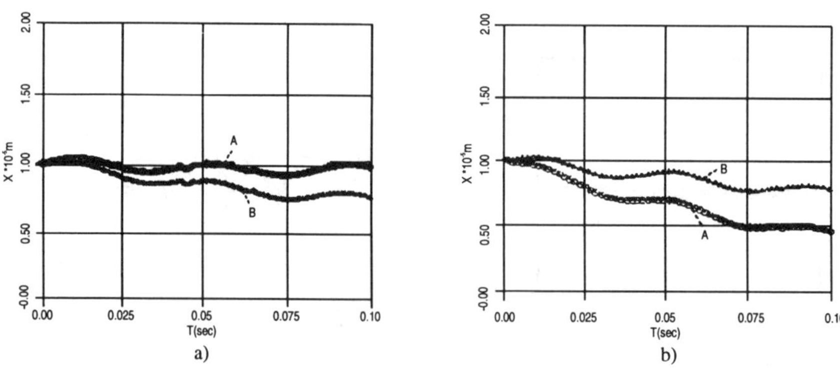

Figure 6.23: Influence of discretization a) A: order 1 ($T_e = 400\ \mu s$), B: order 0 ($T_e = 200\ \mu s$). b) ($T_e = 200 \mu s$) A: order 1, B: order 0.

Figure 6.23 shows, in simulation, the influence of the discretization. Figures 6.23a and 6.23b compare the behavior of the output in the case where the output x reaches a zero position, in the case of order 0 (B) and in the case of order 1 (A). We can see that the control is unstable for $T_e > 300\mu s$ in the case of a zero-order discretization control while in the case of a first-order discretization the control is unstable for $T_e > 450\mu s$. When the system is stable, performances are always better with a first-order discretization (Fig. 6.23b).

6.4 Bibliography

[1] ANDERSEN, B.W. (1976), *The analysis and design of pneumatic systems*, NewYork, John Willey and Sons.

[2] BURROWS, C.R. (1972), *Fluid power servo mechanisms*, Van Nostrand Company, London.

[3] CHARARA, A. (1992), Contribution à la modélisation et à la commande d'un palier magnétique actif en fonctionnement non linéaire, PhD thesis, Université de Savoie.

[4] COMOLET, R., (1979), *Mécanique expérimentale des fluides. T1 Statique et dynamique des fluides non visqueux*, Paris, Masson.

[5] GUSTAFSON, T.K. and WALLER, K.V. (1983), Dynamic modelling and reaction invariant control of pH, *Chem. Eng. Sci.*, **38**, 3, 389–398.

[6] GUSTAFSON, T.K. and WALLER, K.V. (1986), Myths about pH and pH control, *AIChe J.*, **32**, 2, 335–337.

[7] HABERMANN, H. and LIARD, G. (1979) *Practical Magnetic Bearings*, IEEE spectrum, 26–30, september.

[8] JEBAR, H.S. (1997), Design of pneumatic actuator systems, PhD Thesis, Nottingham.

[9] JUTILA, A. (1983), An application of adaptive control algorithms based on physico-chemical modelling in a chemical waste water treatment plant, *Int. J. Control*, **38**, 3, 639–655.

[10] KAIN, J. and WARTELLE, C. (1973), *Étude de la dynamique des vérins pneumatiques*, Mémoires Techniques du Cetim, **17**.

[11] MC CLOY, D. and MARTIN, H.R. (1980), *Control of Fluid Power: Analysis and Design*, Chichester, Ellis Harwood.

[12] MASLEN, E., HERMANN, P., SCOTT, M. and HUMPHRIS, R. R. (1989), Practical limits to the performance of Magnetic Bearings: peak force, slew rate and displacement sensitivity, *Journal of Tribology*, **III**, 331–336.

[13] MATSUMURA, F. and NAKAGAWA, K. (1986), Theory and experiment of magnetic bearing for radial and thrust control, *Electrical Engineering in Japan*, **106**, 2, 123–130.

[14] PASZKIEWICZ, D. (1987), Commande des procédés non linéaires en temps continu et en temps discret, Application à un réacteur chimique de neutralisation, PhD thesis, Institut National des Sciences Appliquées de Lyon.

[15] RICHARD, E. (1990), De la commande linéaire et non linéaire en position des systèmes électropneumatiques, PhD thesis, Institut National des Sciences Appliquées de Lyon.

[16] SHEARER, J.L. (1956), Study of Pneumatic Processes in the Continuous Control of Motion With Compressed Air , Parts I and II, *Trans. Am. Soc. Mech. Eng.*, **78**, 233–249.

[17] YONNET, J.P. (1980), *Étude des paliers magnétiques passifs* , Thèse d'état de l'INP Grenoble.

[18] ZALMANZON, L.A. (1965), *Components for pneumatic control instruments,* Oxford, Pergamon Press.

CHAPTER 7

Approximate closed-loop optimization of nonlinear systems by singular perturbation technique

A.J. FOSSARD, J. FOISNEAU, T. HUN HUYNH

7.1 Introduction

A singularly perturbed dynamic system is a system where state equations can be written in the form

$$\begin{aligned} \dot{X}_1 &= f_1(X_1, X_2, u, \varepsilon) \\ \varepsilon\dot{X}_2 &= f_2(X_1, X_2, u, \varepsilon) \end{aligned} \qquad (7.1)$$

where $X_1 \in R^{n_1}, X_2 \in R^{n_2}, f_1, f_2$ are regular vectorial functions in X_1, X_2, ε, u, and where ε is a small nonnegative parameter.

This form (7.1), called a *standard* form, which will be used in the remainder of this chapter, has two advantages:

1. the origin of the name "singular" because the dimension of system (7.1), which equals $n_1 + n_2$ for every $\varepsilon \neq 0$, degenerates to a reduced order system with only dimension n_1 for $\varepsilon = 0$, defined by the equation $\dot{X}_1 = f_1(X_1, X_2, u, 0)$ under the constraint $f_2(X_1, X_2, u, 0) = 0$

2. If ε is small enough, the system has two time-scales, the variation of velocity then being very different between X_1 and X_2. If $\varepsilon \rightarrow 0$, and if $f_2(X_1, X_2, u, \varepsilon) \neq 0$, the velocity of X_2 tends towards infinity while the velocity of X_1 remains restricted. Rewriting the equations with the time-scale $\tau = t/\varepsilon$, for the purpose analysing of fast variations in X_2, gives

$$\begin{aligned} \frac{dX_1}{d\tau} &= \varepsilon f_1(X_1, X_2, u, \varepsilon) \\ \frac{dX_2}{d\tau} &= f_2(X_1, X_2, u, \varepsilon) \end{aligned} \qquad (7.2)$$

when $\varepsilon \longrightarrow 0$, the velocity of X_1 tends to 0 while the one of X_2 remains limited.

Taking into account the existence of two, or more, time-scales will allow, as can be seen later, a considerable simplification of analysis, simulation and control of a system such as that defined by (7.1). This is particularly true for nonlinear problems in optimal control which will be treated in the second part of this chapter. One will see that, with a moderate loss of optimality, it is possible to get approximate solutions allowing a considerable reduction of computational efforts and for which a practical set up is possible in a closed-loop form, thanks to its expression in terms of state variables.

First, some general remarks, which will be recalled below:

• To say that in equations (7.1) ε is a small positive parameter may be interpreted as saying that one is able to give a numerical value to it. This can be achieved in a certain number of cases where this parameter can be linked to some fixed physical characteristics of the process itself and can then be identified to the ratio of mass, of inertia, *etc.* to the inverse of the stiffness of the transmission *etc.* However, in many of situations, even when it is possible to evidence the parameter ε, the latter will only have a qualitative value and will be linked both to the characteristics of the system and to the scenario of the control under consideration.

• Even in the case where the perturbation parameter ε has a physical meaning, its introduction into the dynamic equations can lead to some more general forms although different from equation (7.1), such that

$$\dot{X} = f(X, u, \varepsilon) \tag{7.3}$$

The recognition of the existence of very different dynamics in (7.3) and the setting in a standard form (7.1) can then be something very complicated.

• Lastly, in several practical cases, on the contrary, it is not possible to define a small parameter even though one knows through experience, or through simulation *etc.* that the system has indeed several dynamics, slow and fast ones. In such a case, the parameter ε will be formally introduced as a "marker", multiplying the derivatives of fast state variables in order to be treated afterwords as a real perturbation parameter. This is then called "forced singular perturbation", which will be broadly used in the following, from a very pragmatic way.

At first the interest of the standard form (7.1) is justified, from the aspect of analysis.

7.2 Multi-time-scales systems and singularly perturbed systems

7.2.1 Dynamic of a singularly perturbed system

7.2.1.1 General

Let us define an autonomous system such that

$$\begin{aligned} \dot{X}_1 &= f_1(X_1, X_2, \varepsilon) \quad X_1(0) = X_{10} \quad X_1 \in R^{n_1} \\ \varepsilon\dot{X}_2 &= f_2(X_1, X_2, \varepsilon) \quad X_2(0) = X_{20} \quad X_2 \in R^{n_2} \end{aligned} \tag{7.4}$$

To find a solution to this problem and assuming ε a very small number, it is natural to set $\varepsilon = 0$ and to check if, in doing so, one can get a reasonable approximate solution for the whole problem. For $\varepsilon = 0$, the system is reduced to

$$\dot{\overline{X}}_1 = f_1\left(\overline{X}_1, \overline{X}_2, 0\right); f_2\left(\overline{X}_1, \overline{X}_2, 0\right) = 0; \overline{X}_1(0) = X_{10} \tag{7.5}$$

where the variables are upper-lined to indicate that the solution obtained is associated with $\varepsilon = 0$ in the t time basis t.

Let S be all the solutions of equation $f_2\left(\overline{X}_1, \overline{X}_2, 0\right) = 0$ and let us assume that at point $\left(X_1^S, X_2^S\right)$ of S the rank of $\dfrac{\partial f_2}{\partial X_2}$ is equal to n_2

According to the theorem of implicit functions, there exists in the vicinity of $\left(X_1^S, X_2^S\right)$ one and only one solution such that

$$\overline{X}_2 = \psi(\overline{X}_1) \tag{7.6}$$

The reduced solution is then represented by

$$\dot{\overline{X}}_1 = f_1(\overline{X}_1, \psi(\overline{X}_1), 0) \quad \overline{X}_1(0) = X_{10} \quad \overline{X}_2 = \psi(\overline{X}_1) \tag{7.7}$$

This solution will be called *reduced* solution or *external* solution. Obviously it verifies the initial conditions on X_1 but not on X_2.

If one performs the change of variables

$$X_2 = \tilde{X}_2 + \psi(X_1) \tag{7.8}$$

the set S is defined by the set of points $\left(X_1^S, X_2^S\right)$ such that $\tilde{X}_2 = 0$ and one obtains in the new coordinates

$$\begin{aligned} \dot{X}_1 &= f_1(X_1, \psi(X_1) + \tilde{X}_2, \varepsilon) = f_1^*(X_1, \tilde{X}_2, \varepsilon) \\ \varepsilon\dot{\tilde{X}}_2 &= f_2(X_1, \tilde{X}_2 + \psi(X_1), \varepsilon) - \varepsilon\frac{\partial \psi}{\partial X_1} f_1(X_1, \tilde{X}_2 + \psi(X_1), \varepsilon) \\ &= f_2^*(X_1, \tilde{X}_2, \varepsilon) \end{aligned} \tag{7.9}$$

$$\text{with } f_2^*(X_1, 0, 0) = 0$$

Within the time-scale $\tau = t/\varepsilon$, one has

$$\frac{\mathrm{d}X_1}{\mathrm{d}\tau} = \varepsilon f_1^*(X_1, \tilde{X}_2, \varepsilon) \qquad \frac{\mathrm{d}\tilde{X}_2}{\mathrm{d}\tau} = f_2^*(X_1, \tilde{X}_2, \varepsilon) \tag{7.10}$$

and as $f_2^*(X_1, 0, 0) = 0$, every point of $S(X_1, 0)$ is an equilibrium point (stable or unstable) for $\varepsilon = 0$. For $\varepsilon = 0$, the behavior of the system is defined by the family of n_2 differential equations

$$\frac{\mathrm{d}\tilde{X}_2}{\mathrm{d}\tau} = f_2^*(X_1^*, \tilde{X}_2, 0) \qquad \tilde{X}_2(0) = X_{20} - \psi(X_{10}) \tag{7.11}$$

where X_1^* can be considered as a constant parameter.

The solution of this equation represents what will be called the (initial) *boundary-layer solution*.

To obtain an approximation of (X_1, X_2), uniformly valid $\forall t$ one will try to superpose the external solution and the boundary-layer solution as follows

$$\begin{aligned} X_1 &= \overline{X}_1(t) \\ X_2 &= \overline{X}_2(t) + \tilde{X}_2(t) \end{aligned} \tag{7.12}$$

The conditions under which such an approximation is valid will be seen through the following examples.

7.2.1.2 Linear example

As a first example, let us consider a linear autonomous system defined by the differential equation

$$\varepsilon y'' + 2y' + y = 0$$

for which one wants to study the trajectory from initial conditions given by y_0, y_0'. By setting $x_1 = y$ and $x_2 = y'$ the equation becomes

$$\begin{bmatrix} \dot{x}_1 \\ \varepsilon \dot{x}_2 \end{bmatrix} = \begin{bmatrix} 0 & 1 \\ -1 & -2 \end{bmatrix} \begin{bmatrix} x_1 \\ x_2 \end{bmatrix} \qquad \begin{aligned} \dot{x}_1 &= f_1(x_1, x_2, \varepsilon) \\ \varepsilon \dot{x}_2 &= f_2(x_1, x_2, \varepsilon) \end{aligned}$$

for which the eigenvalues are given by

$$\lambda_i = \frac{-1 \pm \sqrt{1-\varepsilon}}{\varepsilon}$$

One can see that if ε tends towards 0, one of them tends to -0.5 when the other one tends to $-\infty$ in $-2/\varepsilon$. One can thus imagine the difficulties encountered for simulation of such a system due to the existence of two very different dynamics; one may also notice the possibility to separate the system into one slow subsystem and one fast subsystem. Within the time-scale t and for $\varepsilon = 0$, the previous equations lead to the solution

$$\begin{cases} \overline{x}_1 = x_{10}\, \mathrm{e}^{-0.5t} \\ \overline{x}_2 = -0.5 \overline{x}_1 \end{cases}$$

These equations represent what one calls the *reduced – or slow – or external* solution.

Obviously, as x_2 is no longer considered as a dynamic variable, initial conditions are not verified on this variable (*cf.* figure 7.1). If one performs the change of variables

$$x_2 = \tilde{x}_2 + \overline{x}_2 = \tilde{x}_2 + h(\overline{x}_1) = \tilde{x}_2 - 0.5\overline{x}_1$$

the equations become then, within the time-scale τ

$$\begin{aligned} \frac{dx_1}{d\tau} &= \varepsilon f_1() = \varepsilon(\tilde{x}_2 - 0.5x_1) \\ \frac{d\tilde{x}_2}{d\tau} &= -2\tilde{x}_2 + 0.5\varepsilon(\tilde{x}_2 - 0.5x_1) \end{aligned}$$

which means, for $\varepsilon = 0$

$$\begin{aligned} \frac{dx_1}{d\tau} &= 0 & x_1 &= \text{constant} = \mathrm{x}_{10} \\ \frac{d\tilde{x}_2}{d\tau} &= -2\tilde{x}_2 & \tilde{x}_2 &= \tilde{x}_{20}\, \mathrm{e}^{-2t/\varepsilon} = (x_{20} + 0.5x_{10})\, \mathrm{e}^{-2t/\varepsilon} \end{aligned}$$

these equations represent the fast solution or solution in the boundary-layer.

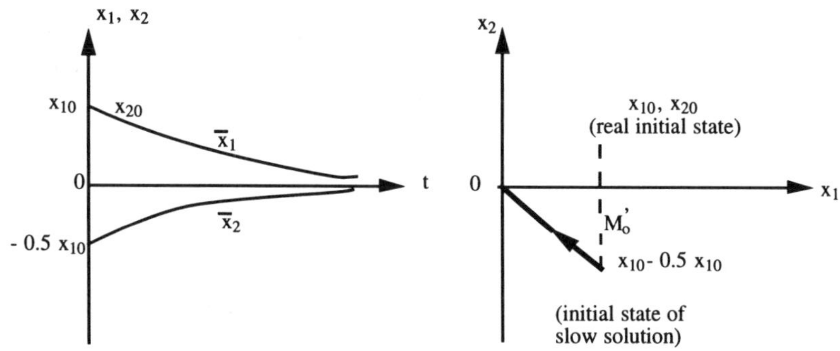

Figure 7.1: Slow motion: time responses and phase trajectory.

As the equilibrium solution of $\tilde{x}$, is equal to 0 and is a stable and attractive position for every $\tilde{x}_{20}$, one can choose the following as an approximate solution (*cf.* figure 7.2)

$$\begin{aligned} x_1 &= \overline{x}_1 = x_{10}\,\mathrm{e}^{-0.5t} \\ x_2 &= \overline{x}_2 + \tilde{x}_2 = -0.5x_{10}\,\mathrm{e}^{-0.5t} + (x_{20} + 0.5x_{10})\,\mathrm{e}^{-2t/\varepsilon} \end{aligned}$$

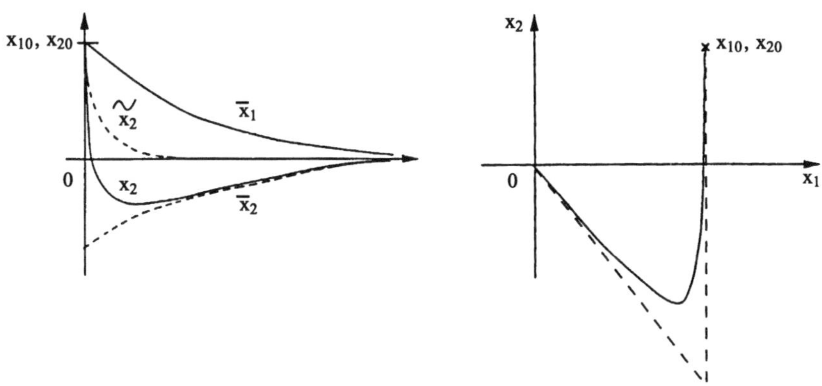

Figure 7.2: Complete motion: time responses and phase trajectory.

Remark 7.1

One can compare the above approximate solutions with the exact solutions given below

$$\begin{aligned} x_1 &= \frac{(1+\sqrt{1-\varepsilon})x_{10} + \varepsilon x_{20}}{2\sqrt{1-\varepsilon}}\,\mathrm{e}^{-\frac{1-\sqrt{1-\varepsilon}}{\varepsilon}t} \\ &\quad - \frac{\varepsilon x_{20} - (1-\sqrt{1-\varepsilon})x_{10}}{2\sqrt{1-\varepsilon}}\,\mathrm{e}^{-\frac{1+\sqrt{1-\varepsilon}}{\varepsilon}t} \\ x_2 &= -\frac{1-\sqrt{1-\varepsilon}}{2\sqrt{1-\varepsilon}}\left[\frac{1+\sqrt{1-\varepsilon}}{\varepsilon}x_{10} - x_{20}\right]\mathrm{e}^{-\frac{1-\sqrt{1-\varepsilon}}{\varepsilon}t} \\ &\quad + \frac{1+\sqrt{1-\varepsilon}}{2\sqrt{1-\varepsilon}}(x_{20} + \frac{1-\sqrt{1-\varepsilon}}{\varepsilon}x_{10})\,\mathrm{e}^{-\frac{1+\sqrt{1-\varepsilon}}{\varepsilon}t} \end{aligned}$$

One can see below the comparison between the approximate and the exact trajectories, in the phase plane for $\varepsilon = 0.01, 0.1$, and temporal evolution of x_1, x_2, showing the fact that, specially for the slow variables, the approximate solution $\overline{x}_1$ provides a good idea of the behavior of the slow variable x_1 (cf. figure 7.3).

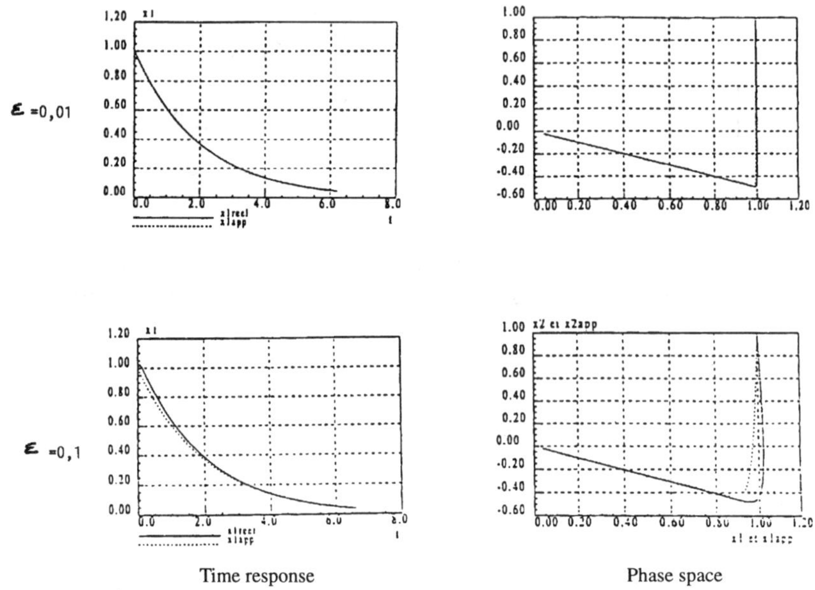

Figure 7.3: Trajectories in phase-plane for various values of ε.

Remark 7.2

One could also, without changing the x_2 variable, write directly equations within a time-scale τ as

$$\frac{\mathrm{d}x_1}{\mathrm{d}\tau} = \varepsilon x_2$$

$$\frac{\mathrm{d}x_2}{\mathrm{d}\tau} = -x_1 - 2x_2$$

which for $\varepsilon = 0$, leads to

$$x_1 = x_{10}$$

$$\hat{x}_2 = (x_{20} + 0.5x_{10})e^{-2t/\varepsilon} - 0.5x_{10}$$

and choose a combination of fast and slow solutions as an approximate solution, valid from $\mathrm{t} = 0$ to $\mathrm{t} = \infty$

$$x_1 = \overline{x}_1 = x_{10}\,\mathrm{e}^{-0.5t}$$

$$x_2 = \overline{x}_2 + \hat{x}_2 - \overline{x}_{20} = (x_{20} + 0.5x_{10})e^{-2t/\varepsilon} - 0.5x_{10}\,\mathrm{e}^{-0.5t}$$

(cf. figure 7.4).

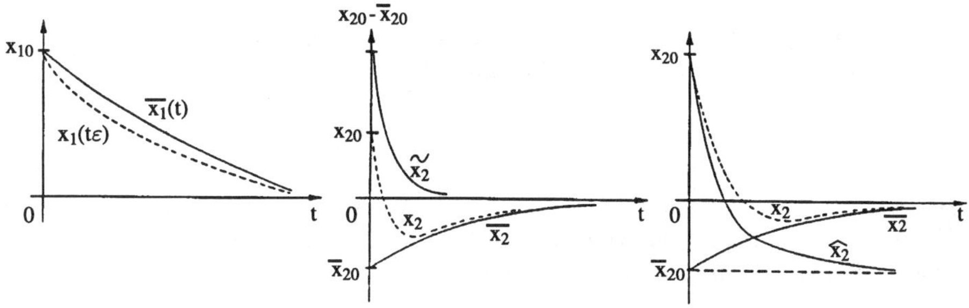

Figure 7.4: Time responses: slow component, fast component, global.

Remark 7.3

This procedure can obviously be extended to any autonomous linear system, in the form

$$\begin{bmatrix} \dot{X}_1 \\ \varepsilon \dot{X}_2 \end{bmatrix} = \begin{bmatrix} A_{11} & A_{12} \\ A_{21} & A_{22} \end{bmatrix} \begin{bmatrix} X_1 \\ X_2 \end{bmatrix}; \quad X_{10}, \quad X_{20}$$

for which the A_{22} matrix is invertible and stable, invertible so that the second group of equations has one, and only one, solution for $\varepsilon = 0$

$$\overline{X}_2 = -A_{22}^{-1} A_{21} \overline{X}_1$$

The reduced system is then defined by

$$\dot{\overline{X}}_1 = \left(A_{11} - A_{12} A_{22}^{-1} A_{21}\right) \overline{X}_1 = Ao\overline{X}_1 \longrightarrow \begin{array}{rcl} \overline{X}_1 & = & e^{A_0 t} X_{10} \\ \overline{X}_2 & = & -A_{22}^{-1} A_{21} \overline{X}_1 \end{array}$$

If, as above one writes

$$X_2 = \tilde{X}_2 - A_{22}^{-1} A_{21} X_1$$

one can find

$$\begin{aligned} \dot{X}_1 &= A_0 X_1 + A_{12} \tilde{X}_2 = f_1^* \left(X_1, \tilde{X}_2\right) \\ \varepsilon \dot{\tilde{X}}_2 &= \left(A_{22} + \varepsilon A_{22}^{-1} A_{21} A_{12}\right) \tilde{X}_2 + \varepsilon A_{22}^{-1} A_{21} A_0 X_1 = f_2^* \left(X_1, \tilde{X}_2, \varepsilon\right) \end{aligned}$$

The slow solution is defined by $\dot{X}_1 = f_1^* (X_1, 0)$ and the fast solution by

$$\varepsilon \frac{\mathrm{d}\tilde{X}_2}{\mathrm{d}t} = A_{22} \tilde{X}_2 = \frac{\mathrm{d}\tilde{X}_2}{\mathrm{d}\tau}$$

It is evident that this implies $\tilde{X}_2 \to 0$ and that the matrix A_{22} must be stable.

7.2.1.3 The nonlinear case: Tihonov's Theorem

One will notice that in the nonlinear case, with respect to the linear one,

1. the equation $f_2\left(\overline{X}_1, \overline{X}_2, 0\right) = 0$ does not usually provide, only one solution
2. the equilibrium solution with $\varepsilon = 0$, can be stable or unstable.

Theorem 7.1

Approximation of system trajectories (7.4) by the approached relations $\overline{X}_1, X_2 = \overline{X}_2 + \tilde{X}_2$ is then provided by Tihonov's theorem. As in (7.4), let

$$\dot{X}_1 = f_1(X_1, X_2, \varepsilon); X_{10}$$
$$\varepsilon\dot{X}_2 = f_2(X_1, X_2, \varepsilon); X_{20}$$

If f_1 and f_2 are continuous functions of X_1, X_2 in an open domain, if $f_2\left(\overline{X}_1, \overline{X}_2, 0\right) = 0$ admits one isolated root $\overline{X}_2 = \psi\left(\overline{X}_1\right)$, if this root is an asymptotic stable equilibrium point for the boundary-layer defined by

$$\frac{\mathrm{d}\tilde{X}_2}{\mathrm{d}\tau} = f_2\left(X_1, \tilde{X}_2, 0\right) \text{ with } X_2 = \tilde{X}_2 + \psi\left(\overline{X}_1\right),$$

if X_{10}, X_{20} remain in the attraction domain of equilibrium $X_{10,}\psi(X_{10})$, if the reduced problem and the complete problem have one, and only one, solution in $[0T]$, then

$$\begin{aligned}\lim_{\varepsilon\to 0} X_1(t,\varepsilon) &= \overline{X}_1(t)\\ \lim_{\varepsilon\to 0} X_2(t,\varepsilon) &= \overline{X}_2(t)\end{aligned} \tag{7.13}$$

Thus one can see, in the nonlinear case, that the "difficulty" may lie in the characterization of the (stable or unstable) equilibrium solution and in the determination of the corresponding attraction domain.

Example 1

For the purpose of illustration, let us consider the system

$$\begin{aligned}\dot{x}_1 &= -2x_1 + x_2 = f_1 \qquad x_{10} > 0\\ \varepsilon\dot{x}_2 &= +x_1^2x_2 - x_2^3 = f_2 \qquad x_{20}\end{aligned}$$

By setting $\varepsilon = 0$ the whole set S of solutions $x_1^2x_2 - x_2^3 = 0$ is given by the three straight lines $x_2 = 0$, $x_2 = x_1$, $x_2 = -x_1$ and $\dfrac{\partial f_2}{\partial x_2}$ is different from 0 at every point of S where $x_1 \neq 0$ (figure 7.5) The reduced system can be then represented in one of the three following forms:

$$\text{a)}\quad \begin{aligned}\overline{x}_2 &= 0 & \overline{x}_1 &= x_{10}\,\mathrm{e}^{-2t}\\ \dot{\overline{x}}_1 &= -2\overline{x}_1 & x_2 &= 0\end{aligned}$$

$$\text{b)}\quad \begin{aligned}\overline{x}_2 &= \overline{x}_1 & \overline{x}_1 &= x_{10}\,\mathrm{e}^{-t}\\ \dot{\overline{x}}_1 &= -\overline{x}_1 & \overline{x}_2 &= x_1\end{aligned}$$

$$\text{c)}\quad \begin{aligned}\overline{x}_2 &= -\overline{x}_1 & \overline{x}_1 &= x_{10}\,\mathrm{e}^{-3t}\\ \dot{\overline{x}}_1 &= -3\overline{x}_1 & \overline{x}_2 &= -x_1\end{aligned}$$

and the fast system by the corresponding forms

$$\text{a}')\quad \frac{\mathrm{d}\tilde{x}_2}{\mathrm{d}\tau} = x_{10}^2\tilde{x}_2 - \tilde{x}_2^3$$

$$\text{b}')\quad \frac{\mathrm{d}\tilde{x}_2}{\mathrm{d}\tau} = -2\tilde{x}_2x_{10}^2 - 3\tilde{x}_2^2x_{10} - \tilde{x}_2^3$$

$$\text{c}')\quad \frac{\mathrm{d}\tilde{x}_2}{\mathrm{d}\tau} = -2\tilde{x}_2x_{10}^2 + 3\tilde{x}_2^2x_{10} - \tilde{x}_2^3$$

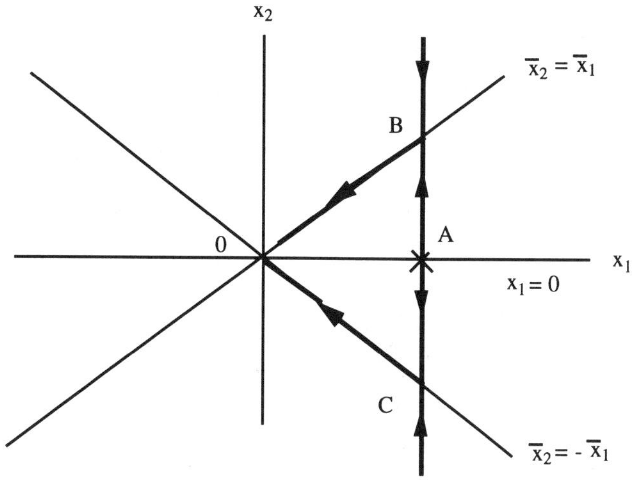

Figure 7.5: Phase trajectories for $\varepsilon = 0$.

One can notice that in the boundary-layer points B and C correspond to stable equilibria and point A corresponds to an unstable equilibrium.

Moreover, the attraction domain of B is upper-half straight line AB and that of C is the lower-half straight line AC .

Therefore, Tihonov's theorem gives

$$\forall x_{10} > 0 \quad x_{20} > 0 \qquad \lim_{\varepsilon \longrightarrow 0} x_1(t, \varepsilon) = x_{10}\, \mathrm{e}^{-t} \qquad \lim_{\varepsilon \longrightarrow 0} x_2(t, \varepsilon) = x_{10}\, \mathrm{e}^{-t}$$

$$\forall x_{10} > 0 \quad x_{20} < 0 \qquad \lim_{\varepsilon \longrightarrow 0} x_1(t, \varepsilon) = x_{10}\, \mathrm{e}^{-3t} \qquad \lim_{\varepsilon \longrightarrow 0} x_2(t, \varepsilon) = -x_{10}\, \mathrm{e}^{-3t}$$

One can find in figure 7-6 the graph of some results for $\varepsilon = 0.01, x_{10} = 1, x_{20} = 1.2, 0.2, -0.2, -1.2$ which shows the attraction phenomenon and two different slow dynamics, according to the initialization, corresponding respectively to the time constants 1 sec ($x_{20} > 0$) and 0.3 sec ($x_{20} < 0$).

7.2.1.4 Links with asymptotic expansions

One will note that the previous approximation corresponds to a zero-order solution provided by the asymptotic expansion method. One recalls that the nth order asymptotic expansion of a function $f(\varepsilon)$ with respect to a parameter ε is given by an expression in the form

$$f(\varepsilon) = \sum_{k=0}^{n} a_k \varepsilon^k + \beta_n(\varepsilon)\, \varepsilon^n$$

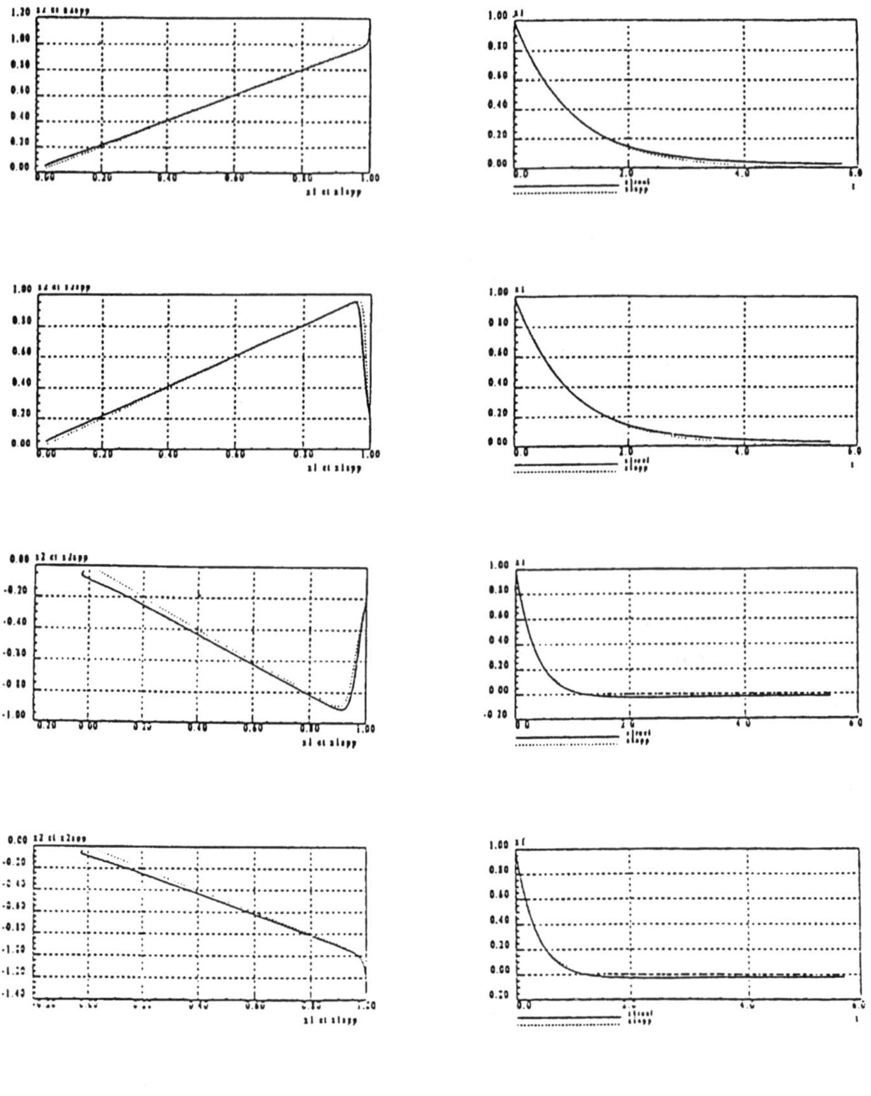

Figure 7.6: Time responses and phase trajectories for various initial conditions.

such that

and

$$\lim_{\varepsilon \to 0} \left[f(\varepsilon) - \sum_{i=0}^{k-1} a_i \varepsilon^i \right] / \varepsilon^k = a_k$$

$$\lim_{\varepsilon \to 0} \beta_n(\varepsilon) = 0$$

The asymptotic expansion techniques have been widely used in numerous domains and in the field of multiple time-scale systems one can quote, for example the matched asymptotic expansion techniques (Ardema and Rajan, 1983) or the multivariable approach of Hoppensteadt (1977).

7.2.1.4.1 Matched asymptotic expansion techniques (Ardema and Rajan, 1983)

In this case, one looks for two asymptotic expansions, independent of each other, the first one for the external (or reduced) solution, the second one for the initial boundary-layer. According to the system (7.4), the reduced solution is then looked for in the form:

$$\begin{aligned}\overline{x}_1(\varepsilon t) &= \overline{x}_{10} + \varepsilon\overline{x}_{11}(t) + \varepsilon^2\overline{x}_{12}(t) + \cdots \\ \overline{x}_2(\varepsilon t) &= \overline{x}_{20} + \varepsilon\overline{x}_{21}(t) + \varepsilon^2\overline{x}_{22}(t) + \cdots\end{aligned}$$

By introducing these expansions into the equations

$$\dot{x}_1 = f_1(x_1x_2) = f_1(\overline{x}_{10}\overline{x}_{20}) + \varepsilon\left[\left.\frac{\partial f_1}{\partial x_1}\right|_{\overline{x}_{10}}\overline{x}_{11} + \left.\frac{\partial f_1}{\partial x_2}\right|_{\overline{x}_{20}}\overline{x}_{21}\right] + \cdots = f_{10} + \varepsilon f_{11} + \cdots$$

$$\varepsilon\dot{x}_2 = f_2(x_1x_2) = f_2(\overline{x}_{10}\overline{x}_{20}) + \varepsilon\left[\left.\frac{\partial f_2}{\partial x_1}\right|_{\overline{x}_{10}}\overline{x}_{11} + \left.\frac{\partial f_2}{\partial x_2}\right|_{\overline{x}_{20}}\overline{x}_{21}\right] + \cdots = f_{20} + \varepsilon f_{21} + \cdots$$

and then by identifying the terms with the same exponents ε one can get a set of differential equations of linear type except for the first one,

$$\begin{array}{ll}\dot{\overline{x}}_{10} = f_{10} & 0 = f_{20} \\ \dot{\overline{x}}_{11} = f_{11} & \dot{\overline{x}}_{20} = f_{21} \\ \vdots & \vdots\end{array}$$

which integration provides the solutions, except for the constants of the integration.

If, one performs an asymptotic expansion for the boundary-layer, within the time-scale τ

$$\begin{aligned}\tilde{x}_1(\varepsilon,\tau) &= \tilde{x}_{10} + \varepsilon\tilde{x}_{11}(\tau) + \cdots \\ \tilde{x}_2(\varepsilon,\tau) &= \tilde{x}_{20} + \varepsilon\tilde{x}_{21}(\tau) + \cdots\end{aligned}$$

one can proceed in the same way, the difference being in that here all the constants of integration are well defined.

The integration constants introduced in the first step are then determined by the matching conditions

$$\lim_{\substack{\varepsilon\to 0\\ t\to 0\\ \tau\to\infty}} \left(\overline{x}_1(\varepsilon,t) - \tilde{x}_1(\varepsilon,\tau)\right) = 0$$

One then obtains a solution which is valid in the external layer and another in the boundary-layer. To get a unique solution, uniformly valid $\forall t$, one can add up both solutions by canceling the common part (pc) as follows:

$$\begin{aligned}x_1(\varepsilon,t) &= \overline{x}_1(\varepsilon,t) + \tilde{x}_1(\varepsilon,\tau) - \mathrm{pcx}_1(\varepsilon,\mathrm{t}) \\ x_2(\varepsilon,t) &= \overline{x}_2(\varepsilon,\tau) + \tilde{x}_2(\varepsilon,\tau) - \mathrm{pcx}_2(\varepsilon,\mathrm{t})\end{aligned}$$

This technique can become very complex if one takes a high-order expansion and this is not further commented here.

7.2.1.4.2 Multivariable approach

In this second approach, one is looking for an asymptotic expansion for each variable x_1, x_2 in the

$$\begin{aligned} x_1(t,\varepsilon) &= \overline{x}_1(t,\varepsilon) + \varepsilon x_1^i(\tau,\varepsilon) \\ x_2(t,\varepsilon) &= \overline{x}_2(t,\varepsilon) + x_2^i(\tau,\varepsilon) - \overline{x}_2(0,\varepsilon) \end{aligned}$$

with

$$\begin{aligned} \overline{x}_1(t,\varepsilon) &= \sum_{k=0}^{n} \overline{x}_{1k}(t)\,\varepsilon^k & x_1^i(\tau,\varepsilon) &= \sum_{k=0}^{n} x_{1k}^i(\tau)\,\varepsilon^k \\ \overline{x}_2(t,\varepsilon) &= \sum_{k=0}^{n} \overline{x}_{2k}(t)\,\varepsilon^k & x_2^i(\tau,\varepsilon) &= \sum_{k=0}^{n} x_{2k}^i(\tau)\,\varepsilon^k \end{aligned}$$

For the expansion of x_1, the term $x_1^i(\tau,\varepsilon)$ is multiplied by ε to take into account that the zero-order solution for the slow variable does not contain any term related to the boundary-layer. To define the problem completely, one also assigns two conditions.

1. $\overline{x}_1(t,\varepsilon), \overline{x}_2(t,\varepsilon)$ fit the system (7.1) at each time,

$$\begin{aligned} \frac{d\overline{x}_1(t,\varepsilon)}{dt} &= f_1(\overline{x}_1(t,\varepsilon), \overline{x}_2(t,\varepsilon)) \\ \varepsilon\frac{d\overline{x}_2(t,\varepsilon)}{dt} &= f_2(\overline{x}_1(t,\varepsilon), \overline{x}_2(t,\varepsilon)) \end{aligned}$$

2. $x_1^i(\tau,\varepsilon), x_2^i(\tau,\varepsilon)$ satisfy

$$\lim_{\tau\longrightarrow\infty} \begin{pmatrix} x_1^i(\tau,\varepsilon) \\ x_2^i(\tau,\varepsilon) - \overline{x}_2(0,\varepsilon) \end{pmatrix} = 0$$

One can then show that from the previous relations $x_1^i(\tau,\varepsilon), x_2^i(\tau,\varepsilon)$ satisfy the differential equations

$$\begin{aligned} \frac{dx_1^i(\tau,\varepsilon)}{d\tau} &= \varepsilon\left[f_1(x_1,x_2) - f_1(\overline{x}_1,\overline{x}_2)\right] \\ \frac{dx_2^i(\tau,\varepsilon)}{d\tau} &= f_2(x_1,x_2) - f_2(\overline{x}_1,\overline{x}_2) \end{aligned}$$

It is clear that the asymptotic expansion at order 0 satisfies the same equations as the external and the boundary-layer solutions, defined in the previous section. One can also note that differential equations associated with higher-order terms are all linear with time dependent coefficients and can be computed step by step in a recurrent way.

7.2.2 Introduction of a perturbation parameter – standard and non standard forms

Up to now, one has assumed that equations were put in a singularly perturbed standard form

$$\begin{aligned} \dot{X}_1 &= f_1(X_1,X_2,\varepsilon) \\ \varepsilon\dot{X}_2 &= f_2(X_1,X_2,\varepsilon) \end{aligned} \tag{7.14}$$

and one could deduce from $f_2(X_1,X_2,0) = 0$ one solution $\overline{X}_2 = \psi\left(\overline{X}_1\right)$, and only one locally. In fact, by writing the equations of a dynamic system, there is little chance that the equations can be expressed directly in the form (1), and also that a perturbation parameter can be introduced *a priori*. In others words, in practice, one will find the following problems:

1 – How to introduce a small parameter ε in the dynamic equations.

2 – Having solved this, if, instead of a standard form (1) one find a non standard form

$$\dot{X} = f(X, \varepsilon) \tag{7.15}$$

how to detect if the system has two time-scales, as if so, how to bring it back to a standard form.

7.2.2.1 Introduction of a physical perturbation parameter

Concerning the first point, for some cases, the parameter ε will be introduced naturally, directly from physical parameters of the process: the ratio of two masses, of two inertias, of two resistances, *etc.* inverse of stiffness of a transmission *etc.* The following examples 2 and 3 illustrate this situation, where it will then be possible to give a "numerical" value to parameter ε. In other cases, the parameter ε, even if linked to the physical properties of the system, has not, strictly speaking, a numerical value (Example 4) and can be linked, not only to the process but also to the scenario in which one wants to perform the evolution and it requires the use of reduced variables (Example 5).

Example 2

Suppose the system in figure 7.7 and let us assume currents i_1, i_2 in inductances L as state variables x_1, x_2. One has then the following equations

$$\begin{pmatrix} \dot{x}_1 \\ \dot{x}_2 \end{pmatrix} = \begin{pmatrix} -\dfrac{r+R}{L} & \dfrac{R}{L} \\ \dfrac{R}{L} & -\dfrac{r+R}{L} \end{pmatrix} \begin{pmatrix} x_1 \\ x_2 \end{pmatrix}$$

Let us assume that between resistances r and R one has such differences that one can set $\dfrac{r}{R} = \varepsilon$, with ε being small. The above equations then become, assuming $R/L = 1$ for the purpose of simplification

$$\begin{pmatrix} \dot{x}_1 \\ \dot{x}_2 \end{pmatrix} = \begin{pmatrix} -(1+\varepsilon) & 1 \\ 1 & -(1+\varepsilon) \end{pmatrix} \begin{pmatrix} x_1 \\ x_2 \end{pmatrix} = Ax = \begin{pmatrix} A_{11} & A_{12} \\ A_{21} & A_{22} \end{pmatrix} x$$

This represents a two time-scale system since in this linear case, eigenvalues are respectively $-\varepsilon$ and $-(2+\varepsilon)$ but the system is not set in its standard form.

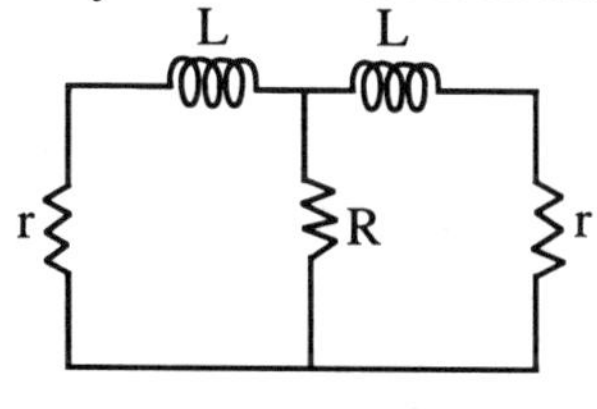

Figure 7.7: Network with two dynamics.

Example 3

Let us consider now the case (Marino-Kokotović, 1988) of a manipulator composed with an inertia arm J_1, with mass M, which is driven by a direct current motor with inertia J_m through a box, the flexibility of which is modeled by a spring with stiffness k. One finds the equations:

$$\begin{aligned} J_1 \frac{\mathrm{d}^2\theta_2}{\mathrm{d}t^2} + Mgl \sin\theta + k(\theta - \theta_m) &= 0 \\ J_m \frac{\mathrm{d}^2\theta_m}{\mathrm{d}t^2} - k(\theta - \theta_m) &= u \end{aligned}$$

where θ, θ_m are respectively the angular positions of the arm and the exit of motor shaft with u the torque applied to the motor shaft.

One then has $x_1 = \theta, \quad x_2 = \dot{\theta}, \quad x_3 = \theta_m, \quad x_4 = \dot{\theta}_m$

$$\begin{aligned} \dot{x}_1 &= x_2 \\ \dot{x}_2 &= -\frac{Mgl}{J_1} \sin x_1 - \frac{k}{J_1}(x_1 - x_3) \\ \dot{x}_3 &= x_4 \\ \dot{x}_4 &= \frac{k}{J_m}(x_1 - x_3) + \frac{1}{J_m} u \end{aligned}$$

In the case where the stiffness of the box is high, k becomes large and by setting $\varepsilon = 1/k$ the equations are written as

$$\begin{aligned} \dot{x}_1 &= x_2 \\ \dot{x}_3 &= x_4 \\ \varepsilon \dot{x}_2 &= -\frac{1}{J_1}(x_1 - x_3) - \frac{\varepsilon Mgl}{J_1} \sin x_1 \\ \varepsilon \dot{x}_4 &= \frac{1}{J_1}(x_1 - x_3) + \frac{\varepsilon}{J_m} u \end{aligned}$$

One can notice that this form, with $X_1 = \begin{bmatrix} x_1 \\ x_3 \end{bmatrix}$ and $X_2 = \begin{bmatrix} x_2 \\ x_4 \end{bmatrix}$

$$\begin{aligned} \dot{X}_1 &= f_1(X_1, X_2) = X_2 \\ \varepsilon \dot{X}_2 &= f_2(X_1, X_2, \varepsilon) = f_2^*(X_1, \varepsilon) \end{aligned}$$

is not in a standard singularly perturbed form, since $\dfrac{\partial f_2^*(X_0)}{\partial X_2} = 0.$

Example 4

In both of the previous cases, the parameter ε was expressed directly from the physical parameters of the process. The situation may be less objective. Let us consider for instance the motion of an aircraft, reduced to a point mass, flying in a vertical plane.

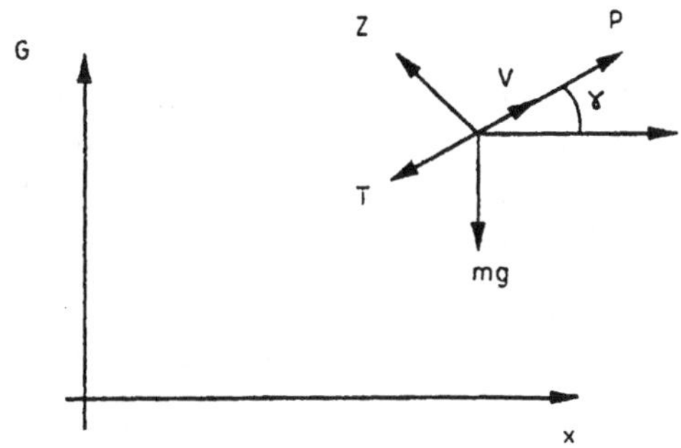

Figure 7.8: Visualization of the forces applied to the system.

If one calls, respectively, P the thrust, T the drag, Z the lift, mg the weight, h the altitude, the equilibrium of forces leads to equations

$$\begin{aligned}\dot{h} &= V\sin\gamma \\ \dot{V} &= \frac{P-T}{m} - g\sin\gamma\end{aligned}$$

where the drag T is linked to the lift Z, thus to attack angle α through the relation

$$Z = \frac{1}{2}\rho S V^2 C_z \quad T = \frac{1}{2}\rho S V^2 C_x \quad C_x = f(C_{x0}, C_z) \quad C_z = g(\alpha) \quad \rho = \rho(h)$$

If one assumes that one has a transport aircraft for which the ratio $\dfrac{P-T}{mg}$ is really small, one can set $\dfrac{P-T}{mg} = \varepsilon f(h, V)$ and rewrite the equations in the form

$$\begin{aligned}\dot{h} &= V\sin\gamma \\ \dot{V} &= g\left[\varepsilon f(h, V) - \sin\gamma\right]\end{aligned}$$

One has introduced a perturbation parameter but one obtains, at this step, a non standard form.

Remark 7.4

One can see, in the case of a missile for which under some conditions one can have a large ratio $(P-T)/m$ at the burning time of the powder propulsion, that one can then have an "inverse" situation with

$$\frac{P-T}{m} = kf(h, V) \qquad k >> 1$$

Setting $\varepsilon = 1/k$, the system would then become

$$\begin{aligned}\dot{h} &= V\sin\gamma \\ \varepsilon\dot{V} &= f(h, V) - \varepsilon g\sin\gamma\end{aligned}$$

which is a standard form.

Example 5

Let us consider a "system" composed of two vehicles in one plane (for example 2 aircraft as in an horizontal plane). Vehicle T moves with constant speed V_T along one axis, used as reference axis. Vehicle M also moves at constant speed V_M, but its direction γ can be changed by applying to it a normal acceleration a.

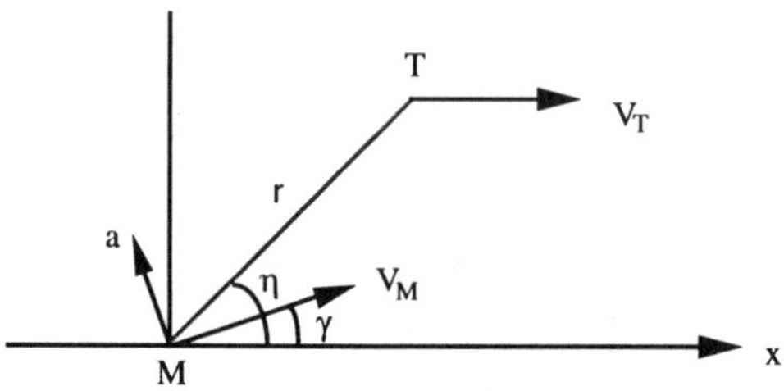

Figure 7.9: Variables used in example 5.

Using the symbols of figure 7.9 the motion equations are

$$\begin{aligned} \dot{r} &= V_T \cos\eta - V_M \cos(\eta - \gamma) \\ r\dot{\eta} &= V_M \sin(\eta - \gamma) - V_T \sin\eta \\ \dot{\gamma} &= \frac{a}{V_M} \end{aligned}$$

Let us assume that the problem, for the vehicle M, consists in joining vehicle T and that the initial distance r_0 between M_0 and T_0 is large with respect to ρ, the minimum turning radius of M. One can then introduce a perturbation parameter $\varepsilon = \dfrac{\rho}{r_0}$.

If R is the turning ratio as a result of the acceleration a, and ρ is its corresponding value related to the maximal acceleration a_M, let us rewrite the last equation in the form

$$\dot{\gamma} = \frac{a}{V_M} = \frac{V_M}{R} = \frac{V_M}{\rho}u \quad \text{with } -1 < u = \frac{a}{a_M} < 1$$

If now one introduces dimensionless quantities

$$r^* = \frac{r}{r_0} \qquad \frac{V_T}{V_M} = \nu < 1$$

the previous equations become

$$\begin{aligned} \dot{r}^* &= \frac{V_M}{r_0}\left[\nu \cos\eta - \cos(\eta - \gamma)\right] \\ \dot{\eta} &= \frac{V_M}{r}\left[\sin(\eta - \gamma) - \nu \sin\eta\right] \\ \varepsilon\dot{\gamma} &= \frac{V_M}{r_0}u \end{aligned}$$

As the second members of all these equations have the same order $\mathrm{O}\left(\dfrac{V_M}{r_0}\right)$, at least at the beginning of the action, the system is set in the standard perturbed form, thus putting forward the slow variables r and η, and one fast variable γ.

7.2.2.2 Standard and non standard forms

Here, in the case where it was possible to introduce a perturbation parameter although the system was not under a standard form, the problem consists in checking if the system has two time-scales, and if so, in trying to bring it back to a standard form.

7.2.2.2.1 Recall of the linear case

In the linear case, the decision whether a system is a two time-scale system is not a major problem, since its dynamic is linked to the eigenvalues of evolution matrix

$$A(\dot{X} = AX + \cdots)$$

which are generally easy to compute.
Knowing these eigenvalues, by increasing modulus, they can be divided into two separate classes such that the maximum of the small eigenvalues is far lower than the minimum of the large eigenvalues (in absolute value).
Having done this, the standard form can be easily obtained through a linear transformation. If n_1 is the number of slow modes, then $n_2 = n - n_1$ is the number of fast modes and as the initial model was divided into the form

$$\begin{pmatrix} \dot{X}_1 \\ \dot{X}_2 \end{pmatrix} = \begin{pmatrix} A_{11} & A_{12} \\ A_{21} & A_{22} \end{pmatrix} \begin{pmatrix} X_1 \\ X_2 \end{pmatrix} \quad \begin{matrix} X_1 \in R^{n_1} \\ X_2 \in R^{n_2} \end{matrix}$$

one block-triangularizes matrix A in the form

$$\begin{pmatrix} \dot{X}_1 \\ \dot{\hat{X}}_2 \end{pmatrix} = \begin{pmatrix} \hat{A}_1 & A_{12} \\ 0 & \hat{A}_2 \end{pmatrix} \begin{pmatrix} X_1 \\ \hat{X}_2 \end{pmatrix}$$

the matrices $\hat{A}_1 \hat{A}_2$ having as eigenvalues, the respectively slow and fast eigenvalues of the initial system. One knows that this can be achieved through the coordinate change

$$\begin{pmatrix} X_1 \\ \hat{X}_2 \end{pmatrix} = \begin{pmatrix} I & 0 \\ L & I \end{pmatrix} \begin{pmatrix} X_1 \\ X_2 \end{pmatrix}$$

where L, is the solution of the Riccati equation

$$A_{21} + LA_{11} - A_{22}L - LA_{12}L = 0$$

and which can also be computed directly by

$$L = -V_{21}V_{11}^{-1}$$

$V_1 = \begin{bmatrix} V_{11} \\ V_{21} \end{bmatrix}$ being the eigenvalue matrix associated with slow modes.
In example 2 of previous paragraph one has for instance $L = -I$ and the transformation leads to

$$\begin{pmatrix} \dot{x}_1 \\ \dot{\hat{x}}_2 \end{pmatrix} = \begin{pmatrix} 1 & 0 \\ -1 & 1 \end{pmatrix} \begin{pmatrix} x_1 \\ x_2 \end{pmatrix} \rightarrow \begin{pmatrix} \dot{x}_1 \\ \dot{\hat{x}}_2 \end{pmatrix} = \begin{pmatrix} -\varepsilon & 1 \\ 0 & -(2+\varepsilon) \end{pmatrix} \begin{pmatrix} x_1 \\ \hat{x}_2 \end{pmatrix}$$

$$= \begin{pmatrix} \hat{A}_1 & A_{12} \\ 0 & \hat{A}_2 \end{pmatrix} \begin{pmatrix} x_1 \\ \hat{x}_2 \end{pmatrix}$$

where $\hat{x}_2$ represents the fast variables.

The system can be then block-diagonalized by the transformation

$$\begin{pmatrix} \hat{x}_1 \\ \hat{x}_2 \end{pmatrix} = \begin{pmatrix} I & -M \\ 0 & I \end{pmatrix} \begin{pmatrix} x_1 \\ \hat{x}_2 \end{pmatrix}$$

where M is solution of Lyapunov's equation

$$\hat{A}_1 M - M\hat{A}_2 + A_{12} = 0$$

which puts the system in a completely decoupled form.

$$\begin{pmatrix} \dot{\hat{x}}_1 \\ \dot{\hat{x}}_2 \end{pmatrix} = \begin{pmatrix} \hat{A}_1 & 0 \\ 0 & \hat{A}_2 \end{pmatrix} \begin{pmatrix} x_1 \\ x_2 \end{pmatrix}$$

With the same example, the Lyapunov's equation then becomes

$$-\varepsilon m + (2+\varepsilon)m + 1 = 0 \qquad m = -0.5$$

and globally the transformation used is

$$\begin{pmatrix} \hat{x}_1 \\ \hat{x}_2 \end{pmatrix} = \begin{pmatrix} 0.5 & 0.5 \\ -1 & 1 \end{pmatrix} \begin{pmatrix} x_1 \\ x_2 \end{pmatrix}$$

which puts the system in the form (related to the standard form within the time-scale $\tau = t/\varepsilon$)

$$\begin{pmatrix} \dot{\hat{x}}_1 \\ \dot{\hat{x}}_2 \end{pmatrix} = \begin{pmatrix} -\varepsilon & 0 \\ 0 & -(2+\varepsilon) \end{pmatrix} \begin{pmatrix} \hat{x}_1 \\ \hat{x}_2 \end{pmatrix}$$

The slow variable is the sum of currents in the inductances and the fast variable is their difference.

Remark 7.5

From initial equations of example 2

$$\begin{pmatrix} \dot{x}_1 \\ \dot{x}_2 \end{pmatrix} = \begin{pmatrix} -(1+\varepsilon) & 1 \\ 1 & -(1+\varepsilon) \end{pmatrix} \begin{pmatrix} x_1 \\ x_2 \end{pmatrix}$$

which becomes, by setting $\varepsilon = 0$

$$\begin{pmatrix} \dot{x}_1 \\ \dot{x}_2 \end{pmatrix} = \begin{pmatrix} -1 & 1 \\ 1 & -1 \end{pmatrix} \begin{pmatrix} x_1 \\ x_2 \end{pmatrix}$$

one can see that:

- *the equilibrium is defined by* $x_2 - x_1 = 0$ *and*
- $\dot{x}_1 + \dot{x}_2 = 0$ *which means that* $x_1 + x_2 =$ constant. *The fact that slow variable is the sum of the currents is not very unusual! (cf. next remark).*

Remark 7.6

In the linear case, it is not always necessary to compute eigenvalues to decide whether the system has two time-scale.

A simple localization can be enough. Indeed there exist several criteria associated with matrix A allowing this localization: Gerschgorine's circles (Dauphin-Tanguy, 1983), Cassini's ovals (Moreigne, 1984). One can also use results from the matrix signatures related to given domains (Dauphin-Tanguy et al., 1982).

7.2.2.2.2 Nonlinear case

In the nonlinear case, the problem is much more complex, with the exception, perhaps, of the nonlinear system of the Lure-Postnikov type. This system consists of a linear part and a static nonlinear one, the equations of which are

$$\begin{aligned} \dot{X} &= AX + Bf(e) \\ y &= Cx \\ e &= r - Cx \end{aligned} \tag{7.16}$$

where $f(e)$ is a function ϕ defined by

$$\phi : \{f \,|f : R \to Rf(e) = k(e)ek_1 < k(e) < k_2\}$$

For more details concerning this special case, one can refer to Dauphin Tanguy *et al.* (1982).
In a general case one encounters two problems: having a system in a non standard form

$$\dot{X} = f(X, u, \varepsilon) \tag{7.17}$$

the question is how to determine if it is a two time-scale system and, if so, how to bring it back to a standard form.
A first usable approach consists in using linear techniques. One should then linearize the equations around a nominal trajectory. The procedure is complex and will not be developed here (Moreigne, 1984).
A second approach, which has been tried by Kelley (1971; 1972) consists in looking directly for nonlinear transformations which separate slow and fast variables in a similar way as in linear cases. Using Kelley's approach, the situation becomes quickly inextricable.
A geometric approach has been suggested, more recently, by R. Marino and P.V. Kokotović (1988). The basic elements of this approach are described below even though this may be very difficult to carry out in real situations.
Let us first notice that if the system is in a standard form

$$\begin{aligned} \dot{X}_1 &= f_1(X_1, X_2, \varepsilon, u) \qquad X_1 \in R^{n_1} \\ \varepsilon\dot{X}_2 &= f_2(X_1, X_2, \varepsilon, u) \qquad X_2 \in R^{n_2} \end{aligned}$$

or within the time-scale τ

$$\begin{aligned} \frac{dX_1}{d\tau} &= \varepsilon f_1(X_1, X_2, \varepsilon, u) \\ \frac{dX_2}{d\tau} &= f_2(X_1, X_2, \varepsilon, u) \end{aligned}$$

the system is reduced, for $\varepsilon = 0$, to

$$\frac{dX_1}{d\tau} = 0 \qquad \frac{dX_2}{d\tau} = f_2(X_1, X_2, 0, u)$$

The first equation shows that, independently of u (within the time-scale τ and for $\varepsilon = 0$), the n_1 states variables X_1 are constant and equal to X_1^* and thus, the system evolves within a variety C with dimension n_2, parametrized by X_1^*, which is called "conservative" by Marino and Kokotović (1988). Within the same time-scales, still for $\varepsilon = 0$, the equilibrium conditions are defined by

$$f_2\left(X_1^*, X_2, 0, u\right) = 0$$

which means a variety E^u of dimension n_1, which can depend on u, and normal to C according to imposed conditions on f_2.

In example 5 (section 7.2.2.1) where the initial equations were in the standard form, C and E^u as represented in figure 7.10.

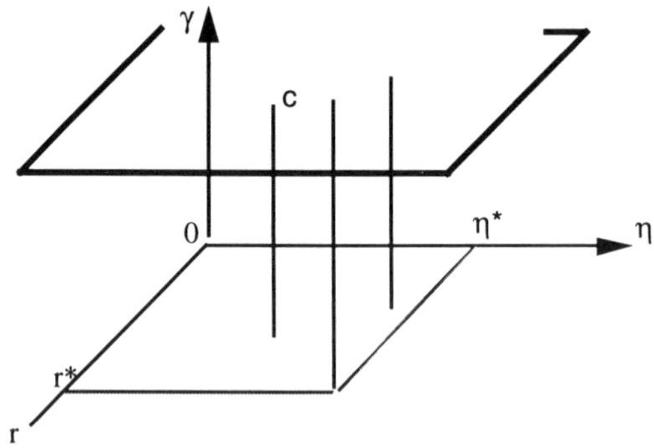

Figure 7.10: Equilibrium and conservation manifolds for example 5.

If one now considers example 4, for $\varepsilon = 0$ the equations become

$$\begin{aligned} \dot{h} &= V \sin\gamma \\ \dot{V} &= -g \sin\gamma \end{aligned} \tag{7.18}$$

thus $g\dot{h} + V\dot{V} = 0$ which leads to $e = gh + \frac{1}{2}V^2 = \text{constant}$. In the (h, V) plane, the conservative variety is defined by the parabolae $e = \text{constant}$.

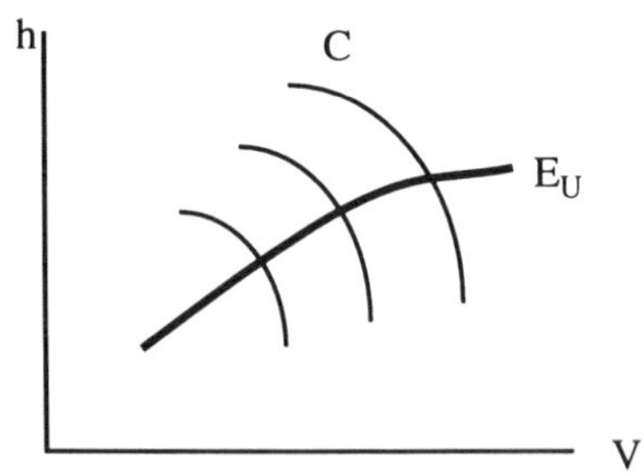

Figure 7.11: Equilibrium and conservation manifold for example 4.

Within the same time-scale, the equilibrium assumes that $\gamma = 0$, *i.e.*

$$mg = Z = \frac{1}{2}\rho(h)\, SV^2 Cz$$

which defines the E_u curve in the same plane.

If one defines the diffeomorphism

$$\xi_1 = gh + \frac{1}{2}V^2$$
$$\xi_2 = h$$

by choosing $\xi_1 = e$ as a slow variable, which seems to be natural since for $\varepsilon = 0$ one has $e =$ constant, the equations (7.18) become

$$\dot{\xi}_1 = \varepsilon f^* \left(\sqrt{\xi_1 - g\xi_2}, \xi_2\right)$$
$$\dot{\xi}_2 = \sqrt{2(\xi_1 - g\xi_2)} \sin\gamma$$

which are now in a standard form. One can notice that as in the linear case of example 2, the slow variable is associated with the conservation property, here $e = gh + 0.5V^2$.
One can also notice that the numbers of slow and fast variables in both of the previous cases are respectively linked to the dimensions of the equilibrium variety and the conservation variety.
Without going further into the details which would involve too much mathematics to be demonstrated here, let us recall the fundamental theorem of Marino (1988):
Consider the system

$$\frac{\mathrm{d}X}{\mathrm{d}\tau} = f(X, \varepsilon, u) \quad X \in R^{n_1+n_2} \tag{7.19}$$

It is a two time-scale system if (but not only if):

(i) for every admissible control, there exists an equilibrium variety with a constant dimension n_1, defined by:

$$E^u : \{X \in S : f(X, 0, u) = 0 \; u \in U\} \tag{7.20}$$

(ii) there exists an involutive distribution with constant rank n_2 such that

$$f(X, 0, u(X0)) \subset \Delta(X) \forall X \in S \text{ et } u \in U \tag{7.21}$$

(iii) for every u, the equilibrium variety E^u is orthogonal to family C of the conservative varieties at every point X which belongs to E^u, *i.e.*, if

$$\Delta(X) \oplus T_x E^u = R^{n_1+n_2} \tag{7.22}$$

Remark 7.7

Condition (i) corresponds to $\dfrac{\partial f_2}{\partial x_2}$ *non singular in the case of a standard form, with symbols used in equation 1.*

Condition (ii) corresponds to a conservative property.

One can notice that the three conditions given above are only sufficient conditions and that one can have a two time-scale system for which these conditions are not satisfied. On the other hand, if and only if the three conditions are verified there exists a local diffeomorphism, which is independent of ε, $\xi_1 = \phi_1(X)$, $\xi_2 = \phi_2(X)$ and which transforms the system $\dot{X} = f(X, \varepsilon, u)$ in

$$\begin{aligned} \dot{\xi}_1 &= \varepsilon F_1(\xi_1, \xi_2, \varepsilon, u) \\ \dot{\xi}_2 &= F_2(\xi_1, \xi_2, \varepsilon, u) \end{aligned} \tag{7.23}$$

where $\xi_1 \in R^{n_1}$ and $\xi_2 \in R^{n_2}$.

Marino and Kokotović (1988) also specify the procedure allowing the determination of the diffeomorphism which sets the system in a standard form: having determined the u set of admissible controls, one chooses $u_0 \in U$ and n_2 functions $\phi_{n_1+1} \ldots \phi_{n_1+n_2}$ such that E^{u_0} is locally described by

$$\phi_i(X) = 0 \qquad i = n_1 + 1, \ldots n_1 + n_2$$

One then computes n_1 functions $\phi_1 \ldots \phi_{n_1}$ such that

$$< d\phi_i, f(x, 0, u) >= 0 \qquad i = 1, \ldots, n_1 \tag{7.24}$$

and checks the conditions

$$\text{rank}\,(\mathrm{d}\phi_1 \ldots \mathrm{d}\phi_{n_1+n_2}) = n_1 + n_2$$

If these conditions are satisfied, the diffeomorphism is defined by

$$\xi_1 = \phi_\ell(X) = \begin{bmatrix} \phi_1 \\ \vdots \\ \vdots \\ \vdots \\ \phi_{n_1} \end{bmatrix} \qquad \xi_2 = \phi_r(X) = \begin{bmatrix} \phi_{n_1+1} \\ \vdots \\ \vdots \\ \vdots \\ \phi_{n_1+n_2} \end{bmatrix}$$

One can notice that this procedure corresponds to what has been performed intuitively in example 4, even if this is not the only possible way, and even if this can lead to an impasse while physical considerations can lead to a more appropriate choice.

For the purpose of illustration, one considers again example 3 (Marino and Kokotović, 1988).

These equations become, within the time-scale $\tau = t/\varepsilon$

$$\begin{aligned} \frac{\mathrm{d}x_1}{\mathrm{d}\tau} &= \varepsilon x_2 & \frac{\mathrm{d}x_3}{\mathrm{d}\tau} &= \varepsilon x_4 \\ \frac{\mathrm{d}x_2}{\mathrm{d}\tau} &= -\frac{1}{J_1}(x_1 - x_3) - \frac{\varepsilon M g l}{J_1}\sin x_1 & \frac{\mathrm{d}x_4}{\mathrm{d}\tau} &= \frac{1}{J_m}(x_1 - x_3) + \frac{\varepsilon}{J_m}u \end{aligned}$$

By setting $\varepsilon = 0$, the equilibrium space is defined by $x_1 - x_3 = 0$, (dimension 3) and according to (7.24) it suggests choosing $\phi_4 = x_1 - x_3$

On the other hand, still with $\varepsilon = 0$, the conservative variety is defined by

$$\frac{\partial \phi}{\partial x_2}\left(-\frac{1}{J_1}(x_1 - x_3)\right) + \frac{\partial \phi}{\partial x_4}\left(\frac{1}{J_m}(x_1 - x_3)\right) = 0$$

which is satisfied for $\phi_1 = J_1 x_2 + J_M x_4$ and for any function which is independent from x_2 and x_4. One could choose

$$\begin{aligned} \phi_2 &= \alpha x_1 + \beta x_3 \\ \phi_3 &= \alpha' x_1 + \beta' x_3 \end{aligned} \qquad \begin{vmatrix} \alpha & \beta \\ \alpha' & \beta' \end{vmatrix} \neq 0$$

One cannot, however, conclude that it is a two-time-scale system with three slow variables and one fast variable because the transversality condition (iii) is not satisfied, since in fact

$$\text{rank}\,(\mathrm{d}\phi_1 \ldots \mathrm{d}\phi_4) < 4$$

7.2.2.3 Forced singular perturbations

As one has already seen, to write a realistic physical system in a standard singularly perturbed form and to separate the system into slow and fast variables can be somewhat complex from a mathematical point of view.
It will be the case, for example, in nonlinear optimization problems in which we are interested hereunder and where the aim is to obtain a control in terms of state variables, in a closed-loop form, easy to implement and providing the "best" compromise computation performances. It will generally be insufficient to define only two classes of dynamics and one will need a multiple-time-scale formulation in the form

$$\begin{aligned} \dot{X}_1 &= f_1(X_1, X_2, \ldots X_k, \varepsilon, u) \\ \varepsilon\dot{X}_2 &= f_2(X_1, X_2, \ldots X_k, \varepsilon, u) \\ \varepsilon^{k-1}\dot{X}_k &= f_k(X_1, X_2, \ldots X_k, \varepsilon, u) \end{aligned} \tag{7.25}$$

In practical situations, even if there is a need to have a mathematical model in order to develop the control law, the designer usually has a long enough experience of the system and of its possible evolution for a given scenario which enables him to know where the slow and the fast variables are and he can thus proceed to a first arrangement.
If necessary, he can always perform simulations with some realistic control schemes, even if they are not optimal, to arrive at a first assessment. He can then define the "velocity" of a variable as the inverse of the time which is used to cover a characteristic domain of variation.

$$v_i = \max_{\substack{u \in \mathcal{D}u \\ x \in \mathcal{D}}} \frac{f_i(X, u)}{\Delta x_i} \tag{7.26}$$

this evaluation being, naturally, performed within the acceptable domain of variations of the controls and states.
Having grouped in the state variables different classes, according to (7.26), the parameter ε is then introduced in the initial equations as a marker of speed (possibly in the form ε^k if there is more than two time-scales), even if it will be used as a real perturbation parameter afterwords.
This technique which is called forced singular perturbations – used in aeronautics for optimal trajectories problems– seems to be the most realistic one. It contains perhaps a degree of empiricism but presents, apart from its simplicity, the advantage of classing the variables, not only in terms of the process itself, but also in terms of the scenario of the control to be used (as will be shown in example of § 7.3.4).

7.2.3 *Application to the decoupling linearized control*

Even from a qualitative point of view, the separation of a global system into one slow system and one (or several) fast systems allows us, from the control aspect, to simplify the control law considerably and, within the scope of the same purpose (optimal or decoupling-linearizing for example) to make a difference between an unrealistic solution, due to its complexity, and a feasible one, even if the initial objective is slightly degraded (due to the initial assumptions, as the dynamic separation can never be performed completely).

The remainder of this chapter will be devoted to applications of optimization of nonlinear systems to multi-time-scale systems. However one first provides an example of a decoupling-linearizing control applied to a system for which the state variables can be divided into two classes, respectively slow and fast.

In chapter 2, one has already seen the principle of a nonlinear decoupling control, based on the linearization between input and output by diffeomorphism and feedback, and the main limitations of this method which are linked to, on the one hand the possibility of getting a subspace of state variables which is not observable due to non-interacting constraints and, on the other hand, the complexity of the control law. These two problems are linked together in so far as, with a system of a fixed order, the dimension of the unobservable variety is linked to characteristic indices and an unobservable subspace, with low dimension, induces large characteristic numbers, thus a complex control law requiring high-order derivations of the outputs.

For the purpose of illustration, one considers the following problem of piloting a missile for which one wishes to control transverse acceleration while keeping a constant roll angle.

Using classical symbols of flight mechanics α the attack angle, β the slide slip angle (assumed to be small), θ the flight path angle, V the velocity, F_X, F_y, F_z, L, M, N the aerodynamic forces and torques along G_x, G_y, G_z, $L_p p$, $L_q q$, $N_r r$ the damping momentum (p, q, r are rolling, pitching and yawing angular speeds), θ, η, ζ equivalent angles of control surfaces equations of motion are:

• Momentum equations

$$\begin{aligned} I_x\dot{p} &= L(\alpha,\beta,\xi) + L_p p \\ I_y\dot{q} + (I_x - I_z)pr &= M(\alpha,\beta,\eta) + M_q q \\ I_z\dot{r} + (I_y - I_x)pq &= N(\alpha,\beta,\zeta) + N_r r \end{aligned}$$

• Force equations

$$\begin{aligned} \dot{\alpha} - q + p\beta &= \frac{1}{mV}\left[F_z(\alpha,\beta) + mg\cos\theta\sin\varphi\right] \\ \dot{\beta} - r - p\alpha &= \frac{1}{mV}\left[F_y(\alpha,\beta) + mg\cos\theta\sin\varphi\right] \end{aligned}$$

(we do not develop G_x). (7.27)

• Kinematic equations

$$\begin{aligned} \dot{\varphi} &= p + \tan\theta\,(q\sin\varphi + r\cos\varphi) \\ \dot{\theta} &= q\cos\varphi - r\sin\varphi \end{aligned}$$

Finally, to control the missile, the output variables to be controlled are:

$$\begin{aligned} y_1 &= \varphi : \text{ rolling angle} \\ y_2 &= \Gamma_y : \text{ acceleration along } G_y \\ y_3 &= \Gamma_z : \text{ acceleration along } G_z \end{aligned}$$

where

$$\begin{aligned} \Gamma_y &= \frac{1}{m}\left[F_y(\alpha,\beta,\zeta) + mg\cos\theta\sin\varphi\right] \\ \Gamma_z &= \frac{1}{m}\left[F_z(\alpha,\beta,\eta) + mg\cos\theta\cos\varphi\right] \end{aligned}$$

As one wishes to control transverse accelerations, especially along the component vertical to the earth, one will notice that flight path angle θ (introduced because $\dot{\varphi}$ is dependent) can be an unstable state. This is not considered to be a drawback because the variations of θ can be further taken into account by guidance laws.

One will also notice that:

- in the force equations, F_z and F_y were assumed to be dependent only on α and β (they are also dependent on η and ζ but in a much smaller extent).

- dynamics of actuators of control surfaces are not taken into account

- one considers in what follows, the input variables

$$u_1 = \dot{\xi} \quad u_2 = \dot{\eta} \quad u_3 = \dot{\zeta}$$

  in order to have a dynamic equation which is linear in terms of control.

Thus the state vector has a dimension

$$X = [p, q, r, \alpha, \beta, \varphi, \theta, \xi, \eta, \zeta]^T$$

and the previous equations become formally

$$\begin{cases} \dot{p} &= f_1(p, \alpha, \beta, \xi) \\ \dot{q} &= f_2(p, q, r, \alpha, \beta, \eta) \\ \dot{r} &= f_3(p, q, r, \alpha, \beta, \zeta) \end{cases}$$
$$\begin{cases} \dot{\alpha} &= f_4(p, q, \alpha, \beta, \varphi, \theta) \\ \dot{\beta} &= f_5(p, r, \alpha, \beta, \varphi, \theta) \end{cases}$$
$$\begin{cases} \dot{\varphi} &= f_6(p, q, r, \varphi, \theta) \\ \dot{\beta} &= f_5(q, r, \varphi) \end{cases} \tag{7.28}$$
$$\begin{cases} \dot{\xi} &= u_1 \\ \dot{\eta} &= u_2 \\ \dot{\zeta} &= u_3 \end{cases}$$

with the output variables

$$\begin{aligned} y_1 &= h_1(X) = \varphi \\ y_2 &= h_2(X) = h_2(\alpha, \beta, \varphi, \theta, \zeta) = \Gamma_y \\ y_3 &= h_3(\alpha, \beta, \varphi, \theta, \eta) = \Gamma_z \end{aligned} \tag{7.29}$$

It is then easy to verify that all characteristic numbers are equal to 3. One has, for example

$$
\begin{aligned}
y_1 &= \varphi \\
\dot{y}_1 &= \dot{\varphi} = f_9\left(p, q, r, \varphi, \theta\right) \\
\ddot{y}_1 &= \sum \frac{\partial f_6}{\partial x_i} f_i \quad i = 1, 2, 3, 6, 7 \\
\dddot{y}_1 &= G\left(\ldots, u_1\right)
\end{aligned}
$$

One encounters the situation where the unobservable induced part has only the dimension one (corresponding to variable θ). From a theoretical point of view, this seems to be a good thing, since θ should rather be considered in this problem as a parameter. In fact, due to the high values of characteristic numbers, one should not apply the theoretical law by using this approach since this would force us to work with third-order derivatives of output variables.

Remark 7.8

One can notice that if one takes into account the fact that F_z and F_y are theoretically dependent on η and ζ, the characteristic numbers should then be equal 2, 0, 0. The unobservable subvariety should then be unstable with dimension 5!

To solve this dilemma, one can take into consideration the fact that a study of the physical process shows that some state variables $(\alpha, \beta, \phi, \theta)$ are slow whereas other variables $(p, q, r, \xi, \eta, \zeta)$ are fast. It can then be possible to divide, to a certain extent, state equations into two groups, in the form of two subsystems connected as shown in figure 7.12.

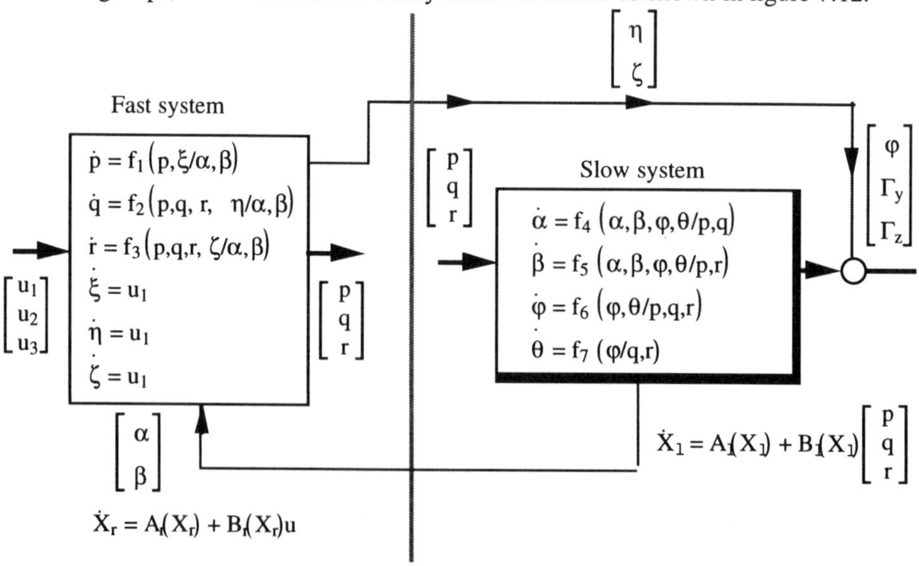

Figure 7.12: Separation of the flight equations into slow and fast subsystems.

Due to this dynamic division, slow variables $\alpha, \beta, \phi, \theta$ are assumed to be constant for the evolution of the fast subsystem and, conversely, fast variables are assumed to be inputs or parameters for the evolution of the slow system.

Slow subsystem

If one considers the slow system, with four state variables $(\alpha, \beta, \phi, \theta)$, where input variables are (p, q, r), where η and ξ are assumed to be constant and where ϕ, Γ_y, Γ_z, are output variables, it is easy to see that all the characteristic numbers are equal to 1. The decoupling matrix Δ_ℓ then becomes

$$\Delta_\ell = \begin{bmatrix} 1 & \tan\theta\sin\varphi & \tan\theta\cos\varphi \\ \frac{1}{\mathrm{m}}\left(-\beta \mathrm{F}_{\mathrm{y}_\alpha} + \alpha \mathrm{F}_{\mathrm{y}_\beta} + \mathrm{mg}\cos\theta\cos\varphi\right) & \frac{1}{\mathrm{m}}\mathrm{F}_{\mathrm{y}_\alpha} & \frac{1}{\mathrm{m}}\left(-\mathrm{F}_{\mathrm{y}_\alpha} + \mathrm{mg}\sin\theta\right) \\ \frac{1}{\mathrm{m}}\left(-\beta \mathrm{F}_{\mathrm{z}_\alpha} - \alpha \mathrm{F}_{\mathrm{z}_\beta} + \mathrm{mg}\cos\theta\cos\varphi\right) & \frac{1}{\mathrm{m}}\left(\mathrm{F}_{\mathrm{z}_\alpha} - \mathrm{mg}\sin\theta\right) & -\frac{1}{\mathrm{m}}\mathrm{F}_{\mathrm{z}_\beta} \end{bmatrix} \tag{7.30}$$

Controlled transverse accelerations Γ_{yc}, Γ_{zc}, correspond to α_c and β_c. If one sets ϕ_c as the controlled roll angle, $|\Delta_\ell|$ is equal to 0 only if

$$1 + \beta_c \sin\varphi_c \tan\theta + \alpha_c \cos\varphi_c \tan\theta = 0$$

which is only true for high values of flight path angle, outside of the variation domain because α_c, β_c usually have small values. Matrix Δ_ℓ is thus invertible in practical cases. It is thus possible to use a linearized decoupled control for the slow subsystem in the form:

$$\begin{bmatrix} p \\ q \\ r \end{bmatrix} = F_\ell + G_\ell \begin{bmatrix} \tilde{\varphi} \\ \tilde{\Gamma}_y \\ \tilde{\Gamma}_z \end{bmatrix} \tag{7.31}$$

where $\tilde{\varphi}$, $\tilde{\Gamma}_y$, $\tilde{\Gamma}_z$ are new input variables in closed-loop (*cf.* figure 7.13) with all characteristic numbers equal to 1.

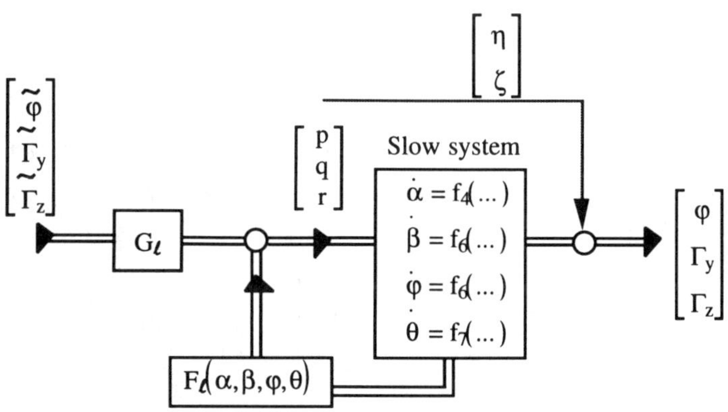

Figure 7.13: Slow subsystem: equations, controls, outputs.

One solution which transforms the system into three chains made of one integrator for each one, is obtained using

$$G_\ell = +\Delta_\ell^{-1} \text{ and } F_\ell = -\Delta_\ell^{-1}\left[L_{A_\ell} C_\ell\right] \tag{7.32}$$

with

$$L_{A_\ell}C_\ell = \begin{bmatrix} 0 \\ \dfrac{1}{mV}\left(\Gamma_z F_{y_\alpha} + \Gamma_y F_{y_\beta}\right) \\ \dfrac{1}{mV}\left(\Gamma_z F_{z_\alpha} + \Gamma_y F_{z_\beta}\right) \end{bmatrix}$$

These three chains can be transformed into three first-order systems with time constants T_1, T_2, T_3 by choosing:

$$G_\ell = +\Delta_\ell^{-1}\begin{bmatrix} 1/T_1 & & \\ & 1/T_2 & \\ & & 1/T_3 \end{bmatrix} \text{ and } F_\ell = -\Delta_\ell^{-1}\left[L_{A_\ell}C_\ell + \begin{vmatrix} \varphi/T_1 \\ \Gamma_y/T_2 \\ \Gamma_z/T_3 \end{vmatrix}\right] \tag{7.33}$$

A state-space representation of the resulting closed-loop subsystem is

$$\begin{aligned} \begin{bmatrix} \dot{\varphi} \\ \dot{\Gamma}_y \\ \dot{\Gamma}_z \end{bmatrix} &= \begin{bmatrix} -\dfrac{1}{T_1} & 0 & 0 \\ 0 & -\dfrac{1}{T_2} & 0 \\ 0 & 0 & -\dfrac{1}{T_3} \end{bmatrix} \begin{bmatrix} \varphi \\ \Gamma_y \\ \Gamma_z \end{bmatrix} + \begin{bmatrix} \dfrac{1}{T_1} & 0 & 0 \\ 0 & \dfrac{1}{T_2} & 0 \\ 0 & 0 & \dfrac{1}{T_3} \end{bmatrix} \begin{bmatrix} \tilde{\varphi} \\ \tilde{\Gamma}_y \\ \tilde{\Gamma}_z \end{bmatrix} \\ Y &= \begin{bmatrix} \varphi \\ \Gamma_y \\ \Gamma_z \end{bmatrix} \end{aligned}$$

Fast subsystem

The fast subsystem, of dimension 6, is in the form

$$\dot{X}_r = A_r\left(X_r\right) + \sum_1^3 u_i \tilde{B}_{r_i}\left(X_r\right)$$

The input of this subsystem is composed by the input $[u_1, u_2, u_3]^T$ of the global system and its outputs should correspond to the inputs $\left[\tilde{\varphi}\tilde{\Gamma}_y\tilde{\Gamma}_z\right]^T$, of the slow closed-loop subsystem; which means, using (7.31)

$$\begin{bmatrix} \tilde{\varphi} \\ \tilde{\Gamma}_y \\ \tilde{\Gamma}_z \end{bmatrix} = G_\ell^{-1}\left[\begin{bmatrix} p \\ q \\ r \end{bmatrix} - F_\ell\right]$$

where the variables $(\alpha, \beta, \phi, \theta)$ of the slow subsystem are assumed to be constant during the evolution of fast subsystem, the matrices F_l and G_l being constant (*cf.* figure 7.14)

As all the characteristic numbers are equal to 2, the decoupling matrix being equal to:

$$\Delta_r = G_\ell^{-1}\begin{bmatrix} L_\xi/I_x & 0 & 0 \\ 0 & L_\eta/I_y & 0 \\ 0 & 0 & L_\zeta/I_z \end{bmatrix} \text{ with } G_\ell^{-1} = \begin{bmatrix} T_1 & & \\ & T_2 & \\ & & T_3 \end{bmatrix}\Delta_\ell$$

and the momentum L, M, N different from zero, no further singularity is introduced.

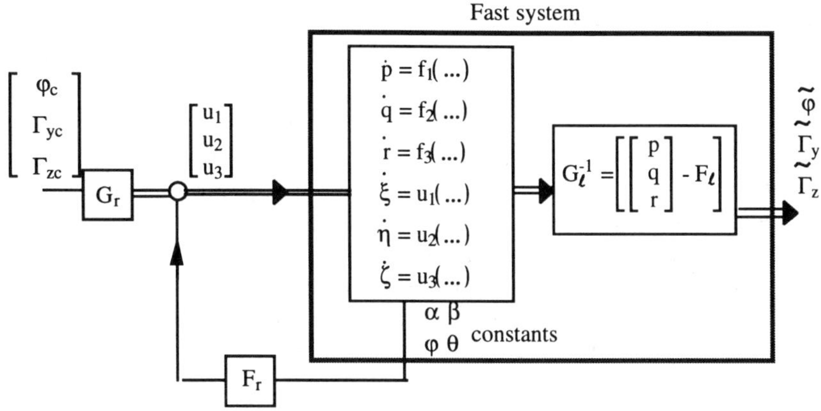

Figure 7.14: Fast subsystem: equations, controls, outputs.

One can thus generate a feedback control law in the form:

$$\begin{bmatrix} u_1 \\ u_2 \\ u_3 \end{bmatrix} = F_r\,[p,q,r,\xi,\eta,\zeta] + G_r = \begin{bmatrix} \varphi_c \\ \Gamma_{yc} \\ \Gamma_{zc} \end{bmatrix}$$

As previously, the matrices F_r and G_r which transform fast subsystems into three chains with two integrators for each of them are defined by

$$G_r = \Delta_r^{-1} F_r = -\Delta_r^{-1}\left[L^2_{A_r} C_r\right]$$

If one wants to get three decoupled chains corresponding to static second-order gains equal to the unity

$$\frac{\omega_i^2}{p^2 + 2\delta_i \omega_i p + \omega_i^2}$$

it is sufficient to take

$$G_r = \Delta_r^{-1} \begin{bmatrix} \omega_1^2 & & \\ & \omega_2^2 & \\ & & \omega_3^2 \end{bmatrix} \qquad F_r = -\Delta_r^{-1}\left[L^2_{A_r} C_r + \left| \begin{matrix} 2\delta_1\omega_1 \dot{\tilde{\varphi}} + \omega_1^2 \tilde{\varphi} \\ 2\delta_2\omega_2 \dot{\tilde{\Gamma}}_y + \omega_2^2 \Gamma_y \\ 2\delta_1\omega_3 \dot{\tilde{\Gamma}}_z + \omega_3^2 \Gamma_z \end{matrix} \right|\right]$$

then, as

$$\begin{bmatrix} \tilde{\varphi} \\ \tilde{\Gamma}_y \\ \tilde{\Gamma}_z \end{bmatrix} = G_\ell^{-1} \left[\begin{bmatrix} p \\ q \\ r \end{bmatrix} - F_\ell \right],$$

one obtains

$$\begin{bmatrix} \dot{\tilde{\varphi}} \\ \dot{\tilde{\Gamma}}_y \\ \dot{\tilde{\Gamma}}_z \end{bmatrix} = G_\ell^{-1} \begin{bmatrix} \dot{p} \\ \dot{q} \\ \dot{r} \end{bmatrix} = G_\ell^{-1} \begin{bmatrix} 1/I_x\,(L + L_p p) \\ \dfrac{1}{I_y}\,(M + M_q q + (I_z - I_x)\,pr) \\ \dfrac{1}{I_z}\,(N + N_r r + (I_z - I_y)\,pq) \end{bmatrix}$$

As for the term $L_{A_r}^{-1}C_r$, the expression of which is not given here, it can be shown that this term only depends on the momentum (and their derivatives) and on the angular velocities p, q, r. F_r and G_r are thus perfectly determined.

Decomposition of the global system into a slow and a fast system thus allows a reduction of characteristic indices, without modifying the unobservable subspace (which remains restricted to θ). Therefore the control law, which was in such a complex form when using the complete model, that carrying out was completely unrealistic becomes much more realistic with this approach. Naturally, as the dynamic separation between slow and fast variables is limited, decoupling and linearization properties are only achieved approximately on the real system. Extensive simulations have shown that the obtained performances were very satisfactory and one may refer to reference Harcaut (1991) for more details.

7.3 Application to the optimization of nonlinear systems

7.3.1 Recall of general resolution method – formulation and difficulties

Let us consider the dynamic system, defined by equations

$$\dot{X} = f(Xu) : \begin{array}{ll} \dot{X}_1 = f_1(X_1, X_2, u) & X_1 \in R^{n_1} \\ \dot{X}_2 = f_2(X_1, X_2, u) & X_2 \in R^{n_2} \end{array} \tag{7.34}$$

where X_1 and X_2 now only represent an arbitrary partition of the state $X \in R^n$, without any other consequence, and f_1, f_2 are vectorial continuous functions with continuous partial derivatives within a domain $X \times U \times t$.

Let us assume that the problem to be solved consists in determining an "optimal" u control, driving the system from an initial state X_{10}, X_{20} to a final state X_{1f}, X_{2f}, which is completely or partially fixed, while minimizing a criterion in the form

$$J = \varphi(X_{1f}X_{2f}) + \int_{t_0}^{t_f} L(X_1, X_2, u)\, dt \tag{7.35}$$

One assumes that the final time is free but that the final state must satisfy the constraints

$$\psi\left(X_{1_f}, X_{2_f}\right) = 0 \tag{7.36}$$

Such a problem is classic and can be solved by two kinds of methods: direct or indirect methods (Sage, 1968).

Direct methods consist in choosing a nominal control law $u(t)$ and in modifying it progressively by small variations $\delta u(t)$ in such a way to satisfy the constraints and to lower the criterion. Sensitivities of cost and constraints are computed through multiple adjoint vectors which are obtained from backward integrations. It is not the topic here to provide more details on these methods, or rather on these computations, which require complex computation codes, are time consuming and often associated with convergence problems, and which, in all the cases, lead to solutions in an open-loop form.

Indirect methods use, in one way or another, the Hamilton-Jacobi equation and allow, theoretically, to write the equations which must verified by the control law, in order to minimize the criterion J.

If one defines the Hamiltonian of the system by

$$H = L(X_1, X_2, u) + \lambda_{X_1}^T f_1(X_1, X_2, u) + \lambda_{X_2}^T f_2(X_1, X_2, u) \tag{7.37}$$

where $\lambda_{X_1}^T$ and $\lambda_{X_2}^T$ are adjoint vectors, one knows that the optimal control u^* minimizes the Hamiltonian in the subspace of allowable controls:

$$u^* = \arg\min_{u \in U} H\left(X_1, X_2, \lambda_{X_1}^T, \lambda_{X_2}^T, u\right) \tag{7.38}$$

For autonomous systems which will be considered hereafter, this condition implies, that the Hamiltonian remains constant along the optimal trajectory. Let us recall that the Hamiltonian must be equal to zero at t_f if this final time is free, and thus also equal to zero along the optimal trajectory if both conditions are present.

It is also clear that the solution (7.38) (which is equivalent to $\dfrac{\partial H}{\partial u} = 0$ if the control is bounded less and the Hamiltonian nonlinear with respect to u^*) allows us to write u* in terms of X_1, X_2 and $\lambda_{X_1}, \lambda_{X_2}$. Now, adjoint variables are given by the adjoint differential equations

$$\begin{aligned} \dot{\lambda}_{X_1} &= -\frac{\partial H}{\partial X_1} = +\frac{\partial L^T}{\partial X_1} - \left(\frac{\partial f_1}{\partial X_1}\right)^T \lambda_{X_1} - \left(\frac{\partial f_2}{\partial X_1}\right)^T \lambda_{X_2} \\ \dot{\lambda}_{X_2} &= -\frac{\partial H}{\partial X_2} = +\frac{\partial L^T}{\partial X_2} - \left(\frac{\partial f_2}{\partial X_2}\right)^T \lambda_{X_1} - \left(\frac{\partial f_2}{\partial X_2}\right)^T \lambda_{X_2} \end{aligned} \tag{7.39}$$

for which the only terminal conditions are fixed

$$\lambda_{X_{if}} = \frac{\partial \varphi}{\partial X_i} + \frac{\partial \psi^T}{\partial X_i} \nu \bigg|_{t_f} \qquad i = 1, 2 \tag{7.40}$$

and where the ν constants must be determined in such a way that they fit the terminal conditions on X_1 and X_2.

In spite of the simplicity of the above formalism, the method is extremely difficult because it requires an explicit resolution of a system of $2n$ equations coupled with optimal control and where the whole consists of a two points boundary value problem because one has both initial conditions on X and final conditions on λ_X .

One will also notice that the solution obtained is not a true control in closed-loop form because u is not given in terms of the state of the system only. Thus if at a given instant the initial state is perturbed, for example by a quantity $\delta X(t_i)$ as a consequence of external perturbations, the resetting of the optimal control is very complex, because one has only two solutions:

1. to compute again the optimal control, starting from new initial conditions.
2. if the perturbation is small enough, to linearize the equations of the motion and the equations of the control around their nominal values, in order to deduce the correction δu^* to be applied, in terms of the deviation δX^* .

Remark 7.9

Correction still remains complex in such a case. It is in the form

$$\delta u^* = -H_{uu}^{-1}\left[H_{ux} + f_u\left(P - SM^{-1}S\right)\right]\delta X^* \tag{7.41}$$

with

$$H_{uu}=\frac{\partial^2 H}{\partial u^2}\quad H_{ux}=\frac{\partial^2 H}{\partial u\partial x}\quad f_u=\frac{\partial f}{\partial u}$$

and the matrices P, S, M are defined by

$$\begin{aligned}\dot{P} &= -PF-F^TP+PRP-Q\\ \dot{S} &= -\left(F^T-PR\right)S\\ \dot{M} &= S^TRS\end{aligned}$$

with

$$\begin{aligned}F &= \frac{\partial f}{\partial x}-\frac{\partial f}{\partial u}\frac{\partial^2 H}{\partial u^2}\frac{\partial^2 H}{\partial u\partial x} & P\left(t_f\right) &= \frac{\partial^2\varphi}{\partial X_f^2}+\frac{\partial}{\partial X_f}\left(\frac{\partial\psi^T}{\partial X_f}\nu\right)\\ R &= \frac{\partial f}{\partial u}H_{uu}^{-1}\frac{\partial f^T}{\partial u} & S\left(t_f\right) &= \frac{\partial\psi_f}{\partial X_f}\\ Q &= H_{xx}-H_{xu}H_{uu}^{-1}H_{ux} & M\left(t_f\right) &= 0\end{aligned}$$

The following example, which looks simple, will demonstrate these difficulties.

7.3.2 Example

Let us consider a planar motion of two vehicles, with constant speed in modulus. The first one, called T, moves with speed v_T, along a trajectory defined by the variation of its direction $\gamma_T(t)$ which is assumed to be known. The orientation of the speed vector v_M of the second, γ_m, is provided through an action on the normal acceleration a, which is assumed to be the control. The aim is to find the control $a(t)$ such that vehicle M encounters vehicle T (the encounter being defined as the crossing at distance R), while minimizing the criterion

$$J=\int_0^{t_f}\left(1+ka^2\right)\mathrm{d}t=\int_o^{t_f}L\mathrm{d}t$$

where t_f is the "encounter" time, and where k is a parameter which allows a weighting between the interception time and the energy of the manoeuvre. Figure 7.15 visualizes geometrical variables of the problem, the equations of which are given by

$$\left.\begin{aligned}\dot{r} &= -v_M\cos\left(\eta-\gamma_m\right)+v_T\cos\left(\eta-\gamma_T\right)\\ \dot{\eta} &= \frac{1}{r}\left[v_M\sin\left(\eta-\gamma_m\right)-v_T\sin\left(\eta-\gamma_T\right)\right]\\ \dot{\gamma}_m &= \frac{a}{v_M}\end{aligned}\right\}\tag{7.42}$$

At first glance, this problem may assumed to be very simple. By comparison with a real "aeronautical" problem, for example, atmosphere, aerodynamics and propulsion effects are neglected. Even in such an extremely simple situation, analytic research of the optimal solution is very complex.

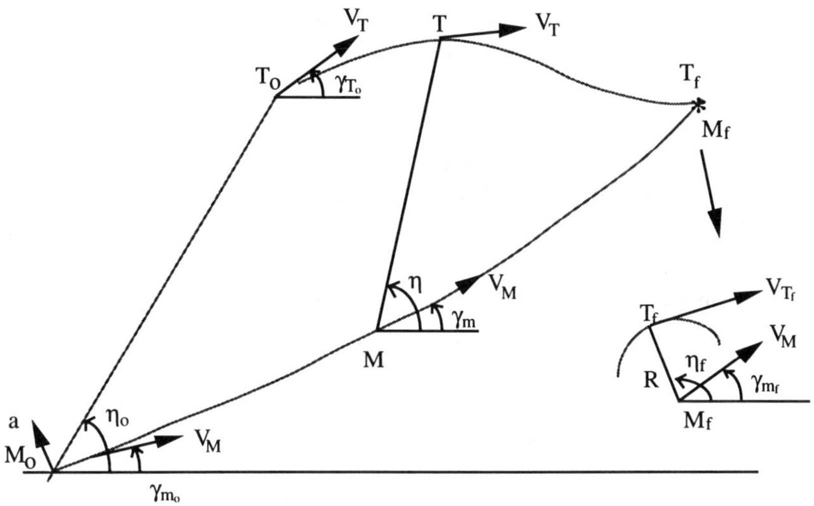

Figure 7.15: The pursuit configuration (T: target, M: missile).

The theoretical solution can be found by using the Hamiltonian and the maximum principle.
Let the Hamiltonian be written in the form:

$$\begin{aligned} H &= -L + \lambda_r \dot{r} + \lambda_\eta \dot{\eta} + \lambda_\gamma \dot{\gamma}_m \\ &= -\left(1 + ka^2\right) + \lambda_r \left[v_T \cos(\eta - \gamma_T) - v_M \cos(\eta - \gamma_m)\right] \\ &\quad + \frac{\lambda_\eta}{r}\left[v_M \sin(\eta - \gamma_m) - v_T \sin(\eta - \gamma_T)\right] + \lambda_\gamma \frac{a}{v_M} \end{aligned}$$

The optimal control a obtained by maximizing H is given by

$$\frac{\partial H}{\partial a} = 0 = -2ka + \frac{\lambda\gamma}{v_M}$$

thus $a = \dfrac{\lambda\gamma}{2kv_M}$. Obviously, one has not solved anything yet because a is defined through the adjoint λ_γ, which is unknown at present time, adjoint variables λ_r, λ_η, λ_γ being given by the equations

$$\begin{aligned} \dot{\lambda}_r &= -\frac{\partial H}{\partial r} = \frac{\lambda_\eta}{r^2}\left[v_M \sin(\eta - \gamma_m) - v_T \sin(\eta - \gamma_T)\right] \\ \dot{\lambda}_\eta &= -\frac{\partial H}{\partial \eta} = -\lambda r \left[v_M \sin(\eta - \gamma_m) - v_T \sin(\eta - \gamma_T)\right] \\ &\qquad -\frac{\lambda_\eta}{r}\left[v_M \cos(\eta - \gamma_m) - v_T \cos(\eta - \gamma_T)\right] \\ \dot{\lambda}_{\gamma m} &= -\frac{\partial H}{\partial \gamma_m} = \lambda_r\, v_M \sin(\eta - \gamma_m) + \lambda_\eta \frac{v_M}{r} \cos(\eta - \gamma_T) \end{aligned} \tag{7.43}$$

To the equation (7.42), the initial conditions are associated: r_0, η_0, γ_{m0}.

For equations (7.43), the initial conditions are unknown and one only knows the final conditions linked to the transversality equations. Particularly, as there are no constraints on η and γ at the final time:

$$\lambda_\eta (t_f) = \lambda_\gamma (t_f) = 0$$

In this very simplified case, the resolution of the equations can be performed analytically (Fossard *et al.*, 1991). As the calculus is complex, the results are here only presented, schematically, in the case where the target moves along a constant direction $(\gamma_T = \text{constant} = 0)$.

One finds a system of three nonlinear equations in the form:

$$\begin{aligned} t_f &= f_1(\gamma_T, \gamma_{m_o}/\gamma_{mf}, \eta_f) \\ f_2(r_o, \eta_o, \gamma_{m_o}, \gamma_T/\gamma_{mf}, \eta_f) &= 0 \\ f_3(r_o, \eta_o, \gamma_{m_o}, \gamma_T/\gamma_{mf}, \eta_f) &= 0 \end{aligned} \tag{7.44}$$

which require on elliptical integrals of first and second type, in the form $F\left(a, \frac{\pi}{2}\right)$, $F(a, u_o)$, $E\left(a, \frac{\pi}{2}\right)$, $E(a, u_o)$ with

$$\begin{aligned} a &= g_1(\eta_f, \gamma_{mf}, \gamma_m) \\ u_o &= g_2(\eta_f, \gamma_{mf}, \gamma_{m_o}) \end{aligned}$$

Even in this case, numerical resolution is difficult and requires a very small discretization step. Indeed, as the theoretical solution corresponds to $\psi_f \longrightarrow 0$, the elliptical integrals $F\left(a, \frac{\pi}{2}\right)$, $E\left(a, \frac{\pi}{2}\right)$ tend to infinity, which does not facilitate the resolution of equations (7.44); particularly any inaccuracy on $\gamma_f = \eta_f$ is reproduced in the whole control and therefore leads to a substantial error.

For the purpose of illustration, a typical case is

$$\begin{aligned} r_o &= 50\,000 \text{ m} & \eta_o &= 60^\circ & \gamma_m &= 0^\circ \\ V_T &= 250 \text{ m/s} & V_M &= 750 \text{ m/s} & k &= 10^{-5} \end{aligned}$$

with a step equal to 5×10^{-4} on ψ_f, final values obtained on ψ_f, η_f $(44^\circ 65)$ lead to a final error of a dozen of meters, while an error of 0.3° on ψ_f linked to a step equal to 10^{-2} on ψ_f leads to a final error of 400 m.

This example, although academic and special, clearly reveals the problems associated with the use of classical optimal control, *i.e.* heaviness of computation, even for the simplest case, sensitivity of accuracy of computation, open-loop character unsuitable with a practical realization.

The next section demonstrates the large amount of simplification which can be achieved through the singular perturbation approach.

7.3.3 *Singular perturbation approach*

Let us assume now that the system of equations (7.34) is set in a standard singularly perturbed form. With respect to paragraph 7.3.1, one assumes that vector X_1 and vector X_2

include respectively slow and fast variables, and therefore equations can be then rewritten under the form:

$$\begin{aligned} \dot{X}_1 &= f_1(X_1, X_2, u) \qquad X_{10} \\ \varepsilon \dot{X}_2 &= f_2(X_1, X_2, u) \qquad X_{20} \end{aligned} \tag{7.45}$$

and the criterion remains unchanged.

"Usually", one should write the Hamiltonian as follows

$$H = -L(X_1, X_2, u) + \lambda_{X_1}^T f_1(X_1, X_2, u) + \lambda_{X_2}^T \frac{f_2(X_1, X_2, u)}{\varepsilon} \tag{7.46}$$

with

$$\begin{aligned} \dot{\lambda}_{X_1} &= -\frac{\partial H}{\partial X_1} = \frac{\partial L^T}{\partial X_1} - \frac{\partial f_1^T}{\partial X_1}\lambda_{X_1} - \frac{1}{\varepsilon}\frac{\partial f_2^T}{\partial X_1}\lambda_{x2} \\ \dot{\lambda}_{X_2} &= -\frac{\partial H}{\partial X_2} = \frac{\partial L^T}{\partial X_2} - \frac{\partial f_1^T}{\partial X_2}\lambda_{X_1} - \frac{1}{\varepsilon}\frac{\partial f_2^T}{\partial X_2}\lambda_{x2} \end{aligned} \tag{7.47}$$

By setting $\frac{\lambda x_2}{\varepsilon} = \overset{*}{\lambda}_{x2}$
the above equations become

$$H = -L(X_1 X_2 u) + \lambda_{X_1}^T f_1 + \overset{*}{\lambda}{}_{X_2}^{T} f_2 \tag{7.48}$$

$$\dot{\lambda}_{X_1} = -\frac{\partial H}{\partial X_1}; \varepsilon \dot{\overset{*}{\lambda}}_{X_2} = -\frac{\partial H}{\partial X_2} \tag{7.49}$$

Therfore these equations show that, by keeping the same form for the Hamiltonian, λ_{X_1} and $\overset{*}{\lambda}_{X_2}$ have the same dynamics as the original state variables, which means that the slow adjoint vector X_1 corresponds to slow variables λ_{X_1} and the fast adjoint vector X_2 corresponds to fast variables $\overset{*}{\lambda}_{X_2}$. Note that for sake of simplification in what follows one will write λ_{X_2} instead of $\overset{*}{\lambda}_{X_2}$.

The form of equations, which then can be distinguished in two groups of $2n_1$ and $2n_2$ variables results in the application of the same procedure as above, shown in section 7.2.1.2 restricting to a closer zero-order approximation which is obtained by superposing the external solution with solutions in the boundary layers of same order.

7.3.3.1 External solution

By setting $\varepsilon = 0$ in the above equations, one obtains:

– concerning the system itself:

$$\dot{\overline{X}}_1 = f_1\left(\overline{X}_1, \overline{X}_2, \overline{u}\right), f_2\left(\overline{X}_1, \overline{X}_2, \overline{u}\right) = 0 \qquad X_1(0) = X_{10}$$

– concerning the adjoint equations:

$$\dot{\overline{\lambda}}_{X_1} = -\frac{\partial H}{\partial X_1} \qquad (15b) \qquad \frac{\partial H}{\partial X_2} = 0$$

with

$$\lambda_{X_1 f} = \left(\frac{\partial \varphi^T}{\partial X_1} + \frac{\partial \psi^T}{\partial X_1}\nu\right)_{tf}$$

It follows that, with respect to the original problem:

– one has to solve a system with dimension n_1 (or $2n_1$) instead of dimension $n_1 + n_2$ (or $2n_1 + 2n_2$)

– from the last adjoint equation $\partial H / \partial X_2 = 0$, the variables X_2 play a role similar to u. They can then be considered as "pseudo-controls", by noticing that the n_1+n_2 "controls" (u and X_2) are linked by n_2 relations

$$f_2\left(\overline{X}_1, \overline{X}_2, u\right) = 0$$

– these n_2 additional constraints $f_2 = 0$, which did not exist in the original problem are "induced constraints". If rank $\left[\dfrac{\partial f_2}{\partial X_2}\right] = n_2$, the equation $f_2\left(\overline{X}_1, \overline{X}_2, u\right) = 0$ can then be solved as

$$\overline{X}_2 = \phi\left(\overline{X}_1, u\right)$$

The Hamiltonian, associated the problem with, is defined by

$$\overline{H} = -L\left(\overline{X}_1, \phi\left(\overline{X}_1 \overline{u}\right), \overline{u}\right) + \lambda_{\overline{X}_1}^T f_1\left(\overline{X}_1, \phi\left(\overline{X}_1 u\right), \overline{u}\right)$$

with

$$\dot{\overline{\lambda}}_{X_1} = -\frac{\partial H}{\partial X_1} = \frac{\partial L^T}{\partial X_1}\left(\overline{X}_1 \overline{u}\right) + \left(\frac{\partial \phi}{\partial X_1}\right)^T \frac{\partial L^T}{\partial \phi} - \left(\frac{\partial f_1}{\partial \overline{X}_1}\right)^T \lambda_{X_1} - \left(\frac{\partial \phi}{\partial \overline{X}_1}\right)^T \left(\frac{\partial f_1}{\partial \phi}\right)^T \lambda_{X_1} \tag{7.50}$$

and

$$\lambda_{X_{1f}} = \left.\left(\frac{\partial \varphi^T}{\partial X_1} + \frac{\partial \psi^T}{\partial X_1}\nu\right)\right|_{t_f}$$

Remark 7.10

If, as will be seen in the following examples, $\dfrac{\partial f_2}{\partial X_2}$ *is not invertible but* rank $\left[\dfrac{\partial f_2}{\partial X_2}, \dfrac{\partial f_2}{\partial u}\right] = n_2$, *an assumption which is usually satisfied in practice, it will always be possible to solve the equation* $f_2\left(\overline{X}_1 \overline{X}_2 u\right) = 0$ *with respect to* n_2 *variables. Thus the problem is not modified except for a change of symbols.*

One will also notice, by rewriting equation (7.50) in the form

$$\dot{\overline{\lambda}}x_1 = \frac{\partial L^T}{\partial X_1}\left(\overline{X}_1, \overline{u}\right) - \left(\frac{\partial f_1}{\partial \overline{X}_1}\right)^T \lambda X_1 + \left(\frac{\partial \phi}{\partial X_1}\right)^T \left[\frac{\partial L^T}{\partial \phi} - \left(\frac{\partial f_1}{\partial \phi}\right)^T \lambda X_1\right]$$

that, formally, the "adjoint variable" associated with $\phi = \overline{X}_2$ *corresponds to the Lagrange multiplier which might appear in the Hamiltonian, as a result of the constraint* $f_2 = 0$.

Remark 7.11 *It is easy to see that both approaches, consisting in either simplifying the state equations, which means by setting directly* $\varepsilon = 0$ *in the equations (7.45) then solving the optimal control of reduced system, or simplifying the optimality equations are identical.*

The optimality equations of the complete system are written, setting $\varepsilon = 0$, as

$$\dot{\overline{X}}_1 = f_1(\overline{X}_1, \overline{X}_2, \overline{u}^*)$$

$$0 = f_2\left(\overline{X}_1, \overline{X}_2, u^*\right) \tag{7.51}$$

$$\dot{\overline{\lambda}}_{X_1} = -\frac{\partial H}{\partial X_1} = \frac{\partial L^T}{\partial X_1} - \left(\frac{\partial f_1}{\partial X_1}\right)^T \lambda_{\overline{X}_1} - \left(\frac{\partial f_2}{\partial X_1}\right)^T \lambda_{\overline{X}_2} \tag{7.52}$$

$$0 = -\frac{\partial H}{\partial X_2} = \frac{\partial L^T}{\partial X_2} - \left(\frac{\partial f_1}{\partial X_2}\right)^T \lambda_{\overline{X}_1} - \left(\frac{\partial f_2}{\partial X_2}\right)^T \lambda_{\overline{X}_2} \tag{7.53}$$

$$\overline{u}^* = \arg\min H\left(\overline{X}_1, \overline{X}_2, \lambda_{\overline{X}_1}, \lambda_{\overline{X}_2}, \overline{u}\right)$$

$$X_1(0) = X_{10} \qquad \lambda_{X_1 f} = \left(\frac{\partial \varphi}{\partial X_1} - \frac{\partial \psi^T}{\partial X_1}\nu\right)\Bigg|_{t_f}$$

$$H = -L\left(\overline{X}_1, \overline{X}_2, u\right) + \lambda_{X_1}^T f_1\left(\overline{X}_1, \overline{X}_2, \overline{u}\right) + \lambda_{X_2}^T f_2\left(\overline{X}_1, \overline{X}_2, \overline{u}\right)$$

If one assumes that $\dfrac{\partial f_2}{\partial \overline{X}_2}$ is invertible along the optimal trajectory, one deduces from (7.51), by using the theorem of implicit functions, that there exists a function ϕ such that:

$$\overline{X}_2 = \phi\left(\overline{X}_1, \overline{u}\right) \quad \text{with} \quad \frac{\partial f_2}{\partial X_2}\frac{\partial \phi}{\partial X_1} + \frac{\partial f_2}{\partial X_1} = 0$$

From (7.53) it follows that

$$\lambda_{\overline{X}_2} = \left(\frac{\partial f_2}{\partial X_2}\right)^{-T}\left[\frac{\partial L^T}{\partial \overline{X}_2} - \left(\frac{\partial f_1}{\partial X_2}\right)^T \lambda_{\overline{X}_1}\right]$$

and, by refering to (7.52)

$$\begin{aligned}
\dot{\overline{\lambda}}_{X_1} &= \frac{\partial L^T}{\partial X_1} - \left(\frac{\partial f_1}{\partial X_1}\right)^T \lambda_{\overline{X}_1} - \left(\frac{\partial f_2}{\partial X_1}\right)^T\left(\frac{\partial f_2}{\partial X_2}\right)^{-T}\left[\frac{\partial L^T}{\partial \overline{x}_2} - \left(\frac{\partial f_1}{\partial X_2}\right)^T \lambda_{\overline{X}_1}\right] \\
\dot{\overline{\lambda}}_{X_1} &= \frac{\partial L^T}{\partial X_1} - -\left(\frac{\partial f_1}{\partial \overline{X}_1}\right)^T \lambda_{\overline{X}_1} - \left(\frac{\partial \phi}{\partial \overline{X}_1}\right)^T \lambda
\end{aligned}$$

with

$$\lambda = \frac{\partial L^T}{\partial \overline{X}_2} - \left(\frac{\partial f_1}{\partial \overline{X}_2}\right)^T \lambda_{\overline{X}_1}$$

If now, on the other hand, one starts from reduced system

$$\begin{aligned}
\dot{X}_1 &= f_1\left(\overline{X}_1, \overline{X}_2, \overline{u}\right) & X_1(0) &= X_{10} \\
0 &= f_2\left(\overline{X}_1, \overline{X}_2, \overline{u}\right) & \overline{X}_2 &= \phi\left(\overline{X}_1, u\right)
\end{aligned}$$

the Hamiltonian then becomes

$$\overline{H} = -L\left(\overline{X}_1, \overline{X}_2, \overline{u}\right) + \overline{\lambda}_{X_1}^T f_1\left(\overline{X}_1, \overline{X}_2, u\right)$$

and

$$\dot{\overline{\lambda}}_{X_1} = -\frac{\partial \overline{H}}{\partial X_1} = \frac{\partial L^T}{\partial X_1} - \left(\frac{\partial f_1}{\partial X_1}\right)^T \lambda_{X_1} + \left(\frac{d\phi}{dX_1}\right)^T\left(\frac{\partial L^T}{\partial X_2} - \left(\frac{\partial f_1}{\partial X_2}\right)^T \lambda_{X_1}\right)$$

which is the same expression as previously obtained.

In the external solution, as it is reduced to a problem where the only dynamic variables of the system are slow variables X_1, it is then clear that, if one satisfies terminal conditions (initial and final) on these variables, the requirements on variables X_2, which are associated with X_1, cannot be fulfilled. The necessary matching will be achieved in the "boundary layers" as can be seen in the following.

7.3.3.2 Solution in the initial boundary layer

In order to fulfill initial conditions on the fast variables X_2 as in section 7.2.1, one must introduce an initial boundary layer.

Within the time-scale $\tau = t/\varepsilon$ equations (7.45) become:

$$\frac{dX_1}{\partial\tau} = \varepsilon f_1(X_1, X_2, u) \qquad \frac{dX_2}{d\tau} = f_2(X_1, X_2, u)$$

and the adjoint equations

$$\frac{d\lambda_{X_1}}{d\tau} = -\varepsilon\frac{\partial H}{\partial X_1} \qquad \frac{d\lambda_{X_2}}{d\tau} = -\frac{\partial H}{\partial X_2}$$

with

$$H = -L(X_1, X_2, u) + \lambda_{X_1}^T f_1(X_1, X_2, u) + \lambda_{X_2}^T f_2(X_1, X_2, u)$$

It follows that, for $\varepsilon = 0$, these equations lead to:

$$\begin{aligned} \frac{d\tilde{X}_1}{d\tau} &= 0 \longrightarrow \tilde{X}_1 = \text{constant} = X_{10} \\ \frac{d\lambda_{\tilde{X}_1}}{d\tau} &= 0 \longrightarrow \lambda_{\tilde{X}_1} = \text{constant} = \lambda_{\overline{X}_1}(0) \end{aligned}$$

Then, one finds the equations

$$\begin{aligned} \frac{d\tilde{X}_2}{d\tau} &= f_2\left(X_{10}, \tilde{X}_2, \tilde{u}\right) \qquad \tilde{X}_{20} = X_{20} \\ \frac{d\tilde{\lambda}_{X_2}}{d\tau} &= -\frac{\partial\tilde{H}}{\partial X_2} \end{aligned} \tag{7.54}$$

with

$$\begin{aligned} \tilde{H} &= -L\left(X_{10}, \tilde{X}_2, \tilde{u}\right) + \lambda_{X_{10}}^T f_1\left(X_{10}, \tilde{X}_2, \tilde{u}\right) + \tilde{\lambda}_{X_2}^{*T} f_2\left(X_{10}, \tilde{X}_2, \tilde{u}\right) \\ \tilde{u}^* &= \arg\max \tilde{H}\left(X_{10}, \lambda_{\overline{X}_{10}}, \tilde{x}_2, \tilde{u}, \lambda_{\tilde{X}_2}\right) \end{aligned} \tag{7.55}$$

One is faced with a problem with dimension n_2, defined by equations (7.54) and (7.55) for which the integral criterion, to be minimized, is given within the time-scale τ

$$\int_0^\infty \left[L\left(X_{10}, \tilde{X}_2, \tilde{u}\right) - \lambda_{X_{10}}^T f_1\left(X_{10}, \tilde{X}_2, \tilde{u}\right)\right] d\tau \tag{7.56}$$

In the same way as in section 7.2.1.3, it is essential to check the convergence of the initial boundary layer towards the external solution, which means that

$$\lim_{\tau\to\infty} \tilde{X}_2(\tau) = \overline{X}_{20} \qquad \lim_{\tau\to\infty} \tilde{u}(\tau) = \overline{u}_o$$

where $\overline{X}_{20}, \overline{u}_0$ are values of X_2 and u at the beginning of the external solution.

No doubt, this is the most critical aspect in the general case because one should check, on the one hand that the singular point $\overline{X}_{20}, \overline{u}_0^*$, is asymptomatically stable and, on the other hand that initial conditions belong to the attraction domain of this point. The stability conditions can theoretically be written down by linearizing the equations in the neighborhood of the external solutions. Indeed the theoretical determination of an attraction domain is usually impossible in complex practical situations when the use of the singular perturbation method becomes necessary. Generally, one only checks whether convergence is indeed achieved, the solution being first computed.

The global control (at this step) will then be given by

$$u = \overline{u}(t) + \tilde{u}(\tau) - \overline{u}_0$$

7.3.3.3 Solution in the final boundary layer

The initial boundary layer has allowed us to fulfil the required initial conditions on the fast variables and to realize the matching with external conditions (if the conditions of Tihonov are satisfied).

If one also has the required final conditions on the fast variables, the matching problem will be set in a similar way and one should consider a terminal boundary layer.

If one sets

$$\sigma = (t_f - t)/\varepsilon = -(t - t_f)/\varepsilon$$

the problem is then defined, by

$$\frac{d\tilde{X}_2}{d\sigma} = -f_2\left(X_{1f}, \hat{X}_2, \hat{u}\right) \qquad \hat{X}_2(\sigma_f) = X_{2f}$$

with the criterion

$$J = \int_0^\infty \left[L\left(X_{1f}, \hat{X}_2, \hat{u}\right) - \lambda_{X_{1f}}^T f_1\left(X_{1f}, \hat{X}_2, \hat{u}\right)\right] d\sigma$$

Convergence towards the external solution is obtained by backward integration of the equation of the boundary layer. As for the initial boundary layer, one must satisfy the conditions of Tihonov's theorem so that:

$$\begin{aligned} \lim_{\sigma\to\infty} \hat{X}_2(\sigma) &= \overline{X}_2(X_{1f}) \\ \lim_{\sigma\to\infty} \hat{u}(\sigma) &= \overline{u}_{t_f} \end{aligned}$$

The global control is then defined by

$$u = \overline{u} + \tilde{u}(\tau) - \overline{u}_0 + \hat{u}(\sigma) - \overline{u}_{t_f}$$

7.3.3.4 Closed-loop control

With respect to the aim which was stated at the beginning of this chapter, it may seen that nothing has been changed on the principle because, basically, the problems which remain to be solved in external solution and in boundary layers are optimization problems of nonlinear dynamic systems (with a finite horizon for the external solution, and infinite horizon for the boundary layers). The only simplification, which seems to be provided, consists of the reduction of the order of optimization problems because one now works with dimensions n_1 and n_2 instead of $n_1 + n_2$. Actually, the possibility to implement a control law in a closed-loop form and in real time will depend on

– the existence or not of terminal conditions on fast variables

– the number of fast variables.

7.3.3.4.1 Case where one does not have terminal conditions on fast variables

If one has no terminal conditions on fast variables, one has seen that calculations to be performed consisted in the determination of the external layer and of only initial layers. The problem becomes especially simple if one has only one scalar fast variable, or, equivalently, if all the fast variables can be ordered in successive time-scales.

One scalar fast variable

Let us assume for example that one has only one scalar fast variable, x_2. Within the initial boundary layer, equation (7.55) then becomes:

$$\begin{aligned}\tilde{H} &= \left[-L(X_{10},\tilde{x}_2,\tilde{u}) + \lambda_{X_{10}}^T f_1(X_{10},\tilde{x}_2,\tilde{u})\right] &&+\tilde{\lambda x_2} f_2(X_{10},\tilde{x}_2,\tilde{u}) \\ &= g(X_{10},\tilde{x}_2,\tilde{u}) &&+\tilde{\lambda x_2} f_2(X_{10},\tilde{x}_2,\tilde{u})\end{aligned}$$

If the problem is stationary, with free end time, the zero value of the Hamiltonian along the optimal trajectory implies that:

$$\lambda_{\tilde{x}_2} = -\frac{g(X_{10},\tilde{x}_2,\tilde{u})}{f_2(X_{10},\tilde{x}_2,\tilde{u})}$$

and minimizing the Hamiltonian $\tilde{H}$ then consists in extremalizing the function $-g/f_2$. Indeed, to minimize $\tilde{H}$ one must write

$$\frac{\partial \tilde{H}}{\partial u} = \frac{\partial g}{\partial u} + \lambda \tilde{x}_2 \frac{\partial f_2}{\partial u} = 0$$

that means, with $g + \lambda \tilde{x}_2 f_2 = 0$

$$f_2 \frac{\partial g}{\partial u} - g \frac{\partial f_2}{\partial u} = 0$$

$$\frac{\partial^2 \tilde{H}}{\partial u^2} = \frac{\partial^2 g}{\partial u^2} + \lambda \tilde{x}_2 \frac{\partial^2 f_2}{\partial u^2} > 0 \quad \longrightarrow \quad \frac{\partial^2 g}{\partial u^2} - \frac{g}{f_2}\frac{\partial^2 f_2}{\partial u_2^2} > 0 \tag{7.57}$$

If one considers the function $\Theta = -\dfrac{g}{f_2}$

$$\frac{\partial \Theta}{\partial u} = -\frac{f_2 \frac{\partial g}{\partial u} - g \frac{\partial f_2}{\partial u}}{f_2^2} = 0 \quad \rightarrow \quad f_2 \frac{\partial g}{\partial u} - g \frac{\partial f_2}{\partial u} = 0$$

$$\left.\frac{\partial^2 \Theta}{\partial u^2}\right|_{\frac{\partial \theta}{\partial u} = 0} = -\frac{f_2\left(\frac{\partial^2 g}{\partial u^2} - \frac{g}{f_2}\frac{\partial^2 f_2}{\partial u^2}\right)}{f_2^2}$$

Condition (7.57) implies

$$\text{if } f_2 > 0 \qquad \frac{\partial^2 \Theta}{\partial u^2} < 0$$

$$\text{if } f_2 < 0 \qquad \frac{\partial^2 \Theta}{\partial u^2} > 0$$

It then results that to minimize $\tilde{H}$ consists of

- minimizing Θ if $f_2 > 0$
- maximizing Θ if $f_2 < 0$

One has thus replaced the dynamic optimization problem by a static optimization problem, much more simple, which can either be solved explicitly or solved first off-line (and the results stored), the control law being then defined in terms of the state.

If the solution in the boundary layer can also be computed explicitly – which limits obviously dimension n_2 of the associated problem – one gets a solution in the compound form:

$$u = \left[\tilde{u}\left(X_{10}, \tilde{X}_2\right) - \overline{u}\left(X_{10}, \overline{X}_{20}\right)\right] + \overline{u}\left[\overline{X}_1, \overline{X}_2\right] \tag{7.58}$$

At instant t_0, the control is a function of the initial state only.

Of course one should keep in mind that in the calculus of the boundary layer the slow variables were assumed to be constant, equal to their initial values. This implies that dynamic separation between slow variables and fast variables is infinite, which is never verified in practice. It follows that, if one applies this control which was computed at instant t_0, the final conditions on X_1 will not be "exactly" fulfilled.

Nevertheless, the simplification of the calculus, obtained by this method, allows anyway

- to perform at least a certain some number k of updating along time-interval $t_0 t_f$ in order to reduce the error (order ε^k), or
- to consider every current instant as an initial instant in order to obtain a control in a closed-loop form, valid all along the trajectory.

The following example in section 7.3.4.1 (which will reconsider the problem in section 7.3.2 in the singular perturbation aspect) will illustrate what may be obtained by this method in this case.
It is clear that, in practical situations, there will be more than one scalar fast variable.

Fast variables dynamically separated

One solution consists of writing down the equations, by separating all the fast variables in as many dynamic classes as scalar variables in the form

$$\begin{aligned} \dot{X}_1 &= f_1\left(X_1, x_2 \ldots x_k\right) \\ \varepsilon \dot{x}_2 &= f_2\left(X_1, x_2 \ldots x_k\right) \\ \varepsilon^k \dot{x}_k &= f_k\left(X_1, x_2 ... x_k\right) \end{aligned}$$

The above procedure is then directly applicable by considering $k - 1$ boundary layers, the problem associated with each of them becomes a one-dimensional problem and thus reduces to a static optimization one. These problems are solved from the external solution towards the solution within the fastest boundary layer and the control which is then applied and updated at every current instant, is given by the control obtained in the last layer.

Non separated fast variables – Linearization of slow layers

The drawback of the above approach is the assumption that all the "fast" variables can be separated into distinct dynamic classes, which means that their variation speed increases significantly. For a large number of situations this assumption is no longer justified.

Another solution, which may be considered, consists in "linearizing" the boundary layer around the nominal solution, defined by the external solution.

If one goes back to equation (7.54)

$$\frac{d\tilde{X}_2}{d\tau} = f_2\left(X_{10}, \tilde{X}_2, \tilde{u}\right)$$

for which the criterion (7.56) was associated

$$\int_0^{\infty} \left[L\left(X_{10}, \tilde{X}_2, \tilde{u}\right) - \lambda_{X_{10}}^T f_1\left(X_{10}, \tilde{X}_2, \tilde{u}\right)\right] d\tau \tag{7.59}$$

around the equilibria values $X_{10}, \overline{X}_2(X_{10}), \overline{u}^*(X_{10}), f_2\left(\overline{X}_1, \overline{X}_2, \overline{u}^*\right) = 0$, one obtains the following system of equations

$$\delta\tilde{X}_2 = \tilde{X}_2(\tau) - \overline{X}_2(X_{10}) \qquad \delta\tilde{u} = \tilde{u} - \overline{u}^*(X_{10})$$

$$\frac{\mathrm{d}}{\mathrm{d}\tau}\delta\tilde{X}_2 = \left.\frac{\partial f_2}{\partial \tilde{X}_2}\right|_0 \delta\tilde{X}_2 + \left.\frac{\partial f_2}{\partial \tilde{u}}\right|_0 \delta\tilde{u}$$

which can be written as

$$\frac{\mathrm{d}}{\mathrm{d}\tau}\delta\tilde{X}_2 = A\delta\tilde{X}_2 + B\delta\tilde{u}$$

since matrices $\left.\frac{\partial f_2}{\partial \tilde{X}_2}\right|_0 \left.\frac{\partial f_2}{\partial \tilde{u}}\right|_0$ evaluated at point $X_{10}, \overline{X}_2(X_{10}), \overline{u}^*(X_{10})$ at $t = 0$ are constant.

Parallel to this, linearization of criterion the leads to replacing (7.59) by

$$\tilde{J} = \int_0^{\infty} \left(\delta X_2^T, \delta u^T\right) \begin{bmatrix} Q & S \\ S^T & R \end{bmatrix} \begin{bmatrix} \delta\tilde{X}_2 \\ \delta\tilde{u} \end{bmatrix} d\tau$$

with

$$Q = \left.\frac{\partial^2 \tilde{H}}{\partial X_2^2}\right| \qquad R = \left.\frac{\partial^2 \tilde{H}}{\partial u^2}\right| \qquad S = \left.\frac{\partial^2 \tilde{H}}{\partial X_2 \partial u}\right|$$

these expressions being computed at $X_{10}, \overline{X}_2(X_{10})$, thus constant.

The optimal control for this linear problem with a quadratic criterion and infinite horizon is given by

$$\delta u^* = -R^{-1}\left(B^T K + S^T\right)\delta\overline{X}_2$$

where K is the solution of the Riccati equation

$$K\hat{A} + \hat{A}^T K + \hat{Q} - KBR^{-1}B^T K = 0$$

with

$$\hat{A} = A - BR^{-1}S^T \quad \hat{Q} = Q - SR^{-1}S^T$$

One then obtains a control law in terms of the state, with the restrictions that

- as for all linearized problem, initial conditions should not be too far away from the equilibrium values of the external solution; and
- in case of an updating of the control, one should possibly solve Riccati's equation at every current time!

Paragraph 7.3.4.2 will illustrate this procedure.

7.3.3.4.2 *Terminal conditions on fast variables*

In the case where one requires terminal conditions on fast variables, the composite control is given by equation (7.58) which is rewritten here:

$$u = \overline{u}\left(\overline{X}_1, \overline{X}_2\right) + \left[\tilde{u}\left(X_{10}, \tilde{X}_2\right) - \overline{u}\left(X_{10}, \overline{X}_{20}\right)\right] + \left[\hat{u}\left(X_{1f}, \hat{X}_2\right) - \overline{u}\left(X_{1f}, \overline{X}_{2f}\right)\right] \tag{7.60}$$

It is composed with three terms relative respectively to the external solution, to the initial and to the final boundary layer solutions. When one stays far enough from the end time, the control corresponding to the last term of expression (7.60) is negligible, as

$$\left|\hat{u}\left(X_{1f}, \hat{X}_2\right) - \overline{u}\left(X_{1f}, \overline{X}_{2f}\right)\right| \leq \overline{e}^{b(t_f - t)/\varepsilon}$$

and one finds the same situation as previously.

On the other hand, when one approaches the end time, the last term can no longer be neglected and one must take into account the terminal layer. Even if the computation of the terminal layer is "similar" to the one used for the initial boundary layer, one should nevertheless notice the differences which seem to be small at first glance from a theoretical point of view but which reveal themselves to be important from a practical point of view.

One will notice particularly two things

1. linearization of equations around the external solution at the end time $\left(\overline{X}_1(t_f)\right)$ results in a linearized optimal control which requires, as for the initial layer, to solve a matrix Riccati equation. One must consider here the definite negative solution.

2. Indeed, the convergence of final boundary layer towards the external solution can only be guaranteed when integration of equations of motion are performed backwards.

For real use, several solutions can therefore be considered:

- The final trajectory is first obtained, as mentioned previously, by integrating off-line equations of the motion backwards, from final conditions, using linearized optimal control of the final boundary layer. The trajectory obtained is then registered and stored in memory, and then used as a nominal guidance track to be followed in direct time.
- You can commute to other control laws, which are either basically different, or coming from the same methodology but through a different splitting of dynamics according to a technique of multi-classification (Fossard and Freitas, 1986).

7.3.4 *Application examples*

7.3.4.1 Interception problem: explicit solution case (one-dimensional fast variable without end condition)

Let us consider again the problem in section 7.3.2 and try to solve this approximately. The only assumption one will make is that M is maneuverable enough such that one can consider the ratio of the minimum turn radius to the total length of trajectory to be small. This implies that γ_m is a fast variable (with definition in section 7.2.2.3) in comparison with variables r and η and one will translate this fact by associating a perturbation parameter ε

with γ_m. Equations are then written in the form: (one assumes $\gamma_T = \text{constant} = 0$)

$$\begin{aligned}
\dot{r} &= -V_M \cos(\eta - \gamma_m) + V_T \cos\eta \\
\dot{\eta} &= \frac{1}{r}\left[V_M \sin(\eta - \gamma_m) - V_T \sin\eta\right] \\
\varepsilon\dot{\gamma}_m &= \frac{a}{V_M} = f_\gamma(a)
\end{aligned}$$

with, by keeping the same expression for the Hamiltonian

$$H = -\left(1 + ka^2\right) + \lambda_r \dot{r} + \lambda_\eta \dot{\eta} + \lambda_\gamma f_\gamma$$

one finds adjoint equations

$$\begin{aligned}
\dot{\lambda r} &= -\frac{\partial H}{\partial r} \\
\dot{\lambda \eta} &= -\frac{\partial H}{\partial \eta} \\
\varepsilon\dot{\lambda\gamma_m} &= -\frac{\partial H}{\partial \gamma_m} = \lambda r V_M \sin(\eta - \gamma_m) + \lambda_\eta \frac{V_M}{r} \cos(\eta - \gamma_m)
\end{aligned}$$

External layer

In the external layer, where one considers only slow variables r and η, the variable γ_m assumed to be fast enough so that it reaches instantaneously any desired value, one gets, by setting directly $\varepsilon = 0$

$$\begin{aligned}
\varepsilon\dot{\gamma}_m &= 0 \quad \text{thus } \overline{a} = 0 \longrightarrow \overline{\gamma}_m = \text{Cte} \\
\varepsilon\lambda\dot{\gamma}_m &= -\frac{\partial H}{\partial \gamma_m} = 0
\end{aligned}$$

One will notice:

- that, in the external layer, a, which was a control for the original system, now becomes a constraint ($a = 0$)
- as $\dfrac{\partial H}{\partial \gamma_m} = 0$, one has to extremalize the Hamiltonian with respect to γ_m which thus becomes a pseudo-control for the external layer.

For this external layer, one must determine the variables which define the slow motion, which means (in addition of $\overline{a} = 0$), $\overline{\gamma}_m, \overline{\eta}$ and the slow adjoint variables $\overline{\lambda}_r, \overline{\lambda}_\eta$ (actually, at this step, one does not need the adjoint variables, but they will be necessary in the boundary layer).

To determine these four variables note the following.

1. One has the following equations: $\dfrac{\partial H}{\partial \gamma_m} = 0$ which provides a first relation

$$\tan(\overline{\eta} - \overline{\gamma}_m) = -\frac{\overline{\lambda}_\eta}{r\overline{\lambda}_r} \tag{7.61}$$

2. One must satisfy terminal conditions on the slow variables, thus

$$r_0 \cos\eta_0 + V_T t_f - V_M t_f \cos\overline{\gamma}_m - R\cos\eta_f = 0 \tag{7.62}$$

$$r_0 \sin \eta_0 - V_M t_f \sin \overline{\gamma}_m - R \sin \eta_f = 0 \tag{7.63}$$

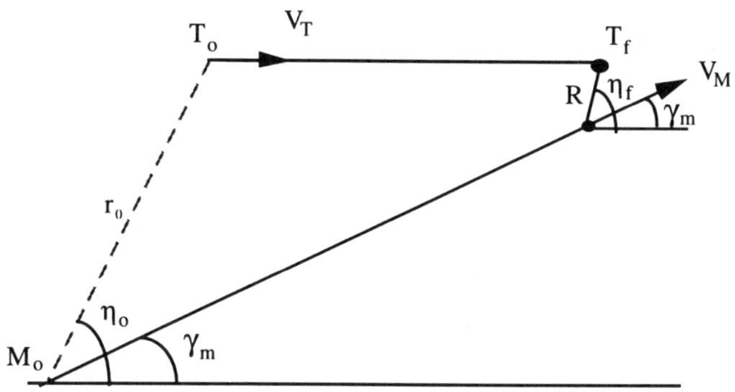

Figure 7.16: Outer solutions.

3. As one has no terminal constraint on η, $\lambda\eta\,(t_f) = 0$ and from (7.61) it follows that

$$\overline{\eta}_f = \overline{\gamma}_m$$

Equations (7.62) and then (7.63) become

$$\begin{aligned} r_0 \cos \eta_0 + V_T t_f - (V_M t_f + R) \cos \overline{\gamma}_m &= 0 \\ r_0 \sin \eta_0 - (V_m t_f + R) \sin \overline{\gamma}_m &= 0 \end{aligned}$$

which provide two equations on $\overline{\gamma}_m$ and t_f and which can be solved easily

$$t_f = \frac{r_0 \sin \eta_0 - R \sin \overline{\gamma}_m}{V_M \sin \overline{\gamma}_m} = \frac{-r_0 \cos \eta_0 + R \cos \overline{\gamma}_m}{V_T - V_M \cos \overline{\gamma}_m}$$

Thus, for $\overline{\gamma}_m$

$$\nu \sin \eta_0 \cos \overline{\gamma}_m - \left(\nu \cos \eta_0 - \frac{R}{r_0} \right) \sin \overline{\gamma}_m - \sin \eta_0 = 0$$

with $\nu = \dfrac{V_M}{V_T}$, an equation which can be solved easily by setting $x = \tan \dfrac{\gamma_m}{2}$

$$(\nu + 1) \sin \eta_0 x^2 + 2 \left(\nu \cos \eta_0 - \frac{R}{r_0} \right) x + (1 - \nu) \sin \eta_0 = 0 \tag{7.64}$$

One has then only one equation of second-order on x, the solution of which is:

$$\overline{\gamma}_m = 2\text{arctan} x$$

4. At last, by setting that, for a stationary problem, the Hamiltonian is zero along optimal trajectory and taking into account (7.61) one finds

$$\overline{\lambda}_r = \frac{\cos(\overline{\eta} - \overline{\gamma}_m)}{V_T \cos \overline{\gamma}_m - V_M} \qquad \overline{\lambda}_\eta = -\frac{r \sin(\overline{\eta} - \overline{\gamma}_m)}{V_T \cos \overline{\gamma}_m - V_M}$$

Initial boundary layer

As the trajectory, in the external layer, is defined by γ_m = constant = $\overline{\gamma}_m$ while the initial condition is γ_{m0}, the aim, within initial boundary layer, consists in determining the Fursning policy which allows the reaching of $\overline{\gamma}_m$ starting from γ_{m_0} .

Within the extended time-scale $\tau = t/\varepsilon$, the equations are written:

$$\frac{\mathrm{d}r}{\mathrm{d}\tau} = \varepsilon\left[-V_M \cos(\eta - \gamma_m) + V_T \cos\eta\right]$$

$$\frac{\mathrm{d}\eta}{\mathrm{d}\tau} = \frac{\varepsilon}{r}\left[V_M \sin(\eta - \gamma_m) - V_T \sin\eta\right]$$

$$\frac{\mathrm{d}\gamma_m}{\mathrm{d}\tau} = \frac{a}{V_M}$$

with

$$\frac{\mathrm{d}\lambda_r}{\mathrm{d}\tau} = -\varepsilon\frac{\partial H}{\partial r} \quad \frac{\mathrm{d}\lambda_\eta}{\mathrm{d}\tau} = -\varepsilon\frac{\partial H}{\partial \eta} \quad \frac{\mathrm{d}\lambda_\gamma}{\mathrm{d}\tau} = -\frac{\partial H}{\partial \gamma_m}$$

If, as previously, one sets $\varepsilon = 0$ it follows that $\dfrac{\mathrm{d}r}{\mathrm{d}\tau} = 0 \quad \dfrac{\mathrm{d}\eta}{\mathrm{d}\tau} = 0$

One assumes, therefore, that during the turn, r and η remain constant which means that during this evolution they remain practically equal to their initial values ρ_0, η_0 . The result is the same for the adjoint variables λ_r and λ_η which remain equal to their initial values $\lambda\overline{r}_0$, $\lambda\overline{\eta}_0$. With those conditions, the Hamiltonian becomes:

$$\tilde{H} = \left\{-\left(1 + ka^2\right) + \overline{\lambda}_r(0)\left[V_T \cos\eta_0 - V_M \cos(\eta_0 - \gamma_m)\right] + \frac{\lambda_{\eta_0}}{r_0}\left[V_M \sin(\eta_0 - \gamma_m) - V_T \sin\eta_0\right]\right\} + \lambda\gamma\frac{a}{V_M} \tag{7.65}$$

with

$$\overline{\lambda}_r(0) = \frac{\cos(\eta_0 - \overline{\gamma}_{m0})}{V_T \cos\overline{\gamma}_{m0} - V_M} \qquad \overline{\lambda}_\eta(0) = -\frac{r_0 \sin(\eta_0 - \overline{\gamma}_{m0})}{V_T \cos\overline{\gamma}_{m0} - V_M}$$

thus

$$\tilde{H} = -\left(1 + ka^2\right) + \frac{V_T \cos\overline{\gamma}_{m0} - V_M \cos(\gamma_m - \overline{\gamma}_{m0})}{V_T \cos\overline{\gamma}_{m0} - V_M} + \lambda\gamma\frac{a}{V_M}$$

The differential equations are then

$$\frac{\mathrm{d}\gamma_m}{\mathrm{d}\tau} = \frac{a}{V_M}$$

$$\frac{\mathrm{d}\lambda\gamma}{\mathrm{d}\tau} = -\frac{\partial\tilde{H}}{\partial\gamma_m} = -\frac{V_M}{V_T \cos\overline{\gamma}_{m0} - V_M}\sin(\gamma_m - \overline{\gamma}_{m0})$$

One will notice that the equilibrium solution of this layer $a = 0$, $\gamma_m = \overline{\gamma}_m$ does correspond to the initial conditions of the external layer.

The fact that only one adjoint variable λ_γ appears in the expression of Hamiltonian allows to solve this equation simply in terms of λ_γ and/or a knowing that

$$\frac{\partial H}{\partial a} = 0 \longrightarrow a = \frac{\lambda\gamma}{2kV_M}$$

and that $H = 0$ along the optimal trajectory

$$\lambda_\gamma^2 = 4kV_M^2 \frac{1-\cos(\gamma_m - \overline{\gamma}_{m0})}{1-\frac{1}{\nu}\cos\overline{\gamma}_{m0}} \qquad \nu = \frac{V_M}{V_T}$$

$$a = \frac{\lambda\gamma}{2kV_M} = -\frac{1}{\sqrt{k}}\sqrt{\frac{1-\cos(\gamma_m - \overline{\gamma}_{m0})}{1-\frac{1}{\nu}\cos\overline{\gamma}_{m0}}}\operatorname{sign}(\overline{\gamma}_{m0} - \gamma_m) \tag{7.66}$$

One will notice that:

- in the expression of $\widetilde{H}$, $\overline{\lambda}_{r(0)}$, $\overline{\lambda}_{\eta(0)}$ being fixed and the only active variable being γ_m, the quantity in square brackets plays, apart from the sign, the role of the new criterion for which one has to search the optimal control
- the control variable a is defined by using only one current state variable γ_m .
- a tends asymptotically (within time-scale τ) towards 0 when $\gamma_m \longrightarrow \overline{\gamma}_m$ and there is thus an asymptotic matching of the solution in the boundary layer towards the solution in the external layer.
  The control a defined by (7.66) is therefore still valid.

Of course, the fact that one sets $\varepsilon = 0$ amounts to the same thing as assuming – after what has been shown in section 7.2.2.1 – that the ratio of the minimum turn radius ρ to the distance to be covered is zero even though it is small. ρ and η, which were assumed to be constant and equal to ρ_0, η_0 in the boundary layer solution, have thus changed and at the end of the turn the vehicle reaches M_1 instead of M_0, resulting in the fact that the target is inevitably lost (*cf.* figure 7.17) since the minimum distance of proximity increases in the same way as $\overline{\gamma}_{m(0)} - \gamma_{m(0)}$ and r_0/ρ .

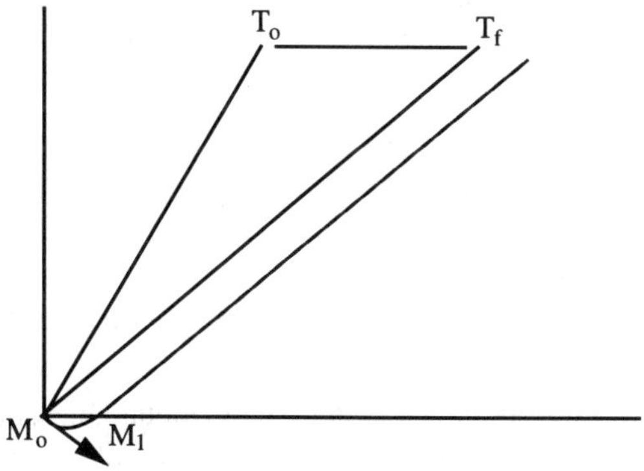

Figure 7.17: Influence of the initial boundary layer.

Two alternative solutions are then possible:

1. When the heading angle $\overline{\gamma}_{m(0)}$ is achieved (with some given accuracy) one resets the problem, starting with values of $r_{(1)}$ $\eta_{(1)}$ $\gamma_{m(1)}$ corresponding to the actual situation at point M_1. As the change in the heading to be performed then becomes small, assumptions are better verified and the process converges rapidly (the procedure is analogous to the one performed with an extension in ε).
2. Every instant is considered as an initial instant. This is possible due to the simplicity

of the calculus required by the singular perturbation solution. This last approach has, moreover, the advantage of not making any assumption on the motion of the target since one can measure at every instant ρ and η (and since evolutions of the target remain small in comparison with those of M).

The problem to be solved is completely illustrated in figure 7.18

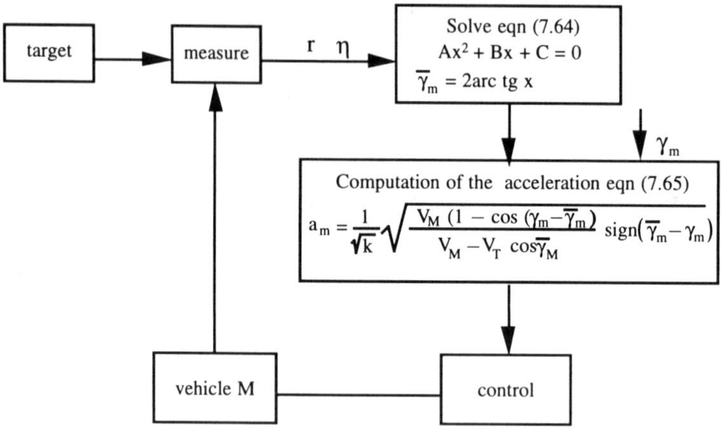

Figure 7.18: Solutions in the initial boundary layer.

By way of illustration, one has considered the same conditions as for the optimal solution. For the computation at the initial instant, one finds in the external layer $\overline{\gamma}_m$ = 43.22°, a crossing distance about 400 m without resetting, as a consequence of the importance of the turn radius (the ratio of turning radius to the length of trajectory is indeed of order of 0.3) (*cf.* figure 7.19).

Figure 7.19 visualizes the results obtained:

- The first diagrams are performances with three updatings. In fact, only the first one, performed at instant about 13 sec, can be seen. The target is then reached (the proximity distance being chosen equal to 15 m).
- The second set of diagrams show a closed-loop form, the calculus being re-computed at every sampling step.

Practically analogous results are obtained in case of frontal attacks and, in the case of a closed-loop updating at every computation step, in the case of a moving target.

7.3.4.2 Optimization of the time to climb for a combat aircraft (several fast variables: linearization and terminal boundary layer)

In this second example, one considers the case of a climb, in a vertical plane, of a combat aircraft. This example can be distinguished from the previous one by the fact that one has two variables in the boundary layer for which the linearization of the equations in this layer is necessary to obtain a solution in a closed-loop form.

Starting from initial conditions (h_0, v_0, γ_0) (altitude, speed, slope) the problem is thus to find a trajectory and a control law leading to fixed end conditions (h_f, v_f) in a minimum time (the final slope γ_f, can be fixed or free).

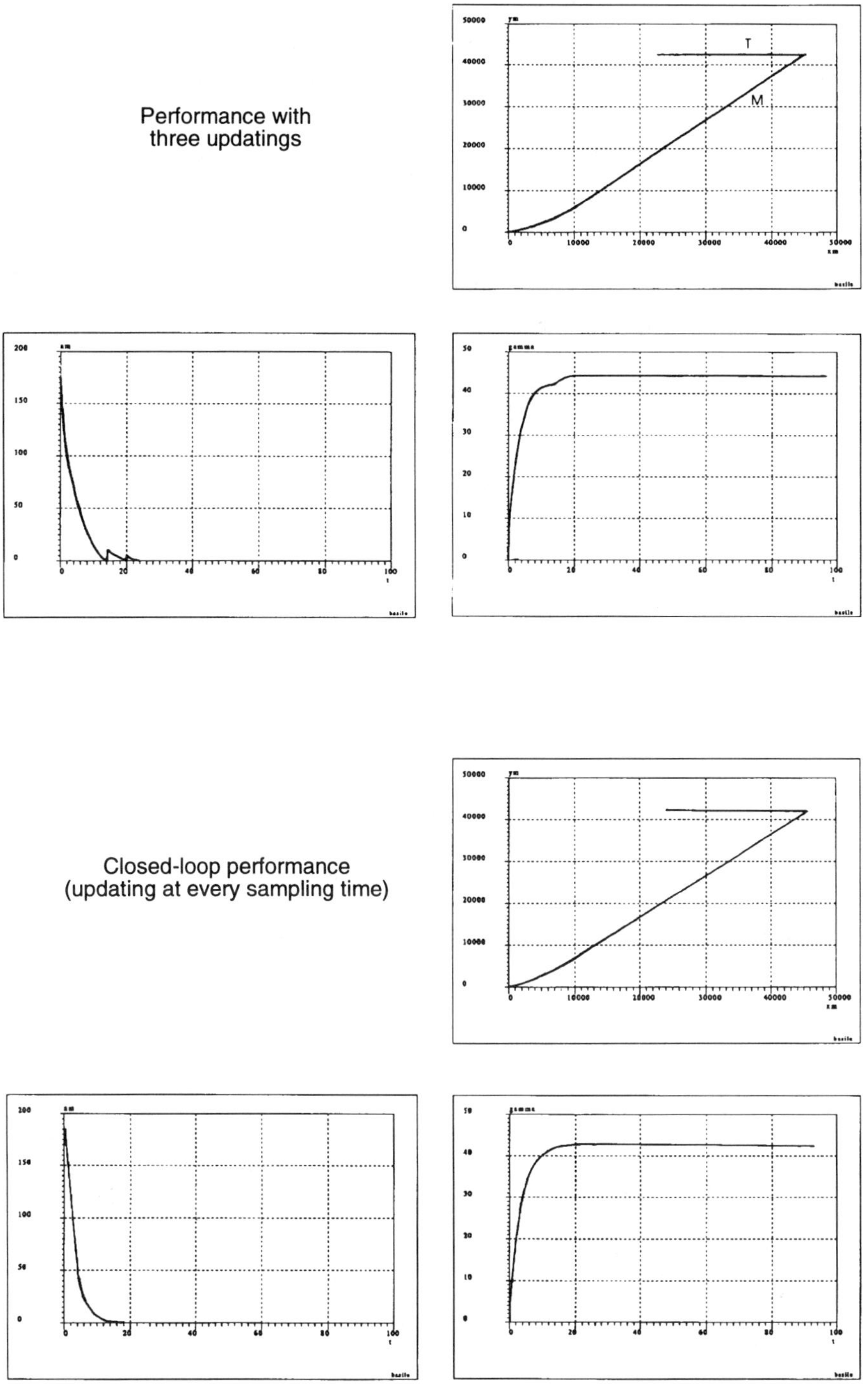

Figure 7.19: Performances with updating.

For the sake of simplicity one assumes that:

- the mass m remains constant during the whole climbing
- the climb is performed with the maximum thrust P_m, which depends on the altitude and the Mach number $P_m\,(M, h)$
- the polar is a parabola of the form

$$C_x = C_{x0}(M) + k\,(M)\,C_z{}^2$$

M being the Mach number $V/a(h)$

- γ and h can be considered as fast variables with respect to the energy $E = h + V^2/2_g$.

With these conditions, the equations of the motion can be written in the following form, which is deduced from figure 7.20, by choosing as a control variable the normal load factor $n_z = \dfrac{Z}{mg}$, Z and T being respectively lift and drag.

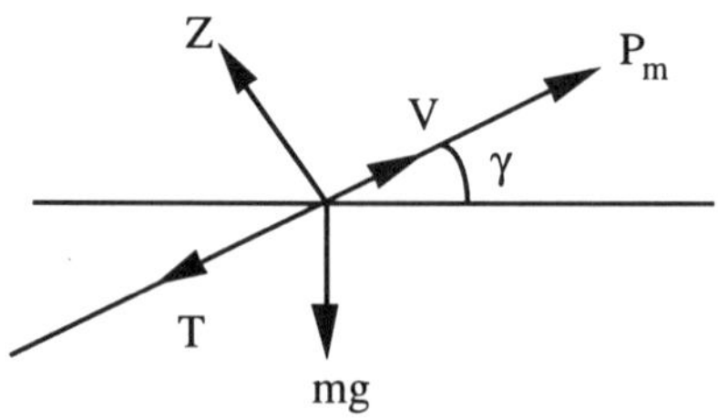

Figure 7.20: Forces acting on the system.

$$\begin{aligned} m\dot{V} &= P_m - T - mg\sin\gamma \\ \dot{h} &= V\sin\gamma \\ \dot{\gamma} &= \frac{g}{V}\,(n_z - \cos\gamma) \end{aligned}$$

n_z, the normal load factor, is given by $n_z = Z/mg = 1/2\rho SV^2C_z/mg$.
In these equations, the drag is given by T

$$T = \frac{1}{2}\rho SV^2C_x$$

with

$$C_x = C_{x0} + kC_z^2$$

thus

$$T = \frac{1}{2}\rho SV^2\left(C_{x_o} + kC_z^2\right) = \frac{1}{2}\rho SV^2C_{x0} + \frac{2km^2g^2}{\rho SV^2}n_z^2 = T_0 + T_1n_z^2$$

If one introduces the variable $E = h + V^2/2g$, by taking into account assumptions on the speed scales, these equations will be rewritten in the form

$$\begin{aligned} \dot{E} &= \frac{V}{mg}\left(P_m - T_0 - T_1n_z^2\right) \\ \varepsilon\dot{h} &= V\sin\gamma \\ \varepsilon\dot{\gamma} &= \frac{g}{V}\,(n_z - \cos\gamma) \end{aligned}$$

If one writes the Hamiltonian in the form

$$H = -1 + \lambda_E \frac{V}{mg}\left(P_m - T_0 - T_1 n_z^2\right) + \lambda_h V \sin\gamma + \lambda_\gamma \frac{g}{V}(n_z - \cos\gamma)$$

with

$$V = V(h, E); T_o = T_0(V, h); T_1 = T_1(V, h)$$

the optimal control n_z satisfies

$$n_z^* = \arg\min H$$

with the adjoint variable defined by

$$\dot{\lambda}_E = -\frac{\partial H}{\partial E}, \quad \varepsilon\dot{\lambda}_h = -\frac{\partial H}{\partial h}, \quad \varepsilon\dot{\lambda}\gamma = -\frac{\partial H}{\partial \gamma},$$

and the transversality conditions corresponding to the final time t_f.
To find a closed-loop control, in terms of state variables, one will use here the method of singular perturbations too.

External layer

The processing of the external layer is absolutely identical to the previous examples. For $\varepsilon = 0$ one obtains:

$$\overline{\gamma} = 0$$

$$n_z = \cos\overline{\gamma} = 1$$

the pseudo control being $\overline{h}$

$$\overline{H} = -1 + \overline{\lambda}_E \frac{V}{mg}(P_m - T_0 - T_1) = 0$$

according to the stationarity, and t_f free.

As the problem in this external layer is of dimension one, one has to solve a static optimization problem and as quantity $(P_m - T_0 - T_1)V$ is always positive it follows that (according to section 7.3.3.4.1)

$$\overline{h}^* = \arg\max_h V(P_m - T_0 - T_1) \tag{7.67}$$

$$\overline{\lambda}_E = +\frac{mg}{\overline{V}\left(\overline{P}_m - \overline{T}_0 - \overline{T}_1\right)^*} > 0$$

For each level of energy E, from E_0 to E_f, equation (7.67) provides the point corresponding to optimal climb-energy profile $\overline{h}^*(E)$. The corresponding values can first be computed off-line and stored in terms of E in a table. Figure 7.21 illustrates the result obtained qualitatively. One will note that one finds $E_f = h_f + {V_f}^2/2g$ at the final instant, which does not mean that one gets h_f and V_f.

Remark 7.12 *One can note a discontinuity on the profile at the proximity of the sound velocity. This is due to the fact that the function $\dot{E}(h, E)$ to be maximized (7.67) presents in the vicinity of Mach 1 two maxima related to two different altitudes. In the subsonic region, the absolute maximum corresponds to the lowest altitude while, in the supersonic region, it corresponds to the highest altitude. There are solutions to bypass this difficulty but this is out of the frame of this study.*

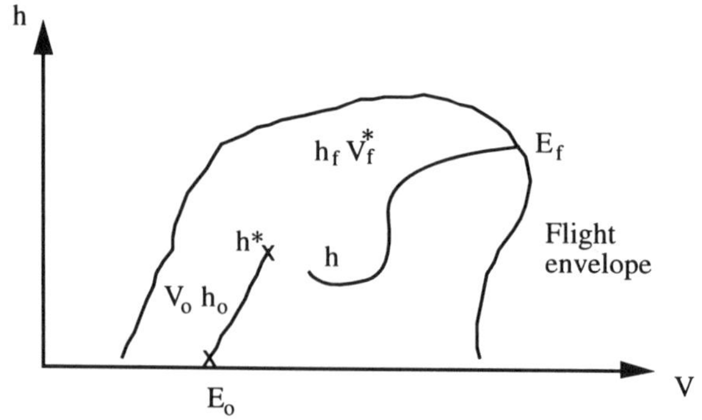

Figure 7.21: Flight profiles (note the discontinuity at sound velocity).

Initial boundary layer

In the initial boundary layer, one must realize variations of altitude and flight path angle which lead, from initial conditions $h_0\ \gamma_0$, to conditions fixed by external layer for $E = E_0$. Within the extended time-scale

$$\begin{aligned} \frac{\mathrm{d}E}{\mathrm{d}\tau} &= \frac{\varepsilon V}{mg}\left(P_m - T_0 - T_1 n_z^2\right) \\ \frac{\mathrm{d}h}{\mathrm{d}\tau} &= V \sin\gamma \qquad (7.68) \\ \frac{\mathrm{d}\gamma}{\mathrm{d}\tau} &= \frac{g}{V}\left(n_z - \cos\gamma\right) \qquad (7.69) \end{aligned}$$

for $\varepsilon = 0$, $E = E_0 =$ constant and $\lambda_E = \lambda_{E(0)}$ The Hamiltonian becomes

$$H = \left[-1 + \overline{\lambda}_{E(0)}\frac{V}{mg}\left(P_m - T_0 - T_1 n_z^2\right)\right] + \lambda_h V \sin\gamma + \lambda_\gamma \frac{g}{V}\left(n_z - \cos\gamma\right)$$

and the optimal control $n_z^* = \arg\min H \Big|_{E = E_0}$ satisfies the equation:

$$\frac{\partial H}{\partial n_z} = -2\overline{\lambda}_{E(0)}\frac{V}{mg}T_1 n_z + \lambda_\gamma \frac{g}{V} = 0$$

that means

$$n_z^* = \frac{mg^2}{2\overline{\lambda}_{E(0)}}\frac{\lambda_\gamma}{V^2 T_1} \qquad (7.70)$$

One will notice that the value given by (7.70) corresponds indeed to the maximum of H since:

$$\frac{\partial^2 H}{\partial n_z^{*2}} = +2\overline{\lambda}_{E(0)}\frac{V T_1}{mg} < 0$$

but it depends, naturally, on λ_γ. As both the adjoint variables λ_h and λ_γ appear simultaneously in the boundary layer, the solution requires to the solution of a two-point boundary problem, nonlinear, with a dimension lower that the one of original problem, but it excludes out any solution in a closed-loop form.

The solution which is used here, and shown in section 7.3.3.4, consists in linearizing the equations of the boundary layer around the solution obtained in the external layer.

$$\overline{h}(0), \overline{\gamma}(0) = 0, \overline{n_z}(0) = 1$$

Equations (7.68) and (7.69), when linearized about external solution at $t = 0$ with

$$\delta h = h - \overline{h}(0) \quad \delta\gamma = \gamma \quad \delta n_z = n_z - 1$$

become, with derivatives computed with respect to τ

$$\begin{bmatrix} \delta\dot{h} \\ \delta\dot{\gamma} \end{bmatrix} = \begin{bmatrix} 0 & V_o \\ 0 & 0 \end{bmatrix} \begin{bmatrix} \delta h \\ \delta\gamma \end{bmatrix} + \begin{bmatrix} 0 \\ g/V_o \end{bmatrix} \delta n_z = AX + B\delta n_z$$

By the same way, linearization of cost criterion associated to optimal control leads to use a criterion of the form, according to section 7.3.3.4

$$\int_o^\infty [\delta h, \delta\gamma, n_z] \begin{bmatrix} Q & S \\ S^T & r \end{bmatrix} \begin{bmatrix} \delta h \\ \delta\gamma \\ \delta n_z \end{bmatrix} \mathrm{d}\tau$$

with

$$Q = \begin{bmatrix} \dfrac{\partial^2 \tilde{H}}{\partial h^2} & \dfrac{\partial^2 \tilde{H}}{\partial h \partial\gamma} \\ \dfrac{\partial^2 \tilde{H}}{\partial\gamma\partial h} & \dfrac{\partial^2 \tilde{H}}{\partial\gamma^2} \end{bmatrix} \qquad S = \begin{bmatrix} \dfrac{\partial^2 \tilde{H}}{\partial h \partial n_z} \\ \dfrac{\partial^2 \tilde{H}}{\partial\gamma\partial n_z} \end{bmatrix} \qquad r = \frac{\partial^2 \tilde{H}}{\partial n_z^2}$$

and

$$\tilde{H} = -1 + \overline{\lambda}_{E_o} \frac{V}{mg} \left[P_m - T_o - T_1 n_z^2\right] + \lambda_h V \sin\gamma + \lambda_\gamma \frac{g}{V} (n_z - \cos\gamma)$$

The derivations are performed while keeping in mind that $E = h + V^2/2g$ remains constant in the boundary layer, and equal to E_o (thus $\mathrm{d}V = -g/V\mathrm{d}h$) and that the obtained expressions are computed at their asymptotic values

$$\gamma = 0 \;\; n_z = 1 \;\; \overline{\lambda}_\gamma = 2\overline{\lambda}_{E_o} \frac{V^2}{mg^2} T_1$$

according to (7.70).

Having performed all the calculations, one finds:

$$Q = \begin{bmatrix} q_1 & 0 \\ 0 & q_2 \end{bmatrix} = \begin{bmatrix} \overline{\lambda}_{EO} \dfrac{\partial^2 \dot{E}}{\partial h^2}\Big|_{\overline{E}} & 0 \\ 0 & 2\overline{\lambda}_{EO} \dfrac{V}{mg} T_1 \end{bmatrix}$$

$$S = \begin{bmatrix} s_1 \\ 0 \end{bmatrix} = \begin{bmatrix} \dfrac{4\overline{\lambda}_{EO} T_1}{mV} - \dfrac{2\overline{\lambda}_{E_o}}{mg} \left[V \dfrac{\partial T_1}{\partial h} - g \dfrac{\partial T_1}{\partial V}\right] \\ 0 \end{bmatrix}$$

$$r = -2\overline{\lambda}_{EO} \frac{V}{mg} T_1$$

The optimal control is then defined by

$$\delta n_z = -R^{-1}\left(B^T K + S^T\right)\begin{pmatrix}\delta h\\ \delta\gamma\end{pmatrix}$$

where K is the solution of the Riccati equation

$$K\hat{A} + \hat{A}^T K + \hat{Q} - KBR^{-1}B^T K = 0$$

with

$$\hat{A} = A - BR^{-1}S^T = \begin{bmatrix} 0 & V_o \\ -\dfrac{gs_1}{rV_o} & 0 \end{bmatrix} = \begin{bmatrix} 0 & V_o \\ \alpha & 0 \end{bmatrix}$$

$$\hat{Q} = Q - SR^{-1}S^T = \begin{bmatrix} q_1 - \dfrac{s_1^2}{r} & 0 \\ 0 & q_2 \end{bmatrix} = \begin{bmatrix} \hat{q}_1 & 0 \\ \mathrm{d} & \hat{q}_2 \end{bmatrix}$$

By setting

$$K = \begin{bmatrix} k_1 & k_2 \\ k_2 & k_3 \end{bmatrix}$$

one finds the equations

$$\frac{g^2}{V_o^2 r}k_2^2 - 2\alpha k_2 - \hat{q}_1 = 0$$

$$\frac{g^2}{V_o^2 r}k_3^2 = 2V_o k_2 + q_2$$

$$k_1 V = \frac{g^2}{V_o^2 r}k_2 k_3 - \alpha k_3$$

then the optimal control

$$\delta n_z = -\frac{1}{r}\left[\left(\frac{gk_2}{V} + s_1\right)\delta h + \frac{gk_3}{V}\delta\gamma\right]$$

Final boundary layer

In the formulation of the problem one had specified final values of h_f and V_f. However, the external solution has resulted in a correct value of energy E_f but the corresponding values of the altitude (and also speed) are not *a priori* equal to the desired final values. As shown in section 7.3.3.4.2 the final trajectory is first obtained by an off-line integration of the equations of motion, in a backward form, from final conditions E_f, h_f, γ_f using the linearized optimal control of the final boundary layer (one takes care to choose the semi definite negative solution of the Riccati equation, which is computed in the same way as for the initial boundary layer by changing only the point used for linearization). The trajectory is then recorded in the form of curves $h_{\mathrm{d}}(E)$ and $\gamma_{\mathrm{d}}(E)$.

In direct time, control n_z is computed by using the form

$$n_z = k_1\left(h - h_{\mathrm{d}}(E)\right) + k_2\left(\gamma - \gamma_{\mathrm{d}}(E)\right)$$

where gains k_1 and k_2 are adjusted in order to obtain a good tracking of prerecorded trajectory.

One will notice that the control n_z which was chosen to fulfill the final condition, is in a same form than the one already used in the initial layer.

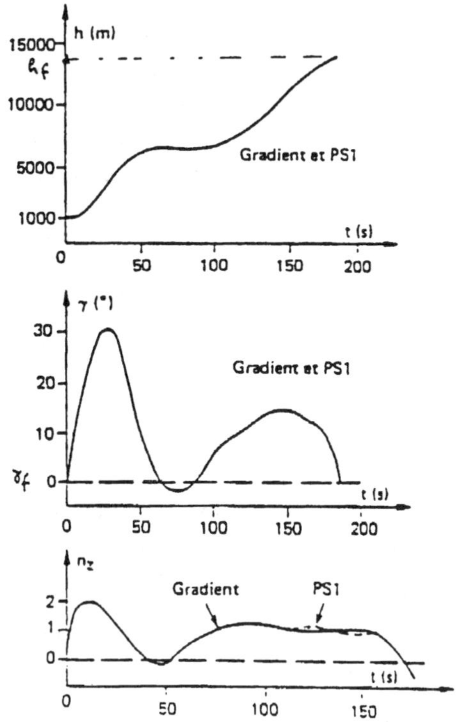

Figure 7.22: Optimal climb (results).

Figure 7.22 visualizes the results obtained (which are compared to those obtained by a classical off line optimization method of a gradient type which is much more complex).

7.4 Conclusion

One has seen, in this chapter, how the recognition of the coexistence of slow and fast dynamics allows, after a formulation of the system in a singularly perturbed form, to simplify analysis and control, both in a theoretical and a practical aspect, at the expense of an approximation which is better when the separation of dynamics is greater.

In the linear case, which has been mentioned here occasionally only for the purpose of understanding, the interest remained essentially in the simplification of algorithms, the conception of a corrector in two steps, the use of different sampling periods for slow and fast parts, since dynamic properties are easy to show and since a control in closed-loop form is always possible.

For more details, one can refer to Fossard (1985; 1982) and to the excellent surveys (Kokotović *et al.*, 1976; Saksena *et al.*, 1984) shown in the bibliography.

In the nonlinear case, the interest is even greater, in the field of analysis (see section 7.2.1.3) and also in the control field. Illustration has been done in the case of an optimal control where classical techniques (dynamic programming, Hamiltonian, *etc.*) generally do not permit us to reach an explicit solution in terms of the state and lead to such complex calculations that any real-time realization becomes a very questionable problem.

One has shown that an approach in a singularly perturbed form allows a considerable simplification of the calculation and an approximation of the optimum solution by a solution in terms of state, which means in a closed-loop form.

Finally, one will note that the interest of recognition of slow and fast parts within a given system lies beyond the limits of analysis and optimal control. For example, the computation of a linearized and decoupled control of a nonlinear system, which has been presented in another chapter, can lead to unfeasible calculations in practice if the characteristic indices of the output are large. The separation into slow and fast systems, which reduces necessarily the associated indices, can there again provide a useful solution. One can refer to Harcaut (1991).

7.5 Bibliography

[1] TIHONOV, A. N. (1952), Systems of differential equations containing small parameters in the derivatives, *Mathem. Sborniki* **31**, 73.

[2] VASILEVA, A. B. (1963), Asymptotic behavior of solutions to certain problems involving nonlinear differential equations containing a small parameter multiplying the highest derivatives, *Russian Math. Surveys*, **18**, 3.

[3] VASILEVA, A. B. (1972), Asymptotic solution of two point boundary value problems for singulary perturbed conditionally stable systems. Singular perturbations order reduction in control system design, *ASME*, NY.

[4] ETDOS, P. and TURAN, P. (1950), On the distribution of roots of polynomials, *Annales of Math.*, **51**, 1, 105–119.

[5] DAUPHIN TANGUY, G. (1983), Sur la représentation multimodèle des systèmes singulièrement perturbés, Thesis Univ. of Lille.

[6] MOREIGNE, O. (1984), Contribution à la modélisation et à la synthèse des systèmes à échelles de temps multiples, Thesis Univ. of Lille.

[7] BOUAYAD, A. (1990), Étude comparative des méthodes d'analyse de systèmes à échelle de temps multiples, Thesis Univ. of Lille.

[8] DAUPHIN-TANGUY, G., BORNE, P. and MEIZEL, D. (1982), *On order reduction of multitime scales systems by singular perturbation and frequency like methods*, IEEE Large scale symposium.

[9] MARINO, R. and KOKOTOVIĆ, P.V. (1988), A geometric approach to nonlinear singularly perturbed control systems, *Automatica*, **211**, 1.

[10] ISIDORI, A. (1989), *Nonlinear control systems*, Springer Verlag.

[11] ARDEMA, M. (1983), Singular perturbations and control, *CISM Courses and lectures*, **280**, Springer-Verlag.

[12] ARDEMA, M. and RAJAN, N. (1983), *Separation of time-scales in aircraft trajectory optimization*, AIAA Atmsph. flight Mechanics conferences, Gatlinburg, USA.

[13] KELLEY, M. J. (1971), Reduced order modeling in aircraft mission analysis, *AIAA Journal*, **9**, 2.

[14] KELLEY, M. J. (1972), State variable selection and singular perturbations. Order reduction in control system design, *ASME*.

[15] FOSSARD, A. J. and UZAN, C. (1989), *Optimisation de trajectoires d'engin par la méthode des perturbations singulières forcées*, Symposium AGARD, Lisbonne.

[16] HUYNH, H. T. (1990), Détermination de lois de guidage quasi optimales en temps réel pour des trajectoires d'avions de combat, *Agardograph A6 301 - Aircraft trajectories computation, prediction, control*, **1**.

[17] SAGE, A. P. (1968), *Optimum systems control*, Prentice Hall.

[18] BRYSON, A. E. and HO Y. C. (1975), *Applied optimal control*, John Wiley.

[19] MIRANDE, M. (1980), *Application of Modern control Theory to the guidance of an air to air dogflight missile*, AGARD conf. Poceedings, **292**.

[20] GUELMAN, M. and SHINAR, J. (1984), Optimal guidance law in the plane, *Journal of guidance*, **7**, 4.

[21] FOSSARD, A. J. and FREITAS, A. (1986), Applications des techniques de perturbations singulières à l'optimisation des trajectoires d'avion, *L'automatique pour l'Aéronautique. SMAI*, Paris.

[22] FOSSARD, A. J., FOISNEAU, J. and HUYNH, H. T. (1991), *Application de la méthodologie perturbations singulières à la détermination de trajectoires optimales en boucle fermée*, Lecture notes of the ENSAE.

[23] FOSSARD, A. J. (1985), *Optimisation de trajectoires en temps réel*, technical report CERT/DERA, 1/7377.

[24] FOSSARD, A. J. (1982), *Systèmes à échelles de temps multiples*, technical report CERT/DERA, 3/7299.

[25] FOSSARD, A. J. (1983), Méthodes de perturbations en automatique. Systèmes linéaires à deux échelles de temps, in *Outils mathématiques pour l'automatique, l'analyse des systèmes et le traitement du signal*, Editions of the CNRS.

[26] KOKOTOVIĆ, P.V., KHAHIL, H.K. and O'REILLY, J. (1986), *Singular perturbation methods in control analysis and design*, New York Academic Press.

[27] KOKOTOVIĆ, P.V. (1994), Applications of singular perturbation techniques to control problems, *SIAM Review*, **26**, 4.

[28] KOKOTOVIĆ, P.V. and SAUER, P.W. (1987), *Integral manifold as a tool for reduced order modeling in nonlinear systems: a synchronous machine case study*, Proceeding of the 26th CDC, Los Angeles.

[29] JANNOU, P.A. and KOKOTOVIĆ, P.V. (1983), Adaptive systems with reduced models, *Lecture notes in Control and Information Sciences*, **47**, Springer, Berlin, NY.

[30] SPONG, M.W., KHORASAM, K. and KOKOTOVIĆ, P.V. (1987), An integral manifold approach to the feedback control of flexible joint robots, *IEEE J. Robot Automat.*, **3**, 4, 291–360.

[31] JUHNG-PERNG, SU, JER-GUANG, HSIECH (1990), Composite feedback control for a class of nonlinear singularly perturbed systems with fast actuators, *Int. Journal Control*, **52**, 3, 571–579.

[32] BARBOT, J.P. and PANTALOS, N. (1991), *Using symbolic calculus for singularly perturbed nonlinear systems*, European Conference on Algebraic Computation in Control, Springer-Verlag, Paris.

[33] SHARKEY, P.M. and O'REILLY, J. (1988), Composite control of nonlinear singularly perturbed systems: a geometric approach, *International J. Control*, **48**, 6, 2991–2506.

[34] HARCAUT, J.PH. (1991), Commande non linéaire. Application au pilotage d'un engin, Doctorate thesis ENSAE.

• Surveys

[35] KOKOTOVIĆ, P.V., O'MALLEY R.E. and SANNUTI P. (1976), Singular perturbations and order reduction in control theory. An overview, *Survey - Automatica*, **12**, 123–132.

[36] SAKSENA, V.R, O'MALLEY, R.E. and KOKOTOVIĆ, P.V. (1984), Singular perturbations and well-scale methods in control theory, *Survey 1976-1984: Automatica*, **20**, 273–293.

[37] HOPPENSTEADT, F. (1977), Assymptotic stability in singular perturbations problems, *Journal of differential equations*.

APPENDIX A

Some differential geometric recalls

J. LÉVINE

A.1 Differentiable manifolds, diffeomorphism

We only sketch here "intuitive" definitions (which are not necessarily easier to present) so that the reader may be able to recognize usual geometric notions. In particular, manifolds are introduced by equations expressed in coordinates. The justification is given through the concepts of diffeomorphism and curvilinear coordinates. The interested reader may refer to [3, 4] for a rigorous presentation.

Definition A.1

*Given a differentiable mapping Φ from $\mathbb{R}^n$ to $\mathbb{R}^{n-p}$ $(0 \leq p < n)$, we assume that there exists at least one x_0 solution of the implicit equation $\Phi(x) = 0$ and that the linear tangent mapping $\mathrm{d}\Phi(x)$ has full rank $(n-p)$ in a neighborhood V of x_0. We call p-*dimensional differentiable manifold *the set X defined by the implicit equation $\Phi(x) = 0$. In other words,*

$$X = \{x \in V | \Phi(x) = 0\} \tag{A.1}$$

If Φ is k times differentiable (analytic), we say that X is a C^k manifold, $k = 1, \ldots, \infty$ (analytic –or C^ω–).

In unambiguous cases, we only speak of manifold.

There are two examples

1. The affine (analytic) manifold: $\{x \in \mathbb{R}^n | Ax - b = 0\}$ is p-dimensional if rank $A = n - p$ and $b \in \mathrm{Im}A$.

2. The sphere of $\mathbb{R}^3$: $\left\{(x, y, z) \in \mathbb{R}^3 | (x-a)^2 + (y-b)^2 + (z-c)^2 - R^2 = 0\right\}$ is an analytic manifold of dimension 2.

The concept of local diffeomorphism is essential to describe manifolds in an intrinsic way (*i.e.* irrespective of any system of coordinates in which the implicit equation $\Phi(x) = 0$ is written).

Definition A.2

Given a C^k mapping (analytic) ϕ, $k \geq 1$, from an open set $U \subset \mathbb{R}^n$ to an open set $V \subset \mathbb{R}^n$, we say that ϕ is a local diffeomorphism *of class C^k (resp. analytic) in a neighborhood $U(x_0)$ of a point x_0 in U if ϕ is invertible from $U(x_0)$ into a neighborhood $V(\phi(x_0))$ of the point $\phi(x_0)$ of V and if ϕ^{-1} is also of class C^k (resp. analytic).*

Homeomorphisms are more general transformations which are useful to analyze topological properties (as opposed to diffeomorphisms that preserve differentiability properties):

Definition A.3

Given a mapping ϕ from an open set $U \subset \mathbb{R}^n$ to an open set $V \subset \mathbb{R}^n$ of class C^k, $k \geq 1$ (resp. analytic), we say that ϕ is a local homeomorphism *of class C^k (resp. analytic) in a neighborhood $U(x_0)$ of a point x_0 in U if ϕ is invertible from $U(x_0)$ into a neighborhood $V(\phi(x_0))$ of the point $\phi(x_0)$ of V and if ϕ^{-1} is continuous.*

Local diffeomorphisms are characterized by the following classic result:

Proposition A.1

A necessary and sufficient condition for ϕ to be a local diffeomorphism in a neighborhood of x_0 is that its tangent linear mapping $\mathrm{d}\phi(x_0)$ is one-to-one.

Curvilinear coordinates provide a particularly important geometric interpretation of diffeomorphisms. In particular, one can find (local) suitable curvilinear coordinates for which the manifold X is expressed as a vector subspace of $\mathbb{R}^p$. It suffices to introduce:

$$y_1 = \Phi_1(x), \ldots, y_{n-p} = \Phi_{n-p}(x), y_{n-p+1} = \Psi_1(x), \ldots, y_n = \Psi_p(x)$$

where the independent functions $\Psi_1, \ldots, \Psi_p$ are such that the mapping

$$x \longrightarrow (\Phi_1(x), \ldots, \Phi_{n-p}(x), \Psi_1(x), \ldots, \Psi_p(x))$$

is a local diffeomorphism. We say that we have "rectified" the coordinates of X since

$$X = \{y | y_1 = \cdots = y_{n-p} = 0\} .$$

A.2 Tangent space, vector field, Lie derivative

Let us recall that the tangent linear mapping $\mathrm{d}\Phi(x)$, expressed in the local system of coordinates $(x_1, \ldots, x_n)$, also called the Jacobian matrix of Φ, is the matrix whose entry at line i and column j is $\frac{\partial \Phi_j}{\partial x_i}(x)$. One can easily check that the normal at the point x to the manifold X is collinear to $\mathrm{d}\Phi(x)$, or, in other words, is a linear combination of columns of $\mathrm{d}\Phi(x)$. Indeed, let $y(t)$ be a differentiable curve contained in X for all $t \geq 0$ small enough, such that $y(0) = x$ (such a curve exists thanks to the implicit function theorem). We thus have $\Phi(y(t)) = 0$ for all $t \in [0, \tau[$ and $\frac{\Phi(y(t)) - \Phi(x)}{t} = 0$. When t tends to

0 then $d\Phi(x).\dot{y}(0) = 0$, where $\dot{y}(0) \stackrel{\text{def}}{=} \frac{dy}{dt}_{|t=0}$, which proves that the vector $\dot{y}(0)$, which is tangent to X at the point x, belongs to the kernel of $d\Phi(x)$. It therefore immediately follows that every element of $d\Phi(x)$ is orthogonal to every tangent vector to X at the point x, Q.E.D.

This motivates the next definition of tangent space to a manifold:

Definition A.4

The tangent space *to X at the point $x \in X$ is the vector space*

$$\mathrm{T}_x X = \mathrm{Ker}\, d\Phi(x) \, .$$

The tangent bundle $\mathrm{T}X$ *is the set* $\mathrm{T}X = \bigcup_{x \in X} (\{x\} \times \mathrm{T}_x X)$.

Note that, according to the fact that $d\Phi(x)$ has rank $n - p$ in the open set V,

$$\dim \mathrm{T}_x X = \dim \mathrm{Ker}\, d\Phi(x) = p \, , \forall x \in V \, .$$

Definition A.5

A (C^k, analytic) vector field f *on X is a (C^k, analytic) mapping sending $x \in X$ into $f(x) \in \mathrm{T}_x X$.*

Definition A.6

An integral curve *of the vector field f is a local solution of the differential equation $\dot{x} = f(x)$.*

Local existence and uniqueness of integral curves of f follow from the fact that f is C^k, $k \geq 1$, and thus locally Lipschitzian.

Integral curves may exist globally. In this case, one speaks of a complete vector field:

Definition A.7

We say that the vector field f is complete if every integral curve of f is defined on $[0, +\infty)$.

Let us consider a system of local coordinates $(x_1, \ldots, x_n)$. The components of the vector field f, in these coordinates, are denoted by $(f_1, \ldots, f_n)^T$. We now show that one can establish a one-to-one correspondence between f and a differential operator called *Lie derivative in the direction f*. Let us denote by $t \longmapsto X_t(x)$ the integral curve of f passing through x at time $t = 0$.

Definition A.8

Let h be a C^1 function of $\mathbb{R}^n$ into $\mathbb{R}$. We call Lie derivative of h in the direction f, *denoted by $L_f h$, the derivative of h along the integral curve of f at $t = 0$, in other words,*

$$L_f h(x) = \frac{d}{dt} h(X_t(x))_{|t=0} = \sum_{i=1}^{n} f_i(x) \frac{\partial h}{\partial x_i}(x) \, .$$

By this formula, an arbitrary vector field f is identified to the linear first-order differential operator $L_f = \sum_{i=1}^{n} f_i(x)\frac{\partial}{\partial x_i}$.

As before, there exist coordinates that *rectify a vector field*. To show this statement, we need the notions of image of a vector field by a diffeomorphism, and of first integral.

To introduce the image of a vector field by a diffeomorphism, let us proceed as follows: let ϕ be a local diffeomorphism, and let $y = \phi(x)$. The image through ϕ of an integral curve of f satisfies

$$\frac{\mathrm{d}}{\mathrm{d}t}y_i(t) = \frac{\mathrm{d}}{\mathrm{d}t}\phi_i(X_t(x)) = L_f\phi_i\left(\phi^{-1}(y(t))\right) \ \forall i = 1, \ldots, n\,.$$

Thus we have the following definition.

Definition A.9

The image through the diffeomorphism ϕ of the vector field f, denoted by $\phi_* f$, is the vector field:

$$\phi_* f = \left(L_f\phi_1\left(\phi^{-1}(y)\right), \ldots, L_f\phi_n\left(\phi^{-1}(y)\right)\right)^T\,.$$

Moreover, we say that f is rectified by diffeomorphism *if*

$$L_f\phi_1\left(\phi^{-1}(y)\right) = 1 \text{ and } L_f\phi_i\left(\phi^{-1}(y)\right) = 0 \quad \forall\, i = 2, \ldots, n\,.$$

As a result of this definition (exercise) the Lie derivative formula is not affected by changes of coordinates: $L_f h(x) = L_{\phi_* f}\left(h \circ \phi^{-1}\right)(y)$.

Definition A.10

A first integral *of f is a function γ satisfying $L_f\gamma = 0$. In other words, γ remains constant along the integral curves of f.*

Proposition A.2

Let x_0 be a point of X such that $f(x_0) \neq 0$. There exists a system of local coordinates $(\xi_1, \ldots, \xi_n)$, which is C^k at x_0, and in which f is rectified, i.e. such that $L_f\xi_1 = 1$ and $L_f\xi_i = 0$ for $i = 2, \ldots, n$. In other words, f admits a system of $n-1$ independent first integrals in a neighborhood of x_0, defined by $L_f\xi_i = 0$ for $i = 2, \ldots, n$.

A.3 Lie bracket

Let us consider, as in the previous section, a vector field f and a regular function h. The Lie derivative operation in the direction f, L_f, can be iterated. One can indeed define $L_f^k h$ for every $k \geq 0$ as follows:

$$L_f^0 h = h \text{ and } L_f^k h = L_f\left(L_f^{k-1}h\right) \ \forall k \geq 1\,.$$

For example, $L_f^2 h = \sum_{i,j=1}^{n}\left(f_i\frac{\partial f_j}{\partial x_i}\frac{\partial h}{\partial x_j} + f_i f_j\frac{\partial^2 h}{\partial x_i \partial x_j}\right)$. The iterated Lie derivative of order k thus defines a linear differential operator of order less than or equal to k.

Accordingly, if $g_1, \ldots, g_k$ are vector fields on X, one can define, as before, the iterated Lie derivative of order $r_1 + \cdots + r_k$.

$$L_{g_1}^{r_1} \cdots L_{g_k}^{r_k} h = L_{g_1}^{r_1} \left(L_{g_2}^{r_2} \cdots L_{g_k}^{r_k} h \right) .$$

Let us now consider two vector fields f and g and let us compute, in the given system of local coordinates, the expression

$$L_f L_g h - L_g L_f h = \sum_{i=1}^{n} \left(\sum_{j=1}^{n} \left(f_j \frac{\partial g_i}{\partial x_j} - g_j \frac{\partial f_i}{\partial x_j} \right) \right) \frac{\partial h}{\partial x_i} .$$

This skew-symmetric expression with respect to f and g, defines a first-order linear differential operator, the second-order terms, which are symmetric, being eliminated. It is thus a new vector field, denoted by $[f, g]$, and called the Lie bracket of f and g.

Definition A.11

The Lie bracket *of the vector fields f and g is the vector field defined by*

$$L_{[f,g]} = L_f L_g - L_g L_f .$$

In local coordinates

$$[f, g] = \sum_{i=1}^{n} \left(\sum_{j=1}^{n} \left(f_j \frac{\partial g_i}{\partial x_j} - g_j \frac{\partial f_i}{\partial x_j} \right) \right) \frac{\partial}{\partial x_i} .$$

The Lie bracket satisfies three properties

1. skew-symmetry : $[f, g] = -[g, f]$
2. $[\alpha f, \beta g] = \alpha\beta[f, g] + (\alpha L_f \beta) g - (\beta L_g \alpha) f$, for every pair of C^∞ functions α and β
3. Jacobi identity: $[f_1, [f_2, f_3]] + [f_2, [f_3, f_1]] + [f_3, [f_1, f_2]] = 0$

The bracket $[f, g]$ has the following geometric interpretation: Denote by $e^{tf}(x)$ the point of the integral curve of f at time t passing through x at time 0 and $e^{tg}(x)$ the point of the integral curve of g obtained in an analogous way. For ϵ small enough, let us compute the expression

$$e^{(-\epsilon g)} \circ e^{(-\epsilon f)} \circ e^{\epsilon g} \circ e^{\epsilon f}(x) .$$

Its Taylor expansion up to the second-order gives (particular case of Baker–Campbell–Hausdorff formula):

$$e^{(-\epsilon g)} \circ e^{(-\epsilon f)} \circ e^{\epsilon g} \circ e^{\epsilon f}(x) = x + \epsilon^2 [f, g](x) + O(\epsilon^3) .$$

Consequently, if for all x in a given neighborhood, one considers the vector space $E(x)$, included in $T_x X$, generated by $f(x)$ and $g(x)$, the bracket $[f, g]$ indicates whether the integral curves of f and g remain tangent to E for t small enough (which happens to be the case if $[f, g] \in E$) or not ($[f, g] \notin E$). In particular, one may expect to find a submanifold Ξ of X that admits $E(x)$ as a tangent space at any point x of the considered neighborhood if $[f, g] \in E$, whereas if $[f, g] \notin E$, such a submanifold cannot exist since the integral curves necessarily leave it.

Before looking at the problem of the existence problem of integral submanifolds, we need to introduce the notion of distribution of vector fields.

A.4 Distribution of vector fields

Definition A.12

A distribution of vector fields D *is a mapping such that to every point* $x \in X$ *corresponds the subspace* $D(x)$ *of the tangent space* T_xX.

Let V *be an open set of* X. *The distribution* D *is* regular with constant rank k *in* V *if there exist regular vector fields* $g_1, \ldots, g_k$ *such that:*

- rank $(g_1(x), \ldots, g_k(x)) = k$ *for all* $x \in V$.
- $D(x) = \text{sp}\{g_1(x), \ldots, g_k(x)\}$ *for all* $x \in V$.

A field of moving orthogonal frames, for example uniformly accelerated, constitutes a classical example of distribution of vector fields. Note that, as for vector fields, the notion of distribution of vector fields does not depend on the choice of coordinates.

Definition A.13

The distribution D *is said to be* involutive *if and only if for every pair of vector fields* f *and* g *of* D *we have* $[f, g] \in D$.

An involutive distribution is therefore characterized by $[D, D] \subset D$.

If D is not involutive, one can define its involutive closure:

Definition A.14

The involutive closure $\overline{D}$ *of a distribution* D *is the smallest involutive distribution containing* D.

Note that $\overline{D}$ can be constructed from the iterated Lie brackets of the fields $g_1, \ldots, g_k$ (see Isidori, 1989).

As an example of non involutive distribution, let us consider the distribution D in $T\mathbb{R}^3$ generated by $g_1 = \dfrac{\partial}{\partial x_1}$ and $g_2 = \dfrac{\partial}{\partial x_2} + x_1 \dfrac{\partial}{\partial x_3}$. Clearly, $[g_1, g_2] = \dfrac{\partial}{\partial x_3}$. The three vectors g_1, g_2 and $[g_1, g_2]$ are linearly independent. Thus D is not involutive and its involutive closure $\overline{D}$ is equal to $T\mathbb{R}^3$.

A.5 Frobenius' theorem

We now focus on the following problem:

"Let D be a regular distribution with constant rank k. Under what condition does there exist a diffeomorphism $\xi = \phi(x)$ such that D becomes $\text{sp}\left\{\dfrac{\partial}{\partial \xi_1}, \ldots, \dfrac{\partial}{\partial \xi_k}\right\}$ in these new coordinates ?"

This problem is often referred to as *rectifiability by diffeomorphism*. The answer is provided by Frobenius' theorem:

Theorem A.1

Let D be a regular distribution with constant rank k. A necessary and sufficient condition for D to be rectifiable by diffeomorphism is that D is involutive.

A.6 Application to the computation of solutions of first-order partial differential equations

We want to solve a system of k first-order partial differential equations in a given open set V of the manifold X:

$$\left\{\begin{array}{lcl} L_{g_1}y & = & 0 \\ \vdots & & \\ L_{g_k}y & = & 0 \end{array}\right. \tag{A.2}$$

where y is the unknown scalar function and where $g_1, \ldots, g_k$ are regular vector fields such that

$$\text{rank } (g_1(x), \ldots, g_k(x)) = k$$

for every $x \in V$.

Introducing the distribution D given by $D(x) = \text{sp}\{g_1(x), \ldots, g_k(x)\}$ for every $x \in V$, if y is a solution of system (A.2), we clearly have $L_g y = 0$ for every $g \in \overline{D}$. Conversely, since $g_1, \ldots, g_k$ are linearly independent, one can find vector fields $g_{k+1}, \ldots, g_r$ such that $g_1, \ldots, g_r$ form a basis of $\overline{D}(x)$ at every point of V and $L_g y = 0$ for every $g \in \overline{D}$ implies in particular that $L_{g_i} y = 0$ for $i = 1, \ldots, k$ and thus that y is solution of system (A.2).

By the Frobenius theorem, a necessary and sufficient condition for D to be rectified in V is that D is involutive. Note that $\overline{D}$ can be rectified whenever its rank is constant in V. Assume momentarily that D is involutive. There exists a diffeomorphism ϕ such that, if we denote $\xi = \phi(x)$, we have

$$D(\xi) = \text{sp}\{\phi_* g_1(\xi), \ldots, \phi_* g_k(\xi)\} = \text{sp}\left\{\frac{\partial}{\partial \xi_1}, \ldots, \frac{\partial}{\partial \xi_k}\right\}$$

in $\phi(V)$. Then, if we assume $z = y \circ \phi^{-1}$ we immediately obtain that $L_g y = 0$ for every $g \in D$, is equivalent to

$$\frac{\partial z}{\partial \xi_1} = 0, \ldots, \frac{\partial z}{\partial \xi_k} = 0$$

or $z(\xi_1, \ldots, \xi_k, \bar{\xi}) =$ constant for every $\bar{\xi} = (\xi_{k+1}, \ldots, \xi_n)$ such that $\xi \in \phi(V)$. The solution y is immediately deduced from z through $y = z \circ \phi$.

Now if D is not involutive and if $\overline{D}$ has constant rank r, following the same lines with $\overline{D}$ in place of D, we find the same result. One should however note that the solution y is not completely determined by system (A.2). Additional equations, generated by the vector fields of $\overline{D}$ which do not belong to D are needed. These equations are often called compatibility equations and the operation of appending these equations to the original system constitutes a particular case of prolongation of the system of PDEs. Thus, one cannot rectify D without rectifying $\overline{D}$ at the same time, which means that $z(\xi_1, \ldots, \xi_r, \bar{\xi}) =$ constant for every $\bar{\xi} = (\xi_{r+1}, \ldots, \xi_n)$ such that $\xi \in \phi(V)$.

Let us also note that the involutive closure of D is quite often equal to the tangent space of the whole $\mathbb{R}^n$, which means that the unique solution y is the null function in V.

We have shown that solving a system of first-order partial differential equations is equivalent to the rectifiability of the corresponding distribution of vector fields (or its involutive closure). Nevertheless, the rectifying diffeomorphism is itself obtained by the resolution of the system $L_{\gamma_i}\phi_j = \delta_{i,j}$ for $i = 1, \ldots, k$, $j = 1, \ldots, n$, with $\{\gamma_1, \ldots, \gamma_k\}$ a basis of $\overline{D}$ and $\delta_{i,j}$ the Kronecker symbol. Consequently, this method is only applicable in practice if a specific basis of $\overline{D}$, for which the equations of the diffeomorphism are simpler, is known. Otherwise, different methods such as the method of characteristics may be utilized. A more detailed overview of this question can be found in [1, 2] for example.

A.7 More on differential forms, duality

The tangent bundle $\mathrm{T}X$ may be seen as a distribution that, to each point x of the manifold X, associates its tangent space T_xX. One immediately verifies that $\mathrm{T}X$ is involutive (by definition of the tangent space) and, by Frobenius' theorem, one can find a local basis of $\mathrm{T}X$ having the form $\left\{\frac{\partial}{\partial x_1}, \ldots, \frac{\partial}{\partial x_n}\right\}$. Let us introduce the dual space of $\mathrm{T}X$, denoted by T^*X, and called the *cotangent bundle* of the manifold X. For that, we define a dual basis $\{\mathrm{d}x_1, \ldots, \mathrm{d}x_n\}$ by the scalar product

$$\left\langle \frac{\partial}{\partial x_i}, \mathrm{d}x_j \right\rangle = \delta_{i,j} \quad , \quad \forall i, j = 1, \ldots, n\,. \tag{A.3}$$

If now ϕ is a C^∞ function from X to $\mathbb{R}$, its differential is defined by $\mathrm{d}\phi = \sum_{i=1}^{n} \frac{\partial \phi}{\partial x_i}\mathrm{d}x_i$. By this formula, the differential of a function may be identified to its tangent linear mapping, which justifies the notations. Thus, $\mathrm{d}\phi$ belongs to T^*X, and if f is a vector field ($f(x) \in \mathrm{T}_xX$ for all x), the scalar product $< \mathrm{d}\phi, f >$ can be deduced from the preceding formula by

$$< \mathrm{d}\phi, f >= \sum_{i=1}^{n} f_i \frac{\partial \phi}{\partial x_i} = L_f\phi\,. \tag{A.4}$$

More generally we have the following definition.

Definition A.15

We call differential form of degree 1, *or 1-form, an element ω of the cotangent bundle T^*X, in other words, a mapping sending $x \in X$ into $\omega(x) \in \mathrm{T}^*_xX$.*

The scalar product of a 1-form, in local coordinates: $\omega = \sum_{i=1}^{n} \omega_i \mathrm{d}x_i$, with a vector field $f = \sum_{i=1}^{n} f_i \frac{\partial}{\partial x_i}$ is given by

$$< \omega, f >= \sum_{i=1}^{n} f_i\omega_i\,. \tag{A.5}$$

Note that a 1-form is not in general the differential of a function. Such a form is called *exact.* A necessary condition of exactness (which is not sufficient!) is that it satisfies (exercise) $\frac{\partial \omega_i}{\partial x_j} = \frac{\partial \omega_j}{\partial x_i}$ for every i and j. A 1-form satisfying this last property is called a *closed form.* One can prove (Poincaré's theorem) that a closed form is exact on a contractile open set.

One can also define on T^*X the dual operation to the Lie bracket, called *Lie derivative of a 1-form.*

We have the formula

$$\mathrm{d}L_f\phi = \sum_{i=1}^{n} \frac{\partial L_f \phi}{\partial x_i} \mathrm{d}x_i = \sum_{i=1}^{n} \left(L_f \left(\frac{\partial \phi}{\partial x_i} \right) + < \mathrm{d}\phi, \frac{\partial f}{\partial x_i} > \right) \mathrm{d}x_i \,. \tag{A.6}$$

The right-hand side of (A.6) defines a new 1-form denoted by $L_f \mathrm{d}\phi$, called Lie derivative of $\mathrm{d}\phi$ along f. If g is another vector field, one can easily check that:

$$\begin{aligned} < L_f \mathrm{d}\phi, g > &= \sum_{i=1}^{n} \left(L_f \left(\frac{\partial \phi}{\partial x_i} \right) + < \mathrm{d}\phi, \frac{\partial f}{\partial x_i} > \right) g_i \\ &= L_f < g, \mathrm{d}\phi > - < [f, g], \mathrm{d}\phi > \,. \end{aligned} \tag{A.7}$$

To define the Lie derivative of a general 1-form, we just extend the latter relation:

Definition A.16

The Lie derivative of the 1-form ω is defined by

$$< L_f \omega, g >= L_f < \omega, g > - < \omega, [f, g] > \,. \tag{A.8}$$

for every pair of vector fields f and g.

An easy calculation shows that, in local coordinates,

$$L_f \omega = \sum_{i=1}^{n} \left(L_f(\omega_i) + \left\langle \omega, \frac{\partial f}{\partial x_i} \right\rangle \right) \mathrm{d}x_i \,. \tag{A.9}$$

If the Lie bracket of f and g is also called Lie derivative of g along f, $L_f g \overset{\mathrm{def}}{=} [f, g]$, and if (A.8) is rewritten as

$$L_f < \omega, g >=< L_f \omega, g > + < \omega, L_f g >$$

we obtain the well-known derivation formula of a scalar product.

If now the 1-form of formula (A.8) is the differential of the iterated Lie derivative of a function ϕ, (A.8) becomes

$$< ad_f g, \mathrm{d}L_f^i \phi >= -\mathrm{d}L_g L_f^{i+1} \phi + L_f(\mathrm{d}L_g L_f^i \phi) \,. \tag{A.10}$$

It suffices indeed to apply (A.8), with $\omega = \mathrm{d}L_f^i \phi$, given $< \mathrm{d}L_f^i \phi, g > = \mathrm{d}L_g L_f^i \phi$. The next proposition is easily derived by induction:

Proposition A.3

The identities

$$L_{ad_f^j g}\phi = 0 \qquad \forall j = 0, \ldots, k \tag{A.11}$$

are equivalent to

$$< ad_f^j g, \mathrm{d}L_f^i \phi >= 0 \qquad \forall\, i + j \le k \tag{A.12}$$

Proof: By assumption, we have $L_{ad_f^j g}\phi = \langle ad_f^j g, \mathrm{d}\phi\rangle = 0\ \forall j \le k$. Assume that $\langle ad_f^j g, \mathrm{d}L_f^i \phi\rangle = 0\ \forall i \le i_0$ and every j such that $i + j \le k$. For $i_0 + 1$, we have, according to (A.8), $\langle ad_f^j g, \mathrm{d}L_f^{i_0+1}\phi\rangle = L_f\langle ad_f^j g, \mathrm{d}L_f^{i_0}\phi\rangle - \langle ad_f^{j+1} g, \mathrm{d}L_f^{i_0}\phi\rangle = 0\ \forall i_0 + j + 1 \le k$, which proves the result.

Conversely, if $< ad_f^j g, \mathrm{d}L_f^i \phi >= 0$ for every $i + j \le k$, this relation holds *a fortiori* for $i = 0$, which gives (A.11).

A.8 Bibliography

[1] ARNOLD,V.I. (1974), *Équations Différentielles Ordinaires*, MIR, Moscow.

[2] ARNOLD, V.I. (1980), *Chapitres Supplémentaires sur la Théorie des Équations Différentielles Ordinaires*, MIR, Moscow.

[3] DEMAZURE, M. (1987), *Géométrie, Catastrophes et Bifurcations*, Presses of l'École Polytechnique, Palaiseau.

[4] GODBILLON, C. (1969), *Géométrie Différentielle et Mécanique Analytique*, Hermann, Paris.

[5] ISIDORI, A. (1989), *Nonlinear Control Systems.* 2nd ed., Springer-Verlag, New York.

APPENDIX B

Mathematical tools and procedures

J. DESCUSSE

This appendix presents the basic theorems for rectifying vector fields or coordinate frames useful to get the change of coordinates leading to the decoupled canonical form reported in chapter 2. The proofs will be omitted; interest readers should read the references quoted at the end of chapter 2, and especially Godbillon (1969), Spivak (1979) and Boothby (1975). We use the same notation and concepts developed in appendix A.

Theorem B.1 (Rectification of vector fields)

Let M be a manifold C^∞. Let X be a vector field C^∞ on M. For all p belonging to M, such that $X_p \neq 0$, there exists a local coordinate chart (U, φ), $p \in U$, such that, $\forall q \in U$, $\varphi(q) = (z_1, ..., z_n)$, we have $X_q = \dfrac{\partial}{\partial z_1}$

In practice, the result being local, we assume that $M = \mathbb{R}^n$ (with the standard coordinates $(x_1, ..., x_n)$) and $p = 0 \in \mathbb{R}^n$. Moreover, without loss of generality, we can always consider, the situation where $X(0) = \left.\dfrac{\partial}{\partial x_1}\right]_0$.

The idea of the proof is backed by the existence, in a neighborhood of zero, of only one integral curve passing through each point $(0, z_2, ..., z_n)$ (Fig. B.1).

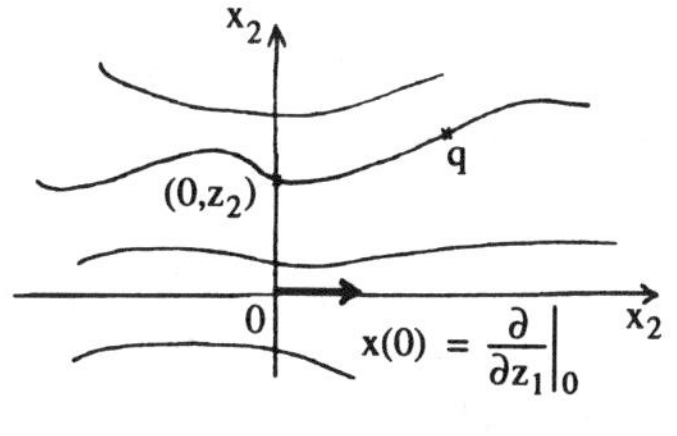

Figure B.1: Representation in the plane.

If q belongs to an integral curve passing through this point, we can locate q by $(z_2, ..., z_n)$ chosen as the last $(n-1)$ coordinates of q and by the time necessary to go from this point to q as the first coordinate of q.

Remark B.1

The numbering is not important. we can take the time as the ith *coordinate of* q *and write* $(z_1, \dots, z_{i-1}, z_{i+1}, \dots, z_n)$ *as the* $(n-1)$ *other coordinates.*

From a practical viewpoint, we calculate the right system of coordinates as follows: we assume that the first component of X is nonzero, we consider Φ_t the flow generated by X and the mapping χ defined in a neighborhood of zero by

$$\chi(z_1, \dots, z_n) = \Phi_{z_1}(0, z_2, \dots, z_n)$$

χ is a diffeomorphism.

Remark B.2

Any nonzero component of X *is suitable for determining this diffeomorphism. If it is the ith, we then choose* z_i *for the "time". Thus,* $z = \chi^{-1}(x)$ *can be used as a local coordinate frame in a neighborhood* U *of* 0.

This system is the desired system. Indeed, in the chart (z, U)*, the vector field* X *is written as:*

$$X = \frac{\partial}{\partial z_1}$$

In both examples given below, the time taken to go from the initial point to q is the last coordinate of q.

Example 1

Consider the vector field

$$X = x_1 \cdot \frac{\partial}{\partial x_1} + \mathrm{e}^{-x_3} \frac{\partial}{x_2} + 1 \cdot \frac{\partial}{\partial x_3} \tag{B.1}$$

with $X(0,0,0) = [0, 1, 1]$, as initial condition.

We search for a new local coordinate frame (z_1, z_2, z_3) such that the vector field X can be written, in a neighborhood of $(0,0,0)$, as: $X(z_1, z_2, z_3) = (0, 0, 1)$.

z_3 is taken as the time to go from p_0 to q along the orbit of p associated with the vector field X. We have the differential equations:

$$\frac{\mathrm{d}x_1}{\mathrm{d}z_3} = x_1 \quad \frac{\mathrm{d}x_2}{\mathrm{d}z_3} = \mathrm{e}^{-z_3} \quad \frac{\mathrm{d}x_3}{\mathrm{d}z_3} = 1$$

Let $x_3 = z_3 + c_3$. Then $x_2 = -\,\mathrm{e}^{-(z_3+c_3)} + c_2$ and $x_1 = c_1 \cdot \mathrm{e}^{z_3}$.

However, when $z_3 = 0$, we must have:

$$\begin{aligned} x_3 &= 0 \\ x_2 &= z_2 \\ x_1 &= z_1 \end{aligned}$$

So:

$$\begin{aligned} x_3 &= z_3 \\ x_2 &= z_2 + 1 - \mathrm{e}^{-z_3} \\ x_1 &= z_1 \cdot \mathrm{e}^{z_3} \end{aligned}$$

Now, from (B.1), we have

$$\dot{x}_1 = x_1$$
$$\dot{x}_2 = \mathrm{e}^{-x_3}$$
$$\dot{x}_3 = 1$$

The system is equivalent to

$$\dot{z}_1\,\mathrm{e}^{z_3} + z_1\dot{z}_3\,\mathrm{e}^{z_3} = z_1\,\mathrm{e}^{z_3}$$
$$\dot{z}_3\,\mathrm{e}^{-z_3} + \dot{z}_2 = \mathrm{e}^{-z_3}$$
$$\dot{z}_3 = 1$$

Thus

$$\begin{aligned} \dot{z}_1 &= 0 \\ \dot{z}_2 &= 0 \\ \dot{z}_3 &= 1 \end{aligned}$$

We actually have $X(z_1, z_2, z_3) = \dfrac{\partial}{\partial z_3}$ in a neighborhood of zero.

Example 2

Consider the vector field

$$X = x_2 \cdot \frac{\partial}{\partial x_1} + \mathrm{e}^{-x_3}\frac{\partial}{\partial x_2} + 1 \cdot \frac{\partial}{\partial x_3}$$

As above, we have the differential equations:

$$\frac{\mathrm{d}x_1}{\mathrm{d}z_3} = x_2 \quad \frac{\mathrm{d}x_2}{\mathrm{d}z_3} = \mathrm{e}^{-x_3} \quad \frac{\mathrm{d}x_3}{\mathrm{d}z_3} = 1$$

and so $x_3 = z_3$ $x_2 = 1 + z_2 - \mathrm{e}^{-z_3}$ $x_1 = z_1 + z_3 + z_2 \cdot z_3 + \mathrm{e}^{-z_3} - 1$

This system is equivalent to

$$-\dot{z}_3\,\mathrm{e}^{-z3} + \dot{z}_3 + z_3\dot{z}_2 + z_2\dot{z}_3 + \dot{z}_1 = z_2 + 1 - \mathrm{e}^{-z_3}$$
$$\dot{z}_3\,\mathrm{e}^{-z_3} + \dot{z}_2 = \mathrm{e}^{-z_3}$$
$$\dot{z}_3 = 1$$

So

$$\dot{z}_3 = 1$$
$$\dot{z}_2 = 0$$
$$\dot{z}_1 = 0$$

Now, we shall pay attention to the rectification of coordinate frames.

Theorem B.2 (Rectification of coordinate frames)

(i) If $(X_1, ..., X_k)$ denote k vector fields C^∞ on M, linearly independent in a neighborhood of p.

(ii) If $[X_\alpha, X_\beta] = 0$ for $1 \leq \alpha, \beta \leq k$, then there exists a local coordinate frame (U, φ), in a neighborhood of p, such that

$$\forall q \in U, \quad \varphi(q) = (z_1, ..., z_n)$$

$$\forall \alpha \in \{1, ..., k\}, X_\alpha = \frac{\partial}{\partial z_\alpha} \text{ on } U.$$

This theorem generalizes the previous one. As above, the result being local, we can assume that $M = \mathbb{R}^n$ and $p = 0$. Let ϕ_α be the flow generated by X_α. Let us define, in a neighborhood of zero, the mapping χ by:

$$\chi(z_1, ..., z_n) = \phi^1_{z_1}\left(\phi^2_{z_2}\left(... \left(\phi^k_{z_k}(0, ...0, z_{k+1}, ..., z_n)\right) ...\right)\right)$$

Therefore, χ is a diffeomorphism.

Then: $z = \chi^{-1}$ can be used as a local coordinate frame, in a neighborhood of zero. It is the right system of coordinates.

Example 3

Consider both of the vector fields

$$X_1 = (x_1, x_2) \qquad X_2 = (-x_2, x_1)$$

We first need to determine the flows ϕ^1_t and ϕ^2_t generated by X_1 and X_2, respectively.

For the first vector field, we have

$$\frac{\mathrm{d}x_1}{\mathrm{d}t} = x_1 \quad \frac{\mathrm{d}x_2}{\mathrm{d}t} = x_2$$

and so $\phi^1_t(a_1, a_2) = (a_1\,\mathrm{e}^t, a_2\,\mathrm{e}^t)$.

X_1 generates on $\mathbb{R}^2$ the flow ϕ^1_t defined by $\phi^1_t(t, x, y) = (x\,\mathrm{e}^t, y\,\mathrm{e}^t)$.

For the second vector field, we have:

$$\frac{\mathrm{d}x_1}{\mathrm{d}t} = -x_2 \quad \frac{\mathrm{d}x_2}{\mathrm{d}t} = x_1$$

and therefore $\phi^2_t(a_1, a_2) = (-a_2 \sin t + a_1 \cos t, a_1 \sin t + a_2 \cos t)$.

X_2 generates on $\mathbb{R}^2$ the flow ϕ^2_t defined by

$$\phi^2_t(t, x, y) = (-y \sin t + x \cos t, x \sin t + y \cos t)$$

We define the mapping $\chi(z_1, z_2)$ by

$$\begin{aligned}\chi(z_1, z_2) &= \phi^2_{z_2}[\phi^1_{z_1}(a_1, a_2)] \\ &= \phi^2_{z_2}[a_1 \,\mathrm{e}^{z_1}, a_2 \,\mathrm{e}^{z_1}] \\ &= [-a_2 \,\mathrm{e}^{z_1} \sin z_2 + a_1 \,\mathrm{e}^{z_1} \cos z_2, a_1 \,\mathrm{e}^{z_1} \sin z_2 + a_2 \,\mathrm{e}^{z_1} \cos z_2]\end{aligned}$$

Let $\chi(z_1, z_2) = \begin{pmatrix} -a_2 \sin z_2 + a_1 \cos z_2 \\ a_1 \sin z_2 + a_2 \cos z_2 \end{pmatrix} \cdot \mathrm{e}^{z_1} = \begin{pmatrix} x_1 \\ x_2 \end{pmatrix}$

We have

$$\begin{pmatrix} \dot{x}_1 \\ \dot{x}_2 \end{pmatrix} \begin{pmatrix} -(a_2 \cos z_2 + a_1 \sin z_2)\dot{z}_2 + (-a_2 \sin z_2 + a_1 \cos z_2)\dot{z}_1 \\ (a_1 \cos z_2 - a_2 \sin z_2)\dot{z}_2 + (a_1 \sin z_2 + a_2 \cos z_2)\dot{z}_1 \end{pmatrix} \cdot \mathrm{e}^{z_1}$$

For the first vector field, we have

$$\begin{pmatrix} \dot{x}_1 \\ \dot{x}_2 \end{pmatrix} = \begin{pmatrix} x_1 \\ x_2 \end{pmatrix} = \chi(z_1, z_2)$$

Then, equalling both elements, we have:

$$\dot{z}_1 = 1$$
$$\dot{z}_2 = 0$$

Thus, we actually have $X_1(z_1, z_2) = \dfrac{\partial}{\partial z_1}$

For the second vector field, we have:

$$\begin{pmatrix} \dot{x}_1 \\ \dot{x}_2 \end{pmatrix} = \begin{pmatrix} -x_2 \\ x_1 \end{pmatrix} = \begin{pmatrix} -(a_1 \sin z_2 + a_2 \cos z_2) \\ -a_2 \sin z_2 + a_1 \cos z_2 \end{pmatrix} \cdot \mathrm{e}^{z_1}$$

Then, equalling both elements, we find

$$\dot{z}_1 = 0$$
$$\dot{z}_2 = 1$$

We actually have $X_2(z_1, z_2) = \dfrac{\partial}{\partial z_2}$

B.1 The Frobenius theorem and the simultaneously integrable distributions

Definition B.1

A distribution $\Delta = \{X_1, \ldots, X_n\}$ on a manifold M with dimension m is said to be involutive if

$$[X_i, X_j] = \sum_{k \in \underline{n}} C^k_{ij}(x) \cdot X_k \quad \forall i, j \in n$$

where C^k_{ij} are functions C^∞ defined on the neighborhood.

Definition B.2

A distribution Δ on a manifold M with dimension m is completely integrable, if $\forall p \in M$, there exists a local coordinate frame (U, ϕ) such that, if $(x_1, ..., x_m)$ denotes the local coordinates, then the n vectors

$$E_i = \phi_*^{-1} \frac{\partial}{\partial(x_i)} \quad i \in \underline{n} \ \dim \Delta = n$$

define a local basis on U for the distribution Δ.

Theorem B.3 (The Frobenius theorem)

A distribution Δ on a manifold M with dimension m is integrable completely $\Longleftrightarrow$ it is involutive.

Remark B.3

A very synthetic proof of this theorem can be found in several books. For instance let us quote the book by Claude Godbillon (1969).

Here, we prefer to give a "constructive" proof of this theorem. It is dual of that one provided by E. Cartan (1945) for exterior differential systems as pointed out to us by P. Rouchon of Automatic Control Center of ENSMP, Fontainebleau, France: E. Cartan considers differential forms, here we work with vectors fields.

Proof

Let $\Delta = \{X_1, ..., X_n\}$ with constant rank, $\dim \Delta = n$. From theorem B.1, about rectification of vector fields, for all p belonging to M, such that $X_1(p) \neq 0$, there exists a local change of coordinates $(U, \Phi), p \in U$, such that:

$$\Phi(q) = (y_1, ..., y_m) \quad \forall q \in U,$$

with $X_1 = \dfrac{\partial}{\partial y_1}$

Let us define the vector fields Y_i by

$$Y_1 = X_1 = \frac{\partial}{\partial y_1}$$
$$Y_i = X_i - X_i(y_1).X_1 \quad \text{for } i = 2, ..., n.$$

These vector fields own the following properties:

(i) Each Y_i is tangent to manifolds such that $y_1 =$ constant. It follows that $[Y_i, Y_j]$ for $2 \leq i, j \leq n$ is tangent to manifolds like $y_1 =$ constant.

(ii) From the involutiveness of Δ, we have

$$[Y_i, Y_j] = \sum_{k \in \underline{n}} C_{ij}^k(y) \cdot Y_k \quad i, j \in n$$

From **(i)** and **(ii)** it follows immediately that $C_{1j}^{\prime k}(y) = 0$

Conclusion

The distribution Δ_1 with dimension $(n-1)$ defined by $\Delta_1 = \{Y_2, ..., Y_n\}$ is involutive and

$$[Y_1, \Delta_1] \subset \Delta_1$$

Let M_1 be a basis matrix of Δ_1. The objective is to find a regular matrix Ω_1 such that we have: $[Y_1, M_1\Omega_1] = 0$.

From (B.2), we already know that there exists a matrix T_1 such that $[M_1, Y_1] = M_1 T_1$.

Therefore, we want to search for Ω_1 such that we obtain: $[Y_1, M_1\Omega_1] = 0$.

We have

$$[M_1\Omega_1, Y_1] = [M_1, Y_1] \cdot \Omega_1 - M_1 \frac{\partial \Omega_1}{\partial y_1}$$

$$\iff [M_1\Omega_1, Y_1] = M_1 \left(T_1\Omega_1 - \frac{\partial \Omega}{\partial y_1} \right)$$

Searching for a regular Ω_1 such that $[Y_1, M_1\Omega_1] = 0$ amounts to solving the system:

$$\frac{\partial \Omega_1}{\partial y_1} = T_1\Omega_1$$

which always admits a solution, in the present situation. In fact, the equation is an ordinary differential equation.

Let Ω_1 be a solution, $M_1.\Omega_1$ is then a basis matrix for Δ_1 such that: $[Y_1, M_1\Omega_1] = 0$.

As Δ_1 is involutive, we can apply the rectification vector fields theorem. We obtain a new local coordinate frame, let $(z_1, z_2, \ldots, z_m)$ with $z_1 = y_1$. After that, we take into account $\Delta_2 = \{Z_3, ..., Z_n\}$ with $Z_i = Y_i - Y_i(z_2) \cdot Y_2$ for $i = 3, ..., n$, exactly as above with Δ_1, ... and the process carries on up to Δ_k such that $\dim \Delta_k = 1$.

Example 4

Let us consider the distribution generated by the three vector fields

$$\begin{pmatrix} 2x_3 \\ -1 \\ 0 \end{pmatrix} \quad \begin{pmatrix} -x_1 \\ -2x_2 \\ x_3 \end{pmatrix} \quad \begin{pmatrix} 0 \\ \frac{1}{2}x_1 \\ x_3 \end{pmatrix}$$

This distribution is regular in a neighborhood of $(0, 1, 1)$. Take z_1 as the time taken for going from the initial point $(0, 0, 0)$ to q.

For the first vector field, we have the differential equations:

$$\frac{\mathrm{d}x_1}{\mathrm{d}z_1} = 2x_3 \quad \frac{\mathrm{d}x_2}{\mathrm{d}z_1} = -1 \quad \frac{\mathrm{d}x_3}{\mathrm{d}z_1} = 0 \tag{B.2}$$

and so $x_3 = z_3 + 1 \;\; x_2 = -z_1 + z_2 + 1 \;\; x_1 = 2x_3z_1 = 2z_1(z_3 + 1)$

When $z_1 = 0$ we have

$$\begin{aligned} x_1 &= 0 \\ x_2(0) &= z_2(0) + 1 \\ x_3(0) &= z_3(0) + 1 \end{aligned}$$

However, the system of differential equations (B.2) is equivalent to:

$$\begin{aligned} 2\dot{z}_1.(z_3+1)+2z_1\dot{z}_3 &= 2\cdot(z_3+1) \iff & \dot{z}_1 &= 1\\ -\dot{z}_1+\dot{z}_2 &= -1 & \dot{z}_2 &= 0\\ \dot{z}_3 &= 0 & \dot{z}_3 &= 0 \end{aligned}$$

For the second vector field, we find

$$\begin{aligned} 2\dot{z}_1\cdot(z_3+1)+2z_1\dot{z}_3 &= -2\cdot z_1(z_3+1)\\ -\dot{z}_1+\dot{z}_2 &= -2(-z_1+z_2+1)\\ \dot{z}_3 &= z_3+1 \end{aligned}$$

$$\iff$$

$$\begin{aligned} \dot{z}_1 &= -2\cdot z1\\ \dot{z}_2 &= -2(z_2+1)\\ \dot{z}_3 &= z_3+1 \end{aligned}$$

At last for the third vector field, we get

$$\begin{aligned} 2\dot{z}_1\cdot(z_3+1)+2z_1\dot{z}_3 &= 0\\ -\dot{z}_1+\dot{z}_2 &= z_1(z_3+1)\\ \dot{z}_3 &= z_3+1 \end{aligned}$$

$$\iff$$

$$\begin{aligned} \dot{z}_1 &= -z_1\\ \dot{z}_2 &= z_1\cdot z_3\\ \dot{z}_3 &= z_3+1 \end{aligned}$$

After rectification of the first vector field, the three vector fields above can be written as

$$\underset{X_1}{\begin{pmatrix}1\\0\\0\end{pmatrix}} \qquad \underset{X_2}{\begin{pmatrix}-2z_1\\-2(z_2+1)\\z_3+1\end{pmatrix}} \qquad \underset{X_3}{\begin{pmatrix}-z_1\\z_1\cdot z_3\\z_3+1\end{pmatrix}}$$

Let us define the vector fields Z_i by

$$\begin{aligned} Z_1 &= X_1 = \frac{\partial}{\partial z_1}\\ Z_i &= X_i - X_i(z_1)\cdot X_1 \quad \text{for } i=2,3. \end{aligned}$$

We then find the three following new expressions

$$\underset{Z_1}{\begin{pmatrix}1\\0\\0\end{pmatrix}} \qquad \underset{Z_2}{\begin{pmatrix}0\\-2(z_2+1)\\z_3+1\end{pmatrix}} \qquad \underset{Z_3}{\begin{pmatrix}0\\z_1\cdot z_3\\z_3+1\end{pmatrix}}$$

and we have:
$$\begin{aligned} [Z_1,Z_2] &= 0\\ [Z_1,Z_3] &= (0,z_3,0)^t \end{aligned}$$

Step 1 Search for T_1 such that $[M_1, Z_1] = M_1 T_1, M_1 = \{Z_2, Z_3\}$. We know that there exists $[Z_1, \Delta_1] \subset \Delta_1$.

Let us write
$$T_1 = \begin{pmatrix} t_1 & t_2 \\ t_3 & t_4 \end{pmatrix}$$

$$\begin{aligned}
M_1 T_1 &= \begin{pmatrix} 0 & 0 \\ -2(z_2+1) & z_1 z_3 \\ z_3+1 & z_3+1 \end{pmatrix} \cdot \begin{pmatrix} t_1 & t_2 \\ t_3 & t_4 \end{pmatrix} \\
&= \begin{pmatrix} 0 & 0 \\ -2(z_2+1)t_1 + z_1 z_3 t_3 & -2(z_2+1)t_2 + z_1 z_3 t_4 \\ (z_3+1)(t_1+t_3) & (z_3+1)(t_2+t_4) \end{pmatrix} \\
&= \begin{pmatrix} 0 & 0 \\ 0 & -z_3 \\ 0 & 0 \end{pmatrix}
\end{aligned}$$

$$\Longrightarrow \qquad \begin{aligned} t_1 &= t_3 = 0 \\ t_2 &= -t_4 = \frac{z_3}{z_1 z_3 + 2(z_2+1)} \end{aligned}$$

Step 2 Search for Ω_1 such that $\dfrac{\partial \Omega_1}{\partial z_1} = T_1 \Omega_1$. This operation amounts to solving the differential system:

$$\begin{aligned}
\frac{\mathrm{d}\omega_1}{\mathrm{d}z_1} &= m_2 \omega_3 \\
\frac{\mathrm{d}\omega_2}{\mathrm{d}z_1} &= m_2 \omega_4 \\
\frac{\mathrm{d}\omega_3}{\mathrm{d}z_1} &= -m_2 \omega_3 \\
\frac{\mathrm{d}\omega_4}{\mathrm{d}z_1} &= -m_2 \omega_4
\end{aligned}$$

$$\Longleftrightarrow \qquad \begin{aligned}
\omega_3 &= \lambda(z_2, z_3)\,[z_1 z_3 + 2(z_2+1)]^{-1} \\
\omega_1 &= -\lambda(z_2, z_3)\,[z_1 z_3 + 2(z_2+1)]^{-1} + \mu(z_2, z_3) \\
\omega_4 &= \nu(z_2, z_3)\,[z_1 z_3 + 2(z_2+1)]^{-1} \\
\omega_2 &= -\nu(z_2, z_3)\,[z_1 z_3 + 2(z_2+1)]^{-1} + \rho(z_2, z_3)
\end{aligned}$$

We have to choose the functions λ, ν, μ, ρ such that Ω_1 is regular on the largest possible domain. For that, it is enough to reckon upon $\Omega_1(z_1 = 0)$.

$$\Omega_1(z_1 = 0) = \begin{pmatrix} -\lambda\,[2(z_2+1)]^{-1} + \mu & -\nu\,[2(z_2+1)]^{-1} + \rho \\ \lambda\,[2(z_2+1)]^{-1} & \nu\,[2(z_2+1)]^{-1} \end{pmatrix}$$

We have $\lambda = 0, \mu = 1,$ because $[Z_1, Z_2] = 0 \quad \Longrightarrow \quad \omega_1 = 1, \omega_3 = 0$

Take, for instance

$$\omega_2(0, z_2, z_3) = 0 \quad \Longrightarrow \quad \nu = 2(z_2+1)$$
$$\omega_4(0, z_2, z_3) = 1 \qquad\qquad \rho = 1$$

Finally, we obtain

$$\begin{aligned} \omega_1 &= 1 \\ \omega_3 &= 0 \\ \omega_2 &= \frac{z_1 z_3}{z_1 z_3 + 2(z_2+1)} \\ \omega_4 &= \frac{2(z_2+1)}{z_1 z_3 + 2(z_2+1)} \end{aligned}$$

or

$$M_1\Omega_1 = \begin{pmatrix} 0 & 0 \\ -2(z_2+1) & 0 \\ (z_3+1) & (z_3+1) \end{pmatrix}$$

The distribution Δ is then generated by the three vector fields

$$\underset{Z_1}{\begin{pmatrix} 1 \\ 0 \\ 0 \end{pmatrix}} \quad \underset{Z_2}{\begin{pmatrix} 0 \\ -2(z_2+1) \\ z_3+1 \end{pmatrix}} \quad \underset{Z_3'}{\begin{pmatrix} 0 \\ 0 \\ z_3+1 \end{pmatrix}}$$

These three vector fields commute, and therefore we can apply theorem B.2 (rectification of coordinate frames).

Let us make the integration of the second vector field with respect to y_2, as variable

$$\frac{dz_2}{dy_2} = -2(z_2+1) \qquad \frac{dz_3}{dy_2} = z_3 + 1$$

so, for $y_2 = 0$

$$z_2(0) = y_2(0) \quad z_3(0) = y_3(0) = 0 \tag{B.3}$$

therefore

$$\begin{aligned} z_2 &= e^{-2y_2} - 1 \\ z_3 &= e^{y_2} - 1 \end{aligned}$$

Let us make the integration of the following third vector field with respect to y_3, as variable

$$\frac{dz_2}{dy_3} = 0 \quad \frac{dz_3}{dy_3} = z_3 + 1$$

so, for $y_3 = 0$

$$z_2(0) = e^{-2y_2} - 1 \quad z_3(0) = e^{y_2} - 1 \tag{B.4}$$

therefore

$$\begin{aligned} z_2 &= \mu = e^{-2y_2} - 1 \\ z_3 &= \lambda e^{y_3} - 1 \qquad \Longrightarrow \lambda = e^{y_2} \end{aligned}$$

Thus, the change of coordinates rectifying Z_2 and Z_3' is

$$\begin{aligned} z_1 &= y_1 \\ z_2 &= e^{-2y_2} - 1 \\ z_3 &= e^{(y_2+y_3)} - 1 \end{aligned}$$

For the second vector field, the system (B.3) then becomes

$$\begin{aligned} \dot{y}_1 &= 0 \\ -2 \cdot \dot{y}_2 \cdot e^{-2y_2} &= -2 \cdot e^{-2y_2} \\ (\dot{y}_2 + \dot{y}_3) \cdot e^{(y_2+y_3)} &= e^{(y_2+y_3)} \end{aligned} \quad \Longleftrightarrow \quad \begin{aligned} \dot{y}_1 &= 0 \\ \dot{y}_2 &= 1 \\ \dot{y}_3 &= 0 \end{aligned}$$

For the third vector field, the system (B.4) then becomes:

$$\begin{aligned} \dot{y}_1 &= 0 \\ -2 \cdot \dot{y}_2 \cdot e^{-2y_2} &= 0 \\ (\dot{y}_2 + \dot{y}_3) \cdot e^{(y_2+y_3)} &= e^{(y_2+y_3)} \end{aligned} \quad \Longleftrightarrow \quad \begin{aligned} \dot{y}_1 &= 0 \\ \dot{y}_2 &= 0 \\ \dot{y}_3 &= 1 \end{aligned}$$

B.2 Some important definitions

Definition B.3

An integral manifold of a distribution Δ *with dimension* n*, on a manifold* M *with dimension* m*, passing through* $q = (a_1, ..., a_m)$ *is the set of points defined by:*

$$x_j = a_j = \text{constant} \quad \textit{for} \quad j = n+1, ..., m.$$

in other words $N = \Phi^{-1}\{x \in \Phi(U); x_j = a_j, j = n+1, \ldots, m\}$. N *is called a leaf of* U.

Definition B.4

Assume that $\dim M = n$ *and let* $\{\Delta_k\}$ *be a family of* k *independent distributions such that* $\dim \Delta_i = \mu_i$. *Then, the* $\{\Delta_k\}$*'s are locally simultaneously integrable if there exists a local coordinate frame* $(x_1, \ldots, x_n)$ *such that the integral manifolds of* $\Delta_j, j \in \underline{k}$*, display* $x_i =$ *constant for* $i \in \underline{n} \backslash \{\mu_{j-1} + 1, \ldots, \mu_j\}$ *with* $\mu_0 = 0$.

The following figure shows the integral manifold of the distribution

$$\Delta = \text{sp}\left\{X = \frac{\partial}{\partial x} + 2x \cdot \frac{\partial}{\partial y};\ Y = \frac{\partial}{\partial y} + 2y \cdot \frac{\partial}{\partial z}\right\}$$

passing through the origin.

Lemma B.1

Let $\Delta_1 \subset \Delta_2 ... \subset \Delta_k$*, be* k *regular involutive distributions on a manifold* M *with dimension* n*, and with* $\dim \Delta_j = \mu_j \forall j \in \underline{k}$. *Then,* $\forall V(x_o) \in V(M)$ *, there exists a local coordinate frame* $(x_1, ..., x_n)$ *such that the integral manifolds of* Δ_j *are written as* $x_i = c_i$ *for* $i \in \{\mu_j + 1, ..., \mu_n\}$.

Proof

We use the Frobenius theorem. Without loss of generality, we restrict our attention to dimension 3, for the sake of shortness.

Consider the three following involutive distributions, such that

$$\Delta_1 \subset \Delta_2 \subset \Delta_3 \quad \text{with} \quad \dim \Delta_i = \mu_j$$

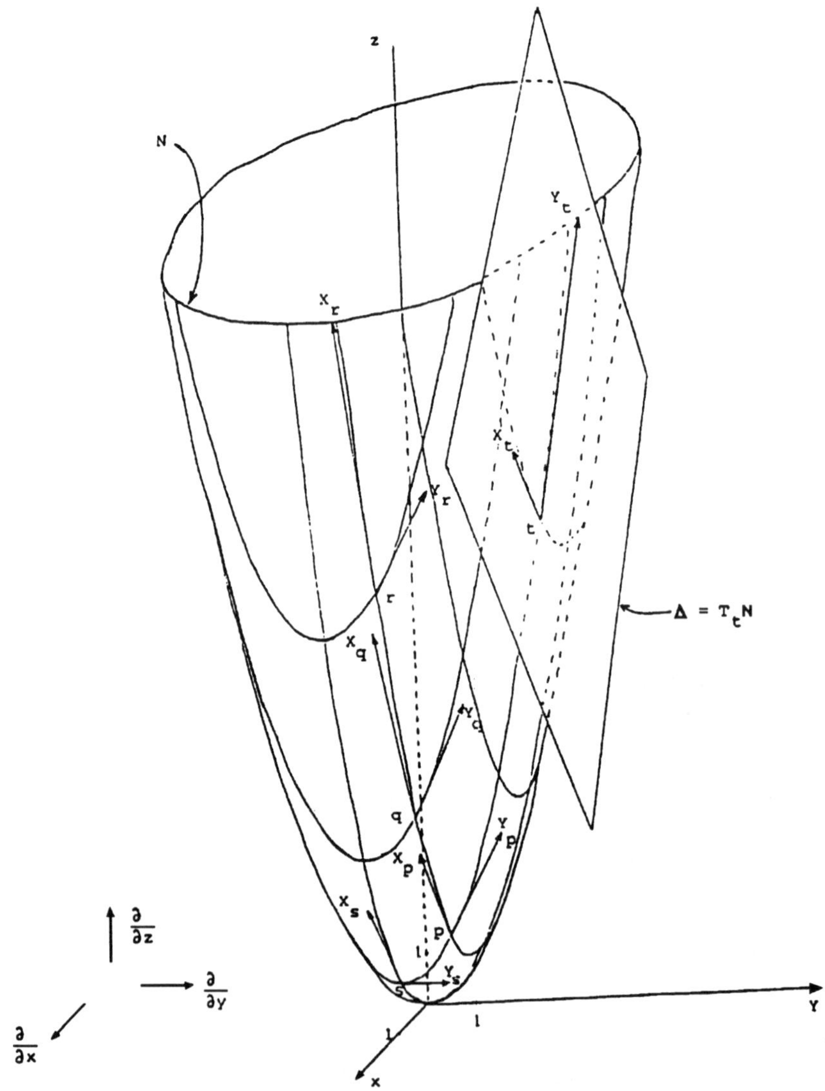

Figure B.2: Integral manifold of the above distribution ($M = \mathbb{R}^3$).

Let the local coordinate frames be

$$\underbrace{\underbrace{\underbrace{(x_1, \ldots, x_{\mu_1},}_{\text{basis of } \Delta_1} \quad x_{\mu_1+1}, \ldots, x_{\mu_2},}_{\text{basis of} \Delta_2} \quad x_{\mu_2+1}, \ldots, x_{\mu_3})}_{\text{basis of } \Delta_3}$$

As Δ_1 is involutive, according to the Frobenius theorem, it is also completely integrable. Then, there exists a local coordinate frame (U, ϕ) such that, if $(y_1, \ldots, y_n)$ denotes the local coordinate frame, $E_i = \phi_*^{-1}\left(\frac{\partial}{\partial y_i}\right)$ $i \leq \mu_1$ are a basis on U for the distribution Δ_1.

Consider the new basis $(y_1, \ldots, y_n, x_{\mu_1+1}, \ldots, x_{\mu_2})$ of Δ_2. The matrix Δ_2 is then written as:

$$\begin{pmatrix} I & \Delta_{12} \\ 0 & \Delta_{22} \end{pmatrix}$$

Applying the change of basis

$$\begin{aligned} Z_i &= \frac{\partial}{\partial y_i} && \text{for} \quad i \leq \mu_1 \\ Z_i &= X_i - \sum_{j \leq \mu_1} X_j(y_i) \cdot X_i && \text{for} \quad \mu_1 < i \leq \mu_2 \end{aligned}$$

In this new basis, Δ_2 takes the form

$$\begin{pmatrix} I & 0 \\ 0 & \Delta_{22} \end{pmatrix}$$

Now, Δ_1 and Δ_2 are involutive $\Longrightarrow$ Δ_{22} is involutive. So, we can apply the Frobenius theorem and then perform the same operations for Δ_3.

Without loss of generality, and to alleviate the writing, we formulate the following theorem for only three distributions.

Theorem B.4

Let $\{\Delta_k\}_3$ be three independent distributions. Then, the two following assertions are equivalent:

(i) *The $\Delta_k, k \in \{1, 2, 3\}$, are locally simultaneouly integrable.*

(ii) $$\begin{aligned} \check{\Delta}_1 &= \Delta_2 + \Delta_3 \\ \check{\Delta}_2 &= \Delta_1 + \Delta_3 \\ \check{\Delta}_3 &= \Delta_1 + \Delta_2 \end{aligned}$$

are integrable and $\Delta = \Delta_1 + \Delta_2 + \Delta_3$ is integrable.

The proof is based upon the Frobenius theorem as well as upon lemma B.1.

Example 5

We consider the distribution generated by the three vector fields

$$\underset{X_1}{\begin{pmatrix} 2x_3 \\ -1 \\ 0 \end{pmatrix}} \quad \underset{X_2}{\begin{pmatrix} -x_1 \\ -2x_2 \\ x_3 \end{pmatrix}} \quad \underset{X_3}{\begin{pmatrix} x_1 \\ 0 \\ x_3 \end{pmatrix}}$$

This distribution is regular in a neighborhood of $(0, 1, 1)$.

As the two first vector fields, being those of example 4, illustrate the Frobenius theorem, we know that the change of coordinates

$$\begin{aligned} x_1 &= 2z_1(z_3+1) \\ x_2 &= -z_1+z_2+1 \\ x_3 &= z_3+1 \end{aligned}$$

rectifies the first vector field.

The second one becomes $\begin{pmatrix} -2z_1 \\ -2(z_2+1) \\ z_3+1 \end{pmatrix}$

For the third vector field, we have

$$\begin{aligned} 2\dot{z}_1 \cdot (z_3+1) + 2z_1\dot{z}_3 &= 2z_1 \cdot (z_3+1) \\ -\dot{z}_1 + \dot{z}_2 &= 0 \\ \dot{z}_3 &= z_3+1 \end{aligned} \iff \begin{aligned} \dot{z}_1 &= 0 \\ \dot{z}_2 &= 0 \\ \dot{z}_3 &= z_3+1 \end{aligned}$$

Applying the change of basis

$$\begin{aligned} Z_1 &= \frac{\partial}{\partial z_1} \\ Z_i &= X_i - X_i(z_1) \cdot X_1 \qquad \text{for } i = 2, 3. \end{aligned}$$

we get the three vector fields

$$\underset{Z1}{\begin{pmatrix} 1 \\ 0 \\ 0 \end{pmatrix}} \quad \underset{Z2}{\begin{pmatrix} 0 \\ -2(z_2+1) \\ z_3+1 \end{pmatrix}} \quad \underset{Z3}{\begin{pmatrix} 0 \\ 0 \\ z_3+1 \end{pmatrix}}$$

with $[Z_i, Z_j] = 0, \quad \forall i, j \in \{1, 2, 3\}$.

We can now apply, for the last two vector fields, theorem B.2 dedicated to the rectification of coordinate fields.

Conclusion

The distributions Δ_k are locally simultaneously integrable.

Now, consider the three distributions defined by:

$$\check{\Delta}_1 = \Delta_2 + \Delta_3 = \left(\begin{pmatrix} -x_1 \\ -2x_2 \\ x_3 \end{pmatrix}, \begin{pmatrix} x_1 \\ 0 \\ x_3 \end{pmatrix} \right)$$

$$\check{\Delta}_2 = \Delta_1 + \Delta_3 = \left(\begin{pmatrix} 2x_3 \\ -1 \\ 0 \end{pmatrix}, \begin{pmatrix} x_1 \\ 0 \\ x_3 \end{pmatrix} \right)$$

$$\check{\Delta}_3 = \Delta_1 + \Delta_2 = \left(\begin{pmatrix} 2x_3 \\ -1 \\ 0 \end{pmatrix}, \begin{pmatrix} -x_1 \\ -2x_2 \\ x_3 \end{pmatrix} \right)$$

A simple calculation shows that the three distributions $\check{\Delta}_1, \check{\Delta}_2, \check{\Delta}_3$ are involutive. Moreover, since the dimension is 3, $\Delta = \Delta_1 + \Delta_2 + \Delta_3$ is involutive.

Lemma B.2

Let $\{\Delta_k\}$ k be independent and involutive distributions, with $\dim \Delta_i = \delta_i$. Assume that Δ_i is generated as follows

$$\Delta_i = \text{sp}\{X_{\rho_{j-1}+1}, ..., X_{\rho_j}\} \qquad i \in \underline{k} \qquad \textit{with} \quad \begin{aligned} \rho_o &= 0 \\ \rho_j &= \sum_{i \in \underline{j}} \delta_i \\ \rho_k &= n \end{aligned}$$

If $\forall X_s \in \Delta_i$, $\forall X_t \in \Delta_j$, $\forall i \neq \forall j$, we have $[X_s, X_t] = 0$, then

(i) *The distributions Δ_j are simultaneously integrable, $j \in \underline{k}$.*

(ii) *The local coordinate frame $(x_1, ..., x_n)$ in which*

$$\Delta_i = \text{sp}\left\{ \frac{\partial}{\partial x_i}, i \in \{\rho_{j-1}+1, ..., \rho_j\}, j \in k \right\}$$

has the following property:

$$X_i = \sum_{h \in J} f_h^i(x_{r_{j-1}+1}, ..., x_{r_j}) \cdot \frac{\partial}{\partial x_h} \qquad \forall j \in \underline{k}, \forall i \in \{\rho_{j-1}+1, \ldots, \rho_j\} = J$$

The proof is based upon the Frobenius theorem.

Example 6

Following the above example, consider the distributions defined by

$$D_1 = \check{\Delta}_3 = \Delta_1 + \Delta_2 = \left(\begin{pmatrix} 2x_3 \\ -1 \\ 0 \end{pmatrix}, \begin{pmatrix} -x_1 \\ -2x_2 \\ x_3 \end{pmatrix} \right)$$

$$D_2 = \Delta_3 = \begin{pmatrix} x_1 \\ 0 \\ x_3 \end{pmatrix}$$

We know that both of these distributions are independent and involutive. From the example 5, we can assert that the three distributions $\Delta_i, i \in \{1,2,3\}$ are simultaneously integrable, but also that $[D_1, D_2] = 0$.

The change of coordinates:

$$\begin{aligned} x_1 &= 2z_1(z_3+1) \\ x_2 &= -z_1 + z_2 + 1 \\ x_3 &= z_3 + 1 \end{aligned} \qquad (B.5)$$

rectifies the first vector field.

The second one becomes $\begin{pmatrix} -2z_1 \\ -2(z_2+1) \\ z_3+1 \end{pmatrix}$

For the third vector field, we have:

$$\begin{aligned} 2\dot{z}_1(z_3+1) + 2z_1\dot{z}_3 &= 0 \\ -\dot{z}_1 + \dot{z}_2 &= 0 \\ \dot{z}_3 &= z_3 + 1 \end{aligned} \iff \begin{aligned} \dot{z}_1 &= 0 \\ \dot{z}_2 &= 0 \\ \dot{z}_3 &= z_3 + 1 \end{aligned}$$

Applying the change of basis

$$\begin{aligned} Z_1 &= \frac{\partial}{\partial z_1} \\ Z_i &= X_i - X_i(z_1) \cdot X_1 \quad \text{for } i = 2, 3. \end{aligned}$$

We then get the three vector fields

$$\underset{Z_1}{\begin{pmatrix} 1 \\ 0 \\ 0 \end{pmatrix}} \qquad \underset{Z_2}{\begin{pmatrix} 0 \\ -2(z_2+1) \\ z_3+1 \end{pmatrix}} \qquad \underset{Z_3}{\begin{pmatrix} 0 \\ 0 \\ z_3+1 \end{pmatrix}}$$

with $[Z_i, Z_j] = 0, \quad \forall i, j \in \{1,2,3\}$.

We can now apply theorem B.2 (rectification of coordinate frames) to the last two vector fields.

The change of coordinates straightening up Z_2 and Z_3 is

$$\begin{aligned} z_1 &= y_1 \\ z_2 &= e^{-2y_2} - 1 \\ z_3 &= e^{(y_2+y_3)} - 1 \end{aligned} \tag{B.6}$$

So, we get the three vector fields

$$\begin{pmatrix} 1 \\ 0 \\ 0 \end{pmatrix} \quad \begin{pmatrix} 0 \\ 1 \\ 0 \end{pmatrix} \quad \begin{pmatrix} 0 \\ 0 \\ 1 \end{pmatrix}$$

Putting (B.6) into (B.5), we obtain the system

$$\begin{aligned} x_1 &= 2y_1 e^{(y_2+y_3)} \\ x_2 &= -y_1 + e^{-2y_2} \\ x_3 &= e^{(y_2+y_3)} \end{aligned}$$

In order to calculate the three initial vector fields as functions of the variables y_i we must first calculate the $\dfrac{\partial}{\partial x_i}$ in function of the $\dfrac{\partial}{\partial y_j}$. We have

$$\left[\frac{\partial}{\partial y}\right] = \begin{pmatrix} 2e^{(y_2+y_3)} & -1 & 0 \\ 2y_1 e^{(y_2+y_3)} & -2e^{-2y_2} & e^{(y_2+y_3)} \\ 2_{y_1} e^{(y_2+y_3)} & 0 & e^{(y_2+y_3)} \end{pmatrix} \cdot \left[\frac{\partial}{\partial x}\right]$$

that is

$$\left[\frac{\partial}{\partial x}\right] = \begin{pmatrix} \frac{1}{2} e^{-(y_2+y_3)} & -\frac{1}{4} e^{(y_2-y_3)} & \frac{1}{4} e^{(y_2-y_3)} \\ 0 & -\frac{1}{2} e^{2y_2} & \frac{1}{2} e^{2y_2} \\ -y_1 e^{-(y_2+y_3)} & \frac{1}{2} y_1 e^{(y_2-y_3)} & e^{-(y_2+y_3)} - \frac{1}{2} y_1 e^{(y_2-y_3)} \end{pmatrix} \cdot \left[\frac{\partial}{\partial y}\right]$$

Calculation of the three initial vectors fields as functions of variables y_i

Replacing the variables x_i into the variables y_j and $\dfrac{\partial}{\partial x_i}$ into $\dfrac{\partial}{\partial y_j}$, we find the expressions of the vector fields in the basis $\left(\dfrac{\partial}{\partial y_1}, \dfrac{\partial}{\partial y_2}, \dfrac{\partial}{\partial y_3}\right)$, which gives

$$\begin{aligned} X_1 &= \frac{\partial}{\partial y_1} \\ X_2 &= -2y_1 \frac{\partial}{\partial y_1} + \frac{\partial}{\partial y_2} \\ X_3 &= \frac{\partial}{\partial y_3} \end{aligned}$$

Conclusion

The local coordinate frame (y_1, y_2, y_3), in which

$$D_1 = \text{sp}\left\{\frac{\partial}{\partial y_i}\right\};\ i \in \{1,2\} \quad \text{and} \quad D_2 = \text{sp}\left\{\frac{\partial}{\partial y_3}\right\}$$

verifies:

$$\begin{aligned} X_1 &= \sum_{i=1 \text{ à } 2} f_i^1(x_1,x_2)\cdot\frac{\partial}{\partial x_i} \\ X_2 &= \sum_{i=1 \text{ à } 2} f_i^2(x_1,x_2)\cdot\frac{\partial}{\partial x_i} \\ X_3 &= f_3^3(x_3)\cdot\frac{\partial}{\partial x_3} \end{aligned}$$

Definition B.5

Let a partition $I_1,\ldots,I_n$, of $\{1,\ldots,m\}$. Let $\{d_0,\ldots,d_\alpha\}$ such that $d_0 = 0$ and $\sum_{i\in\underline{\alpha}} d_i = n$.

We say that the system $\dot{x}(t) = f(x(t)) + \sum_{i\in m} g_i(x(t))\cdot u_i(t)$ admits a parallel decomposition in the local coordinate frame $(x_1,\ldots,x_n)$ if it takes the form

$$\begin{aligned} &\dot{x}^1(t) = f^1(x^1) + \sum_{l_1\in I_1} g_{l_1}(x^1)u_{l_1} \\ &\vdots \\ &\dot{x}^\alpha(t) = f^\alpha(x^\alpha) + \sum_{l_\alpha\in I_\alpha} g_{l_\alpha}(x^\alpha)u_{l_\alpha} \end{aligned}$$

with

x^1 is built up with the d_1 first coordinates $(x_1,\ldots,x_{d_1})$ of x.

x^2 is built up with the d_2 following coordinates $(x_{d_1+1},\ldots,x_{d_1+d_2})$ of x.

$\vdots$

x^α is built up with the d_α last coordinates of x.

Then, the vector fields look like:

$$\begin{aligned} f^i(x^i) &= \sum_{j\in(h_i-h_{i-1})} a_j(x^i)\frac{\partial}{\partial x_j} \\ g_{l_i}(x^i) &= \sum_{j\in(h_i-h_{i-1})} g_{ji}(x^i)\frac{\partial}{\partial x_j} \end{aligned}$$

and

$$f(x) = \begin{pmatrix} f^1(x^1) \\ . \\ . \\ . \\ f^\alpha(x^\alpha) \end{pmatrix} \qquad g(x) = \begin{pmatrix} g_{l_1}(x^1) & . & . & . \\ . & g_{l_2}(x^2) & . & . \\ . & . & . & . \\ . & . & . & . \\ . & . & . & g_{l_\alpha}(x^\alpha) \end{pmatrix}$$

Definition B.6

We call L, the accessibility distribution of the system, the set defined by

$$L = \{ad_f^q \cdot g_i, i \in \underline{m},\ q \in \mathbb{N}\}_{\text{L.A}}$$

Property

If for all $x \in M$, $L(x) = T_x M$, then the system is strongly accessible. In other words

$$L_{oj} = \{ad_f^q \cdot g_i, i \in I_j, q \in \mathbb{N}\}_{\text{L.A.}} \quad \text{with} \quad d_j(x) = \dim(L_{oj}(x))$$

Theorem B.5

If the system is strongly accessible, then it admits a parallel decomposition $\Longleftrightarrow$ *there exists a partition* $(I_1, ..., I_a)$ *of* $\{1, ..., m\}$ *such that the following conditions are fulfilled:*

(i) $\sum_{j \in \underline{\alpha}} d_j(x) = n.$

(ii) $\left[ad_f^q \cdot g_s, ad_f^r \cdot g_t\right] = 0 \quad \forall s \in I_j,\ \forall t \in I_i,\ \forall r, q \in \mathbb{N},\ \forall i, j \in \underline{\alpha},\ i \neq j$

Example 7

Let the system be defined by $\dot{x} = f(x) + \sum_{i \in (1;2)} g_i(x) \cdot u_i$ with

$$f(x) = \begin{pmatrix} 2x_3 \\ -1 \\ 0 \end{pmatrix}, \quad g_1(x) = \begin{pmatrix} -x_1 \\ -2x_2 \\ x_3 \end{pmatrix}, \quad g_2(x) = \begin{pmatrix} x_1 \\ 0 \\ x_3 \end{pmatrix}$$

It is strongly accessible, because $[f, g_1] = -2f$

Moreover, we have $[f, g_2] = 0$ and $[g_1, g_2] = 0$ (B.7)

So

$$L_{01} = \{g_1, [f, g_1]\} \tag{B.8}$$
$$L_{02} = g_2$$

(B.8) $\Longrightarrow$ condition **(i)** of theorem B.5 is verified.

(B.7) $\Longrightarrow$ condition **(ii)** of theorem B.5 is verified.

Consider the coordinate frame already determined for the above example

$$\begin{aligned} x_1 &= 2y_1\, \mathrm{e}^{(y_2+y_3)} \\ x_2 &= -y_1 + \mathrm{e}^{-2y_2} \\ x_3 &= \mathrm{e}^{(y_2+y_3)} \end{aligned}$$

We know that when replacing the variables x_i into the variables y_j and the $\dfrac{\partial}{\partial x_i}$ into the $\dfrac{\partial}{\partial y_j}$, we obtain the expressions of the vector fields in the basis $\left(\dfrac{\partial}{\partial y_1}, \dfrac{\partial}{\partial y_2}, \dfrac{\partial}{\partial y_3}\right)$, that is:

$$\begin{aligned} f &= \frac{\partial}{\partial y_1} \\ g_1 &= -2y_1 \frac{\partial}{\partial y_1} + \frac{\partial}{\partial y_2} \\ g_2 &= \frac{\partial}{\partial y_3} \end{aligned}$$

In fact, this coordinates frame provides a block diagonalization. Indeed, we find the system:

$$\begin{pmatrix} \dot{y}_1 \\ \dot{y}_2 \\ \dot{y}_3 \end{pmatrix} = \begin{pmatrix} 1 \\ 0 \\ 0 \end{pmatrix} + \begin{pmatrix} -2y_1 & 0 \\ 1 & 0 \\ 0 & 1 \end{pmatrix} \cdot \begin{pmatrix} u_1 \\ u_2 \end{pmatrix}$$

B.3 Practical difficulties

Rectifying a vector field, which is the basic operation for obtaining a canonical form, requires the formal integration of a differential system. It is clear that the latter integration is not always achievable, even with classical analytic functions.

The use of the Frobenius theorem implies the formal integration of ordinary differential equations with an almost arbitrary choice of initial conditions, which does not facilitate the practical solution.

We hope that this mathematical appendix will provide a convenient help for users wishing to implement input–output linearization techniques. However, they must keep in mind that, beyond a pure theoretical context, saying or writing "there exists a local coordinate frame such that", is a very optimistic short-cut, since the calculation of the coordinate frame under consideration is not at all easy, and is even, in some situations, impossible!

B.4 Bibliography

[1] CARTAN, E. (1945), *Les systèmes différentiels extérieurs et leurs applications géométriques*, Hermann Paris.

[2] GODBILLON, C. (1969), *Géométrie différentielle et mécanique analytique*, Hermann Paris - Collection Méthodes.

[3] SPIVAK, M. (1979), *Differential geometry*, Publish or Perish, Washington, USA.

[4] BOOTHBY, WILLIAM M. (1975), *An introduction to differentiable manifolds and Riemannian geometry*, Academic Press, New-York. USA.

Index

585327 - (I) - (1,1) - OSBN 80° - PUB

MASSON Éditeur
120, boulevard Saint-Germain
75280 Paris Cedex 06
Dépôt légal : août 1997

Achevé d'imprimer sur les presses de la
SNEL S.A.
Rue Saint-Vincent 12 – B-4020 Liège
tél. 32(0)4 343 76 91 - fax 32(0)4 343 77 50
juillet 1997 - 8108